Natural Computing Series

Series Editors: G. Rozenberg
Th. Bäck   A.E. Eiben   J.N. Kok   H.P. Spaink

Leiden Center for Natural Computing

More information about this series at http://www.springer.com/series/4190

Anthony Brabazon • Seán McGarraghy

Foraging-Inspired Optimisation Algorithms

 Springer

Anthony Brabazon
School of Business
University College Dublin
Dublin, Ireland

Seán McGarraghy
UCD Centre for Business Analytics
University College Dublin
Dublin, Ireland

ISSN 1619-7127
Natural Computing Series
ISBN 978-3-030-09640-3   ISBN 978-3-319-59156-8   (eBook)
https://doi.org/10.1007/978-3-319-59156-8

This Springer imprint is published by the registered company Springer Nature Switzerland AG
The registered company address is: Gewerbestrasse 11, 6330 Cham, Switzerland

To Maria, my mother Rose, and to the memory of my father Kevin
Tony

To Milena, Martin and Alex
Seán

Preface

In recent times there has been growing interest in *biomimicry*, or 'learning from nature', with many disciplines turning to natural phenomena for inspiration as to how to solve particular problems in their field. Examples include the development of pharmaceutical products based on substances found in plants, and inspiration for engineering designs based on structures and materials found in nature.

Another strand of 'learning from nature' concerns the development of powerful computational algorithms whose design is inspired by natural processes which implicitly embed computation. *Biologically inspired algorithms* take inspiration from a wide array of natural processes, including evolution, the workings of the central nervous system and the workings of the immune system, in order to develop algorithms for optimisation, classification, function approximation and other purposes. The foraging activities of animals and other organisms are an additional source of inspiration for the design of computational algorithms. Foraging behaviours, along with some of the algorithms that have been developed to date by drawing on these behaviours, form the focus of this book.

The aim of foraging is to acquire valuable resources such as food, shelter and mates. A practical problem is that in general the forager does not know in advance and with certainty the location of resources in the environment. Good foraging strategies, therefore, need to embed a robust search process based on sparse and noisy information. A rich literature based on the foraging strategies of various organisms has been developed including ant colony algorithms, honeybee algorithms and bacterial foraging algorithms, amongst others.

As yet there is no unifying text which provides comprehensive coverage across these algorithms. This book closes that gap. In addition to overviewing the main families of optimisation algorithms which stem from a foraging metaphor, we contextualise these algorithms by introducing key concepts from foraging and related literatures, and also identify open research opportunities.

The book is divided into seven parts. Part I presents a series of perspectives from the literatures on foraging, sensory ecology and social learning which are relevant for algorithmic design. Part II provides a framework for later chapters, by introducing a number of taxonomies and a general metaframework which help categorise the large literature on foraging-inspired optimisation algorithms. Parts III to V introduce a range of algorithms whose inspiration is drawn from the foraging activities of vertebrates, invertebrates and nonneuronal organisms respectively. In Part VI, a number of algorithms are introduced whose inspiration is drawn from formal models of foraging outlined in Part I. In the final chapter of the book, we outline some open research opportunities.

We hope that this book will be of interest to academics, students and practitioners who are seeking a detailed discussion of the current state of the art in foraging-inspired algorithms. Particular target audiences include those interested in informatics, data science and management science. The book is written so as to be accessible to a wide audience and no prior knowledge of foraging-inspired algorithms is assumed.

The rich complexity of foraging in nature, and its capability to inspire algorithmic design, is a truly fascinating subject of study. We hope that you enjoy your journey through this book.

Anthony Brabazon
Seán McGarraghy
Dublin, July 2018

Acknowledgements

The inspiration for this text arose during the preparation of our last book, *Natural Computing Algorithms* (also published by Springer), in which we provided coverage of a wide array of computational algorithms whose metaphorical roots come from natural phenomena in biology, chemistry and physics. During that project it became apparent that while foraging-inspired algorithms are a significant field of research in their own right, no unifying text existed concerning them. As a result, the idea for this book emerged.

As with all book projects, multiple people have contributed. We thank our research colleagues in the field for their generous sharing of ideas through their publications and through discussions we have had with them at conferences and academic meetings. We would also like to acknowledge with thanks the contribution of members (past and present) of the Natural Computing Research & Applications Group at University College Dublin (http://ncra.ucd.ie). Discussions with our undergraduate and postgraduate students, across a range of modules we have taught, have also helped to mould the material in this book.

We extend our thanks to Ronan Nugent, Senior Editor at Springer. His encouragement of this project from its earliest stages helped ensure it moved beyond an 'interesting idea' to reality. Ronan's invaluable advice on early drafts of the manuscript has resulted in a far stronger final book. We would also like to note Ronan's significant contribution to the field of natural computing. His knowledge of key themes and emerging trends in the field, combined with his encouragement of multiple authors, has helped shape the dialogue which exists in the field today.

Most importantly, we each extend a special thank you to our families. You bear the 'cost' in terms of the late nights and weekends which were devoted to the writing of this book. We each thank our families for your love and understanding. Without your support this book would never have been written.

Anthony Brabazon
Seán McGarraghy

Contents

1

Introduction

The vital ingredient for life is energy. Every animal must forage for the resources it needs to generate this energy in order to sustain life, grow and reproduce. The observation that active foraging requires organisms to undertake a search process in order to locate resource items has clear parallels with optimisation, where one is searching for a solution, a structure or a vector of model parameters, in a possibly high dimensional space. As nature has tuned the efficiency and effectiveness of foraging strategies over many millions of years, exploration of these can potentially provide insights for the design of powerful optimisation algorithms. Foraging and the design of optimisation algorithms which have been inspired by foraging form the focus of this book.

1.1 What Does This Book Cover?

The study of foraging is a fascinating and highly multidisciplinary field of research with implications far beyond the confines of biology. Foraging has been a canonical setting for the study of search, reward seeking and information processing [308]. These and related themes have wide impact in fields such as biology, economics, robotics and computer science, to name but a few.

Questions such as where and when to forage, and what to forage for, implicitly embed computation. A large research literature has arisen concerning various families of foraging-inspired algorithms, including ant colony optimisation algorithms, honeybee optimisation algorithms and bacterial foraging optimisation algorithms. In each case, metaphorical inspiration has been taken from aspects of the real world foraging behaviour of a specific organism to design algorithms for discrete or real-valued optimisation. These algorithms have been successfully applied to a multiplicity of real world problems. It is notable that there is no unifying text which provides comprehensive coverage across the domain of foraging-inspired algorithms. This book closes that gap.

In writing this book we have been mindful of the commentary of Kenneth Sörensen [539] and others that natural computing [77] needs to adopt a more critical

© Springer Nature Switzerland AG 2018
A. Brabazon, S. McGarraghy, *Foraging-Inspired Optimisation Algorithms*,
Natural Computing Series, https://doi.org/10.1007/978-3-319-59156-8_1

perspective when assessing the worth of new metaheuristics. Accordingly, we seek to provide a strong context for the exploration of algorithms deriving from a foraging metaphor.

Even a casual browsing of research studies concerning foraging behaviours quickly reveals that the field is highly multidisciplinary. A text which aims to develop an informed understanding of both the state of the art in the field of foraging-inspired algorithms and open areas for future research in this area must provide a context for these algorithms by introducing key concepts from foraging and related literatures. In this book, we initially provide an introduction to selected aspects from these literatures of which all users and designers of foraging-inspired algorithms should be aware, covering formal models of foraging, and elements from the literatures on sensory ecology and learning.

A notable characteristic of the literature concerning foraging-inspired algorithms is that little attempt is typically made to rigorously identify the similarities and differences between newly proposed algorithms and existing ones. An important tool in sharpening our understanding of existing and proposed algorithms is the development of taxonomies which allow us to tease out these similarities and differences. In this book we introduce a number of exemplar taxonomies which can be used for this purpose. We also illustrate that most foraging algorithms can be encapsulated in a high-level metaframework, with differing operationalisations of elements of this framework giving rise to alternative algorithms with distinct search characteristics.

In addition, this book provides a comprehensive overview of a wide span of foraging-inspired optimisation algorithms. While including detailed coverage of the more popular algorithms, we supplement this with coverage of a number of lesser-known algorithms. In some cases, these algorithms are quite new and further research will be required to assess whether they can scale to higher-dimensional problems, and to determine whether their results are competitive in terms of quality and computational effort required against those of other heuristics. Nonetheless, their inclusion serves to illustrate the breadth of research in this field. We conclude this book with suggestions for future research.

Outline of This Chapter

Whilst every living thing must access food resources, it does so in a specific context. This context comprises a physical environment and a complex ecological niche of conspecifics, predators and prey. Unsurprisingly, given the diversity of these contexts, the range of foraging strategies found in nature is truly enormous.

In this introductory chapter we begin our exploration of foraging by emphasising the diversity and complexity of life on earth and by defining what we mean by foraging (Sects. 1.2 and 1.3). We then explore the general factors which impact on an organism's choice of foraging strategy (Sect. 1.4). This naturally leads to a discussion of how the combination of these factors and the choice of foraging strategy produces a distribution of foraging 'payoffs' (Sect. 1.5). Of course, not all food capture strategies entail an active search for food items, and in Sect. 1.6, for comparison purposes, we introduce two alternative approaches to obtaining food, namely, agriculture and food kleptoparasitism (food theft). We conclude the chapter in Sect. 1.7 by providing an overview of the structure of the rest of the book.

1.2 The Diversity of Life

Life on earth is estimated to have arisen some 3.7–4 billion years ago [151, 439] and consists of an incredible array of organisms, ranging from tiny unicellular creatures which are invisible to the naked eye to the plants and animals with which we are familiar on a daily basis. All nonviral life can be classified into three primary domains (Fig. 1.1): *Archaea*, *Bacteria* and *Eukaryota*. Archaea and bacteria comprise

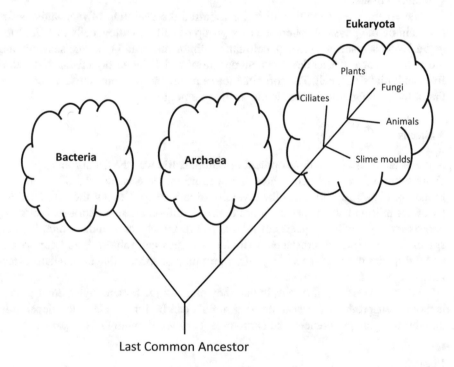

Fig. 1.1. Stylised tree of life with a subset of the Eukaryota shown. All branches of the tree originated from a common ancestor

the *prokaryotes*, simple unicellular life forms without a membrane bound nucleus, mitochondria or other organelles. The third domain of life, the eukaryotes, first arose approximately 2 billion years ago [381]. Eukaryotes possess a more complex cellular structure, with a well-defined membrane-bound nucleus (with chromosomal DNA) and organelles. All animal and plant life belongs to the eukaryotic domain. Given that the number of microbial species alone is estimated to exceed one trillion (10^{11}–10^{12}) [368], it is evident that the diversity of living organisms is unimaginably large.

Multicellular Life

The emergence of multicellular eukaryotes was a key landmark in the development of life, as multicellularity allows an increase in both organism size and structural complexity. Unicellular creatures encounter physical size limits, as individual cells have a decreasing surface-to-volume ratio as they get larger. This makes it more difficult to absorb and internally transport sufficient resources from the environment to support their energy requirements. In contrast, multicellular organisms can generate structural complexity via cellular differentiation, leading to the development of specialised tissue.

The emergence of multicellular life required the evolution of mechanisms for maintaining the physical coherence of a group of cells (to allow cells to 'stick' together), mechanisms to facilitate cellular communication and resource sharing, and the evolution of a 'master development program' which allows organisms to develop from an initial single cell. The consequence of having to overcome these hurdles was that it took a long time for multicellular life to emerge.

Animals

Another landmark in the development of complex life was the Cambrian explosion, about 540 million years ago, which saw a rapid increase in the size of the animal kingdom. Figure 1.2 illustrates the diversity of animal species. Of the estimated total of 1.4 million known species, more than 1.3 million are invertebrates, including approximately 1 million species of insects [628]. Insects, arachnids (which include spiders and scorpions), crustaceans (including crabs and lobsters), and centipedes and millipedes make up the *arthropods*, a grouping of species that have a hard external skeleton and jointed limbs.

All animals are *heterotrophs*, in that they must forage for and ingest other organisms or their products for sustenance. Animals can be further classified depending on their foraging preferences into carnivores, herbivores, omnivores or parasites.

Plants

Plants, another sizeable grouping of multicellular life (the grouping of vascular plants alone consists of some 400,000 distinct species [623]), are *autotrophic* in that they generally produce their food requirements from carbon dioxide and water using energy from sunlight to generate carbohydrates. Although plants are autotrophs, they can still be considered as foraging for sunlight and other requirements such as carbon dioxide, water and nitrogen.

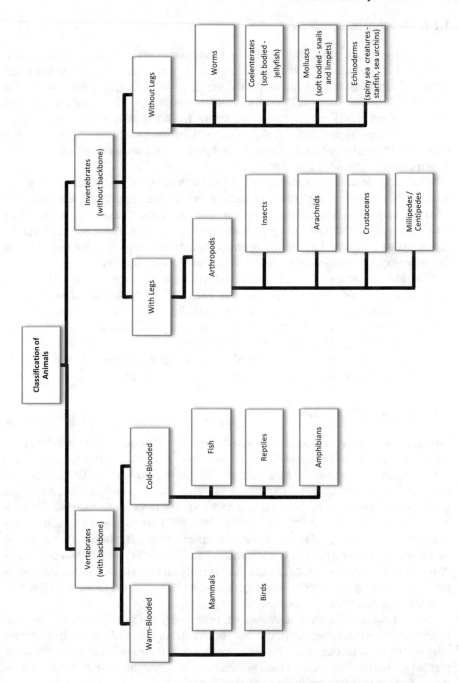

Fig. 1.2. Classification of animals (not all animal groups shown). Only about 4.8% of all animal species are vertebrates [628]

1.2.1 Foraging Interactions

As noted by Yeakle and Dunne [652], the behaviour of all evolutionarily successful organisms is constrained by two requirements. They must pass on their genetic material and they must acquire the necessary energy to do so. Foraging behaviours play a key role in determining evolutionary success. Organisms do not forage in isolation from one another; rather, their activities interact. All organisms are part of a *food web*. Adaptation by any individual participant in a food web can impact on many others, including its prey, its predators, conspecifics and members of other species which consume the same food items.

One example of interactive effects in a food web is a *trophic cascade*. This occurs when (for example) an apex predator reduces the number of, or alters the behaviour of, its direct prey, in turn reducing the degree of predation by those animals on their own prey in the next (lower) trophic level in the food web. These cascades can be top-down, as would be the case where an apex predator is removed from a food web, or bottom-up, as could occur where a primary producer is removed.

Trophic cascades can extend all the way from an apex predator to soil nutrient levels. An illustration of this is provided in [413], which found that changes in the population of dingoes (an apex predator in Australia) impacted on the abundance of kangaroos (which are herbivores), in turn impacting on their consumption of vegetation, ultimately impacting on vegetation growth and soil nutrient levels. Food webs are complex and changes in participants' abundance or behaviour can have multiple and time-varying impacts.

Coevolution and Foraging

Foraging takes place in a competitive environment, with competition occurring between individuals and between species. The direct competition between predators and their prey leads to an *antagonistic* coevolutionary arms race. The fitnesses of the entities in an antagonistic coevolutionary arms race are competitive, with both predator and prey seeking to adapt and counteradapt in order to increase their fitness at the expense of the other. For example, development of even a rudimentary visual capability by a predator will place a nonsighted prey organism at a significant disadvantage, unless it too can produce mutant varieties with enhanced sensory capabilities which allow it to evade capture. As virtually all organisms are simultaneously predators and prey (of other organisms), there are a host of coevolutionary influences continually acting on every species.

Not all coevolutionary adaptations are necessarily competitive [265]. Coevolutionary changes can be *mutualistic*, where both entities benefit from complementary adaptations such as can occur between flowers and their pollinators. In this case, plants provide food rewards to their pollinators, as they contribute to the plant's reproductive success by distributing its pollen.

1.3 What Is Foraging?

All living things need to acquire a variety of resources over time, such as food, shelter, nesting sites and mates. In this book we primarily focus attention on foraging for food resources, and references to foraging can therefore be considered in this context unless otherwise indicated.

The definition of foraging adopted in this book is taken from Stephens and Krebs [545] as 'a repeated sequence of actions: search, encounter, decide'. Under this definition, *search* can encompass a strategy of waiting in place as well as active traversal of the environment in an effort to find resources. An *encounter* occurs when a food item is located, and the organism must then *decide* whether to attempt to appropriate the resource. Following this, a *foraging strategy* can be broadly considered as a strategy for searching an environment in order to encounter and appropriate food resources. Hence, a complete strategy will cover the operationalisation of a search process, encounter behaviours and choices as to which items are considered as prey.

Perhaps the most widely known foraging behaviours in nature are those of animals which we can easily observe, such as large predators actively searching an environment for prey items. Complex life forms such as animals employ an array of strategies to capture the resources they require. However, even the simplest creatures need to 'earn a living'. Unicellular organisms can capture nutrients by engulfing and absorbing macromolecules or other substances (this process is called *endocytosis*) or by active or passive transport of nutrients across their cell membrane.

In all cases, the basic problem faced by a mobile forager is how best to search an environment for resources when their location is not known with certainty a priori.

1.4 Choice of Foraging Strategy

Multiple factors influence the choice of foraging strategy as illustrated in Fig. 1.3 [10, 48, 49, 147, 168, 521]. In turn these factors impact on the payoffs of any specific foraging strategy.

Many of the influences on foraging are genetically determined, as they relate to the morphology and the cognitive capabilities of an organism. Genetic influences can also produce subtle effects on the foraging activity of an organism. For example, the fruit fly *Drosophila melanogaster* has a specific gene (nicknamed the *for* gene) which encodes for an enzyme cGMP PKG. Depending on the level of expression of this enzyme, larvae of the fruit fly exhibit different foraging behaviours. Some exhibit *rover* behaviour and wander around in search of food. Others exhibit *sitter* behaviour and tend to remain in a limited area. Rovers possess one allele of the *for* gene, and this produces higher levels of PKG expression than the allele possessed by sitters [132]. The same gene in the honeybee *Apis mellifera* plays a critical role in its life cycle, with increased production of PKG triggering the start of foraging behaviours in older female honeybees.

A full discussion of all of the influences in Fig. 1.3, and the complex interactions that occur between them in a foraging context, would require a substantial text in

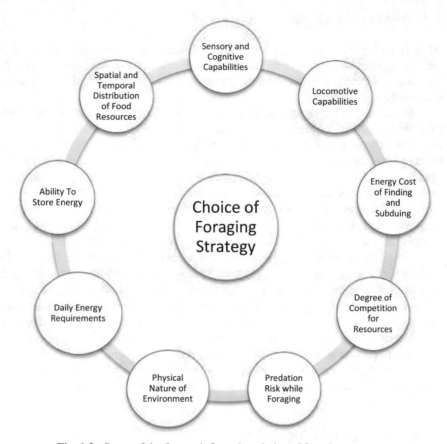

Fig. 1.3. Some of the factors influencing choice of foraging strategy

itself. In the following chapters we focus attention on a subset of themes from the foraging and related literatures which will be most useful in enabling us to place the existing range of foraging-inspired optimisation algorithms into context, and which will provide insights into other, as yet unexplored, avenues for the design of novel algorithms.

In this chapter we therefore restrict our attention to a short discussion of the influence of the physical environment, the spatial and temporal distribution of resources, daily energy requirements, and the costs of foraging. The following subsections underscore the complexity of the influences which impact on the choice of foraging strategy for even simple organisms and also highlight the strong context dependence of any particular foraging strategy.

Physical Environment and Spatial Distribution of Resources

The physical environment faced by an organism plays a significant role in deter-mining what sensory, cognitive and morphological attributes will be required for successful foraging. The distinct physical characteristics of an environment will im-pact on issues such as food availability and on the ease with which the terrain can be traversed. Consider, for instance, the distinctions between the challenges posed to a forager in a high-temperature desert setting with limited prey availability compared with the challenges that a forager will face in a tropical jungle environment. As the challenges differ, we can expect foraging strategies also to differ.

Aquatic environments provide a good example of an environment with character-istics which strongly influence foraging behaviours. Organisms which inhabit these environments are challenged to maintain high internal temperatures as cold water leaches heat from a body. Consequently, most fish are cold-blooded and maintain a body temperature which is close to that of the surrounding water. Active marine predators tend to live at shallower (warmer), depths where they can maintain the higher core temperature which is needed to generate the necessary levels of muscle temperature to engage in active hunting. For fish living in colder water at greater depths, typical resulting behaviours include slow rates of movement, and ambush rather than active foraging strategies. An interesting exception to this general rule is provided by the opah or *moonfish* (one of the only warm blooded fish currently known to exist), which lives at depths of 45–300 metres and which maintains its temperature at 5°C above that of the surrounding water [615]. The ability to main-tain a higher internal temperature allows it to be an active predator of prey such as squid.

Spatial and Temporal Dimensions of Foraging

The spatial distribution of resources can vary notably between differing environments and this distribution can play a key role in determining whether individuals forage alone (*solitary foraging*) or as part of a group. Environments faced by organisms can range from resource-rich environments, where foraging may require little effort, to resource-poor environments, where food is only found in ephemeral clusters. When food resources are distributed in a patchy fashion the key limitation in foraging is often the determination of the location of each patch, and, as a result, species that feed on food resources distributed in this manner are more likely to live in groups and share information on the location of food resources. Group foraging may also arise owing to the benefits of being able to attack and capture larger prey items.

Resources in the environment are not just distributed across space (the landscape) but are also distributed across time. This implies that the foraging problem faced by organisms is both *spatial* and *temporal*, with [119, p. 2271] commenting that '...many problems faced by biological systems can be reduced to the need to gather resources, distributed across space and time, as efficiently as possible.'

Frequently, the location and quality of food resources can change over short pe-riods of time as a result of factors such as consumption and degradation via environ-mental influences. Good food-foraging strategies in these environments will need to be adaptive to changing conditions and to feedback based on their degree of past success. In other words, an ability to learn from past experience or from social learning, such as occurs when a parent demonstrates foraging techniques to its young, will enhance foraging success and survival. Evolutionary adaptations which enhance the ability of organisms to learn and communicate can therefore be very useful.

Apart from short-term fluctuations in food availability, in many regions of the world food resources are seasonal. This raises the question as to what strategies an organism can adopt to deal with the limited availability or nonavailability of preferred food resources (including water) at certain times of the year.

One option is seasonal migration to a new location. In cases where the food supply lessens but at least limited food amounts are still available locally, another option is to change dietary habits in line with seasonal food availability. Alternatively, an animal may seek to *insure* itself against the risk of starvation by accumulating energy reserves which may be stored externally (as a hoard of food) or internally (as fat). The latter strategy may involve trade-offs such as a reduced locomotion capability (due to increased body mass) and may in turn increase the animal's 'mass-dependent predation risk' [11]. Another strategy to deal with limited food availability is to reduce energy consumption by entering into a dormant state such as hibernation.

Daily Energy Requirements

The physical size of an organism is a key driver of its energy requirements with larger organisms generally requiring a higher daily energy intake. Lifestyle also plays a key role in determining its energy requirements, with higher rates of locomotion and higher metabolic rates (*metabolism* describes the network of chemical reactions that enable organisms to generate energy and other molecules that they need to survive) leading to increased energy requirements.

Another factor which impacts on the energy needs of organisms is whether they are *ectothermic* or *endothermic*. Ectotherms rely on the environment to achieve their optimal body heating (examples include amphibians, reptiles, most fish species, and invertebrates) whereas endotherms maintain an elevated body temperature based on heat arising from cellular or mitochondrial metabolism. Endotherms have higher energy demands and therefore require greater energy intake than ectotherms of similar mass [593].

The vast majority of animal species are ectothermic and thus are very sensitive to changes in the ambient temperature [22]. The differing levels of energy requirements give rise to differing movement and foraging strategies between the two classes of organism, with ectotherms generally displaying less movement and less active foraging strategies than endotherms.

Energy Costs of Foraging

Alternative foraging strategies have differing costs. Broadly speaking, there are two primary modes of foraging. One is *sit-and-wait* (also referred to as ambush foraging), where the predator invests little time in searching for prey and typically only attacks a prey item once it moves well within sensory detection range. In essence, the predator sits and lets the environment 'flow' past it [47]. In contrast, *active foraging* arises when the predator expends energy in searching the environment for prey.

The payoff of each mode of foraging depends on the nature of the environment faced by a predator and the nature of the prey items, and is influenced by multiple interlinked physiological and behavioural traits.

The optimal sensory system and physical morphology for each generic strategy differ. A sit-and-wait predator might have an acute visual sense, be able to blend into the background (making it difficult for prey to detect) and be capable of short bursts of rapid movement to enable prey capture [593]. In contrast, active-search predators may use multiple senses, such as visual and chemotactic senses, for prey detection (chemotactic senses would facilitate the detection of nonmoving prey) and have a physiology which is capable of supporting significant movement around the environment [593]. However, the two modes of foraging are not necessarily mutually exclusive and some animals will switch between them based on environmental context such as prey abundance, hunger level and body condition [152].

A taxonomy of foraging costs is provided by *Griffiths' triangle* (Fig. 1.4) [228]. This positions predatory species according to the relative energy costs they incur in

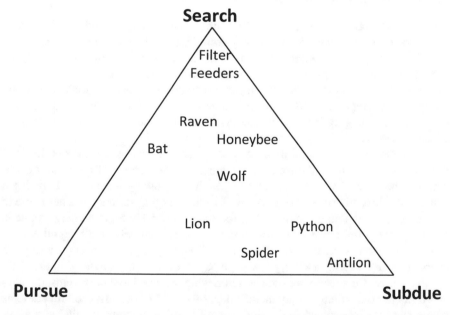

Fig. 1.4. Griffiths' triangle (adapted from [228]). Predators are classified in Griffiths' taxonomy based on their relative costs of foraging over three phases of predation: (i) search for prey, (ii) pursuit of prey and (iii) subduing and digestion of prey

each of the three phases of predation, namely:

i. search for and location of prey;
ii. pursuit and attack; and
iii. handling and digestion.

In Fig. 1.4, we illustrate in stylised form the relative positioning of a number of foraging animals. In the case of ambush foragers such as constricting snakes, for example pythons and boas, most of the energy investment in the foraging process is expended in subduing their prey. Lions tend to live in environments where they are surrounded by prey, therefore their search costs are low and the main energy expenditure during foraging concerns the pursuit and subduing of their prey. In contrast, wolves tend to patrol large territories and incur higher search costs than lions. Filter feeders, organisms such as clams, krill, sponges, and many fish which feed by straining food particles from water, expend little energy in pursuing or subduing prey. Their main energy expenditure during foraging occurs during the filtering (or search) phase.

This taxonomy reminds us that there are many ways to earn a living and not all foraging strategies entail an active search for potential prey items.

1.5 Payoffs of Foraging Strategies

Environments vary over time in terms of the relative density and capabilities of predators, prey and competitors that they contain, as well as in terms of their physical characteristics. As foraging environments are dynamic, the application of any specific foraging strategy will not produce the same outcome on each occasion it is implemented. A foraging strategy will generate a distribution of payoffs or 'returns' if implemented repetitively. The choice between individual foraging strategies, or indeed between foraging-inspired algorithms for optimisation purposes, is therefore a choice between probability distributions of outcomes [545].

The shape of the probability density function underlying this distribution will depend on the nature of the foraging strategy, the capabilities of the foraging animal and the state of the environment in which the strategy is employed. Taking a simplified, illustrative example, the payoffs to the foraging strategy of a herbivore in a specific location that consumes plant food with relatively limited energy value is likely to have a right-skewed distribution (i.e. many relatively small payoff events, where the payoff is measured as being the net energy payoff after considering the energy cost of foraging, and very few large payoff events), as illustrated in Fig. 1.5.

In contrast, for a carnivore such as a lion which chases large prey items, the payoffs of its hunting attempts may include negative returns (i.e. unsuccessful attempts where energy is consumed in the hunt but no positive outcome results) and a relatively small number of 'wins' with significant payoffs when a large prey item is successfully obtained (Fig. 1.6).

The first moment of the probability density function corresponds to the expected payoff of a foraging strategy per time period, with the higher moments of the distribution (variance, skewness and kurtosis) describing how payoffs vary around this

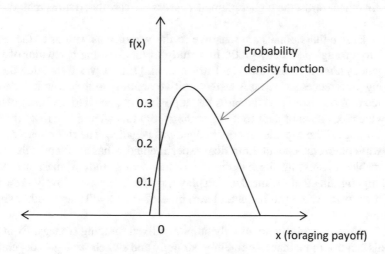

Fig. 1.5. Illustration of a probability density function for a herbivore in a location with plentiful food supplies where most foraging events produce a small positive net energy outcome

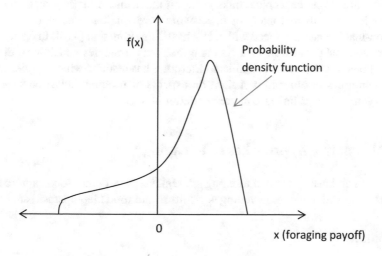

Fig. 1.6. Illustration of a probability density function for a carnivore showing a left skew as many hunting attempts produce a negative energy outcome

value. In other words, the higher moments assess the risk that returns may vary from expected.

It is known that animals are sensitive to the variance as well as to the expected payoff to a foraging strategy [560]. In a study of the foraging behaviour of yellow-eyed juncos *(Junco phaeonotus)* by Caraco et al. [100] it was found that the birds' foraging preferences for rewards were sensitive not only to the mean but also to the variance of food rewards. If the birds had a positive expected (daily) energy budget, they were risk averse in their foraging strategy, but this altered to a risky (high variance) strategy if the expected energy budget was negative. The risk preference of the birds was context dependent. Empirical experimentation has also shown that animals are capable of assessing skewness in outcomes, with a study of macaque monkeys [207] highlighting that the monkeys display a preference for positively skewed payoffs in settings where the mean and variance of the payoffs were otherwise held constant.

Of course, animals do not usually employ a fixed foraging strategy. Even simple animals can use more than one foraging strategy and switch strategies depending on their internal state and the state of the environment. An increasing level of hunger may trigger an animal to employ a 'riskier' foraging strategy which leaves it open to a higher chance of predation, with a predation event producing the ultimate negative return on foraging behaviour for the animal which is consumed!

There are strong conceptual links between the decision-making scenarios faced by foraging animals and many of those which are studied in economics where a decision-making agent is faced with a stochastic environment [474]. In each setting, the decision maker is faced with choosing between strategies which have differing distributions of outcomes. A significant literature has resulted which applies models from economics in order to cast light on aspects of foraging behaviours. We will discuss some of these linkages in greater detail in Sect. 2.1.

1.6 Alternative Approaches to Foraging

Apart from ambush and active foraging strategies, several other food capture strategies are observed in nature including agriculture and food kleptoparasitism.

Agriculture

An alternative to a hunter-gatherer existence is to engage in the domestication of animals and plants. This step in human evolution occurred about 12,000 years ago, initially in the *fertile crescent*, the region around south-eastern Turkey, Iraq and western Iran [256]. Only a limited number of species of plants and animals can be domesticated, and agriculture started in regions where these species were initially found. Wheat was domesticated by approximately 9000 BC, peas and lentils by about 8000 BC, olive trees by about 5000 BC, horses by about 4000 BC, and grapevines around 3500 BC. To this day, approximately 90% of the calories consumed by humans come from plants and animals domesticated by 3500 BC. Remarkably, no significant new plant or animal has been domesticated in the last 2000 years [239].

One of the benefits of the 'agricultural revolution' was the creation of a larger and more predictable food supply. This in turn led to a significant increase in the human population. It is estimated that there were some 5–8 million nomadic foragers in 10,000 BC. By the first century AD, global population had increased to approximately 250 million. The growth of food production, and in particular the creation of food surpluses, led in turn to the first large urban centres and the development of more sophisticated systems of rule, bureaucracy and division of labour. It also resulted in the first significant modifications of the physical environment by humans due to land clearing and irrigation projects.

Nonhuman Examples of Agriculture

Humans are not the only organism that has engaged in agriculture; nor, indeed, were they the first to do so. A number of species of ants, termites, beetles and worms display advanced agriculture that involves seeding, cultivation and harvesting of fungi, referred to as *fungiculture*.

These three insect lineages independently evolved the characteristic of cultivating fungi for nutrition about 40–60 million years ago [527], with the behaviour evolving once in each of ants and termites but at least seven times in beetles. In the case of the relevant species of ants, this represented an evolutionary transition from the life of a hunter-gatherer of arthropod prey, nectar and other plant juices to the life of a farmer subsisting on cultivated fungi [416], with the ants harvesting leaves and other biomass to feed *fungal gardens* in their underground nests. The gardens are also tended by the ants in order to keep them free of parasites and moulds. The fungus is typically used to feed ant larvae, with adult ants feeding on leaf sap [327]. Today, some 210 species of ants (of the tribe Attini) engage in various forms of fungiculture [547].

Termites of the tribe Macrotermitidae engage in similar behaviours, cultivating a fungus garden in their nests, feeding the fungus through the continuous provision of predigested plant material. The nests have a carefully designed architecture which provides an environment with a controlled temperature and humidity suitable for fungus growth. These termite societies (comprising some 330 species) domesticated African Termitomyces mushrooms (fungi) more than 30 million years ago and have become dependent on it as a source of food [1, 547].

An alternative strategy is adopted by the common ragworm species *Hediste diversicolor*, which engages in gardening behaviour to supplement its diet. During this behaviour, it creates a garden by burying cordgrass seeds, consuming the shoots once the seeds sprout [665].

A number of examples of more primitive instances of agriculture (more akin to husbandry) are displayed by some other organisms. The marine snails *Littoraria irrorata* have been found to farm fungi, providing biomass substrate to promote fungal growth whilst also weeding the resulting growth by consuming invasive fungi [527].

An obligate cultivation mutualism between damselfish and an alga in a coral reef ecosystem has been observed [241], whereby the damselfish manages the algal farm through territorial defence against invading grazers and by the weeding of unpalatable algae. The algae are harvested and consumed by the damselfish as a staple food.

As noted in [665], instances of gardening or farming by invertebrates are more likely to be an adaptive feeding strategy to obtain nourishment. There is no evidence that such animals are consciously or intentionally engaging in these behaviours.

Food Kleptoparasitism

Food kleptoparasitism is a foraging strategy in which an animal obtains food resources by stealing them from other animals [83]. Food theft is one of the most common forms of exploitation in nature and has been found in species groupings as diverse as marine invertebrates, insects, fish, reptiles, birds and mammals [211, 411]. The behaviour may vary in form from a large mammal such as a lion or a wolf pack driving away a smaller predator from a kill, to stealing from a store of food cached by another animal.

Kleptoparasitism is divided into two categories: *obligate*, where the kleptoparasite obtains all its food resources from theft; and *facultative*, where the animal obtains a portion of its food resources from theft. Some kleptoparasites only steal from their own species (intraspecific kleptoparasitism), and others steal from members of other species (interspecific kleptoparasitism).

One interesting example of the latter is provided by some ants of the species *Ectatomma ruidum*, termed *thief ants*, which specialise in stealing food items from the nests of neighbouring ant colonies of different species. In contrast to the normal foraging behaviour of *Ectatomma ruidum* ants, the thief caste (which live side by side in the same colony with nonthief ants) only steal food from colonies that are outside their colony's own home range. Although there are no morphological differences, nor any differences in diet, between normal and thief foragers, the thief caste displays a number of distinct behavioural characteristics including avoidance of conspecifics whilst en route to and from the nest of a neighbouring colony, an increased tendency to drop their stolen food items when grabbed, and faster movement whilst in the home range of the target colony than in the home range of their own colony [396]. It is unknown whether these behaviours are learned or innate.

Food theft can also occur amongst plants. One of the best-known examples of this is the dodder plant (or *Cuscuta pentagona*). The dodder is unusual as it has no leaves and lacks chlorophyll, which is necessary for photosynthesis. Therefore, the plant is incapable of generating its own food supply. Instead, this vine-like plant finds a host plant and attacks it by burrowing into its vascular system and draining off the nutrients it requires. The dodder plant identifies potential targets by detecting volatile chemicals released by its favoured host targets (particularly the tomato vine), in a manner akin to 'smelling' the host target [104]. When the dodder's behaviour is examined using time lapse photography, the dodder shoot tip is seen to move in a circular motion, initially randomly, and then growing and rotating in the direction of the nearest preferred host.

In summary, while most exemplars of foraging-inspired algorithms draw metaphorical inspiration from strategies which entail an active search for food items by individual organisms or groups of organisms, there are many alternative resource capture strategies in nature, including agriculture and food kleptoparasitism. Potentially, some of these strategies could also inspire the design of computational algorithms. In later chapters, a number of algorithms inspired by the foraging strategies of sit-and-wait foragers will be introduced.

1.7 Structure of Book

The remainder of this book is structured as follows. In Part I a series of perspectives on foraging which are relevant to algorithmic design and analysis are presented. Chapter 2 discusses formal models from the foraging literature and also some concepts from the growing field of movement ecology. A key input into the foraging activity of animals is information captured by their senses. Chapter 3 provides an introduction to themes from sensory ecology, highlighting the primary sensory modalities found in nature. Past foraging experience and socially transmitted information from other animals are also important in a foraging context. Chapter 4 provides an overview of some concepts from both individual and social learning. It also covers important concepts from the social foraging literature.

Part II consists of a single chapter (Chap. 5) which provides an introduction to the literature on foraging-inspired optimisation algorithms. In order to make this literature more accessible, a number of taxonomies of the various algorithms are presented. The chapter also outlines a metaframework within which most foraging-inspired algorithms can be conceptualised.

The remainder of the book is devoted to the exposition of a wide array of foraging-inspired algorithms which have been developed for optimisation purposes. These are classified into three categories, with algorithms derived from the foraging activities of vertebrates being discussed in Part III, algorithms inspired by the foraging activities of invertebrates being discussed in Part IV and algorithms derived from the activities of nonneuronal organisms being introduced in Part V.

In Part VI, we consider whether ideas from some of the formal models of foraging outlined in Chap. 2 can be used as a source of design inspiration. Exemplars of a number of such algorithms are discussed in Chap. 17. Part VII (Chap. 18) introduces the idea of using an evolutionary algorithm to create a foraging strategy or foraging-inspired algorithm. Finally, in Chap. 19 we make some concluding comments and describe open research opportunities.

Part I

Perspectives on Foraging

In the next three chapters we present a series of perspectives on foraging which are relevant to algorithmic design and analysis. Chapter 2 discusses formal models from the foraging literature and also some concepts from the growing field of movement ecology. A key input into the foraging activity of animals is information captured by their senses. Chapter 3 provides an introduction to themes from sensory ecology, highlighting the primary sensory modalities found in nature. Past foraging experience and socially transmitted information from other animals are also important in a foraging context. Chapter 4 provides an overview of some concepts from both individual and social learning, as well as covering important concepts from the social foraging literature.

Formal Models of Foraging

The study of foraging behaviour has generated a voluminous literature with multiple tributaries, including theoretical models of foraging behaviour, empirical studies of the foraging behaviours of specific species, and contributions from neuroscience which consider how information is captured via the senses and processed by the nervous system, producing behavioural outputs. Discussion of elements across each of these streams of research is required in order to underpin a deep understanding of foraging-inspired algorithms.

In this chapter we introduce a number of theoretical models of foraging, namely:

i. optimal foraging theory (OFT);
ii. the ideal free distribution (IFD);
iii. game theory; and
iv. predator–prey models.

Following a discussion of each of these models, we overview the field of *movement ecology*, which studies the movement behaviours of organisms, and conclude the chapter by introducing some concepts from the study of networks. Apart from providing insight into aspects of foraging behaviours, the topics introduced in this chapter have directly inspired the design of a number of optimisation algorithms, which are covered later in this book.

2.1 Optimal Foraging Theory

Many problems in economics naturally fall within a constrained optimisation framework. Taking the basic economic analysis of consumer demand, this models the problem facing the consumer as how best to allocate their limited budget across potential purchases of goods and services, where each item purchased generates a quantity of 'utility' (a notional construct which seeks to measure the absolute or relative satisfaction received from consuming a good or service), in order to maximise total utility. The decision variables are the amounts spent on each product or service, and the consumer is seeking to determine the best strategy (i.e. a vector of those spends over all

© Springer Nature Switzerland AG 2018
A. Brabazon, S. McGarraghy, *Foraging-Inspired Optimisation Algorithms*,
Natural Computing Series, https://doi.org/10.1007/978-3-319-59156-8_2

possible goods and services). This formulation produces a constrained optimisation problem, namely, to maximise total expected utility subject to a budget constraint.

Frequently, the outcome of a decision will be stochastic. In the case of finance, future returns on investments are uncertain. A common problem facing investors (who are typically risk-averse) is how to construct a portfolio of investments so as to maximise expected returns for a given level of market risk. One approach to this is provided by *modern portfolio theory* (MPT), where the problem is cast as being the maximisation of expected return for a given level of risk, defined in MPT as a variance.

Both human and nonhuman decision makers face similar challenges, including having to make decisions in stochastic environments and having to make decisions under constraints. Unsurprisingly, the theoretical literature concerning foraging has borrowed some conceptual frameworks and perspectives from economics [296, 474, 545]. One of the best-developed applications is *optimal foraging theory* (OFT) [168, 379].

OFT seeks to predict how solitary organisms will behave during foraging. The cornerstone of the theory is the assumption that foraging can be viewed as a constrained optimisation problem, where the aim is to maximise an objective function payoff such as net energy gained per unit of time, or, more comprehensively, a utility function which considers the risk associated with the expected payoffs. The optimal solution to this problem (i.e. the optimal value of the decision variables in the foraging strategy) is the organism's best foraging strategy. This strategy maximises the objective function payoff, subject to the various physical and environmental constraints which preclude the animal from obtaining an unlimited payoff. OFT also assumes that evolution will optimise the foraging strategies of organisms through preferential selection of organisms with better-quality strategies.

The theory argues that it should be possible, using OFT, to predict the actual behaviour that an optimal forager would exhibit under assumptions as to the information available to it and the constraints that the forager is operating within. Departures from predicted behaviour may highlight previously unknown physical or cognitive constraints on an animal, previously unrecognised features concerning the environment within which foraging is taking place, or that the choice of foraging currency (see the next subsection) is incorrect.

2.1.1 Operationalising OFT

Operationalising the OFT model requires the definition of three elements: the currency, constraints and decision variables (Fig. 2.1) [545].

Currency

In the context of OFT, the currency is defined as the item that is being optimised by the foraging animal. For example, individual items of food will produce different energy payoffs when consumed but there may be differing energy costs in acquiring each food item. Hence a focus on net energy gain (the energy obtained from food

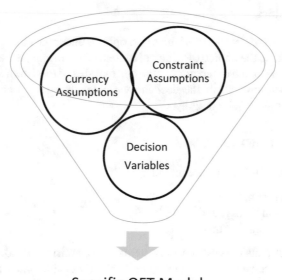

Specific OFT Model

Fig. 2.1. Operationalising an OFT model requires a choice of decision variables and assumptions as to currency and constraints

less the energy expended in finding, subduing and consuming it) is a more plausible foraging currency than the gross energy obtained from consumption of a food item. Food items may also vary in the time taken to acquire them, so another plausible currency unit is net energy gain per unit time.

The correct choice of currency will be context dependent. In the case of honeybees, carrying heavy loads when foraging shortens their lifespan, which in turn reduces the lifetime contribution of a foraging bee to the food stores in its colony. Experimental evidence from observations of real honeybees indicates that the currency being optimised in this setting is energy gained per unit of energy expended rather than net energy gain per unit of time [626].

Constraints

The constraints on foraging behaviours are the factors which limit an animal's feasible foraging choices and, consequently, the foraging payoffs which could arise. These constraints may result from the physiology of an animal, such as its ability to carry food, or limits on its speed of movement. Constraints may also arise owing to features of the environment, such as the terrain in which an animal is foraging or the availability of edible prey in the environment.

Decision Variables

Possible examples of decision variables (i.e. options under the control of the foraging organism) could be the choice of which prey items to consume, the amount of time a forager spends in a patch of territory or the optimal size of food item an animal should feed on. The listing of all of the decision variables describes a complete foraging strategy and could be quite lengthy.

2.1.2 Strands of OFT Literature

Key shoots of the OFT literature include:

- the optimal diet model;
- patch selection theory; and
- central place foraging theory.

The first model considers which prey items an organism should consume. Patch selection theory considers how an organism should decide when to leave an area (patch) in which it is currently foraging; a special case of this is central place foraging theory, where the forager must return to a particular place such as a nest or a den. In each case, the decision variables differ. For example, in deciding which prey to consume, the individual decision variables may correspond to the probability that a forager will pursue a specific prey type after encountering it. In the case of when to leave a patch, the decision variable may be patch residence time [545].

Optimal Diet Model

The *optimal diet model* [169, 379], also known as the *prey choice model*, analyses the behaviour of a forager that can encounter different types of prey and is then faced with the choice of which to consume. Each type of prey may have a differing cost of finding, capturing and digesting, and may provide a differing balance of nutrients and toxins. It is known that many animals can self-select diet nutrient composition in order to maximise their fitness [283]. The optimal diet model considers whether a predator should consume a prey item that it encounters during foraging (i.e. after it has been found) or whether it should ignore it and continue to search for a more profitable item.

Apart from foraging for foodstuffs based on their nutrient content, there are also examples of animals foraging for plant and other substances in order to self-medicate in the case of illness or in order to prevent illness. This behaviour is also known as *zoopharmacognosy*. One of the early authors to refer to this behaviour was Janzen [280] who commented that 'plant-eating vertebrates may do it on occasion as a way of writing their own prescriptions.'

In the thirty years since Janzen's paper, self-medicating behaviours have been identified in multiple species. Huffman and Seifu (1989) [270] noted that wild chimps chewed 'bitter' piths and swallowed selected leaves of otherwise low nutritional value in order to fight off intestinal worm infections and provide relief from related gastrointestinal upsets. Evidence of self-medication in red colobus monkeys who preferentially eat tree barks when infected with parasites has been uncovered in [208]. Medicating behaviours have also been identified in a number of social insects, such as the black ant *Formica fusca*, which changes its dietary preference when exposed to fungal pathogens by preferentially selecting a potentially harmful substance, a weak hydrogen peroxide solution, for consumption [70]. This chemical, which can be found in plant leaves, can help overcome fungal infections in ants but is deadly to them in higher doses. Healthy ants which had not been exposed to the fungus avoided this chemical.

In summary, diet choices are driven by several factors and these may transcend the apparent nutritional content of the items consumed.

Patch Selection Theory

Patch selection theory is concerned with the situation where prey are concentrated in small, separated patches in the environment. The theory addresses the question as to when a forager should leave a patch in which resources are being depleted owing to consumption if it is to maximise its long-term resource capture [260].

A key concept in this strand of literature is the *marginal value theorem* (MVT) [107]. As a forager spends more time within a specific patch, prey resources become depleted and the level of energy gain per unit of time decreases. Leaving the patch to find a new foraging territory incurs an energy cost. The theorem states that it is optimal to leave a patch when the marginal rate of gain in a patch falls to the overall average rate of gain for the environment [260]. Hence, foragers will stay longer in a patch if the patches are far apart and thus there is a higher cost of travel between patches, or if the current patch is rich in resources.

Central place foraging theory is a version of the patch model which seeks to describe the behaviour of a forager that must return to a specific place (a nest or a den) to consume food, to feed a mate or offspring, or to hoard food.

2.1.3 Critiques of OFT

While OFT provides a useful perspective from which to theorise about foraging behaviour, it is also recognised by most researchers that the basic OFT model is a simplification of foraging behaviour. Real-world objective functions and constraints can be complex and difficult to determine with accuracy.

It can also be argued that the assumption that natural selection will produce optimal foraging strategies (as defined by OFT) is questionable. As noted in [294], it is more plausible to think of behaviours as being selected and shaped based on their contribution to overall evolutionary success rather than being selected as being optimal for a single function. While foraging success is important in ensuring survival for reproduction, there are many other factors which contribute to this as well. For example, female lions seek to balance foraging for food, defending their territory and protecting their young. Genes which encode elements contributing to foraging may also encode other physical and behavioural traits. Maximising the contribution of one behaviour to survival may trade off against another factor.

Foraging usually involves trade-offs such as between predation risk and feeding. For example, foraging close to cover may reduce predation risk but will reduce the range of foraging options available. Similarly, spending more time watching for predators will reduce the time available for feeding. Given the potential consequence of being identified by a predator while foraging, it is plausible to assume that predation risks have a significant effect on the foraging strategies of organisms.

Another notable shortcoming of OFT is that the basic models consider only the behaviour of individuals and do not consider foraging as a *game* (discussed in Sect. 2.3) between predators and prey, or the consequences of animals foraging as a group.

2.2 Ideal Free Distribution

The *ideal free distribution* (IFD) model [190] seeks to explain the way that animals distribute themselves across patches of resources in the environment and argues that they will aggregate in patches in proportion to the amount of resources in each, as this distribution will minimise resource competition and maximise fitness for all individuals. If the resources in patch A are twice those in patch B, all other things being equal, the IFD would suggest that there should be twice as many foragers in patch A (Fig. 2.2).

The theory makes several assumptions: that animals are aware of each patch's quality; that they choose to forage in the patch with the highest quality; and that they are freely able to move unhindered and at no cost from one patch to another. All animals are assumed to be competitively equal (equally able to forage).

Under these idealised conditions, the outcome is a Nash equilibrium with all individuals ending up in patches in proportion to each patch's resources. It should be noted that the model assumes that each animal acts individually and that foraging location decisions are not group based.

While the assumptions underlying the theory are quite strong, there is empirical evidence that the foraging distributions of many animals and plants approximate the IFD. As with OFT, identification of empirical deviations from the IFD can help researchers to better understand the specific information-processing or environmental limitations faced by foragers. Amongst other insights arising from the IFD is that the outcomes of the foraging decisions of individual organisms are interdependent, with the payoffs to an individual depending on the foraging strategies of others.

2.3 Foraging as a Game

Game theory seeks to understand how rational individuals make decisions when the outcomes of these decisions are mutually interdependent [492]. Although initially developed in economics, the realisation that many decision scenarios in the natural

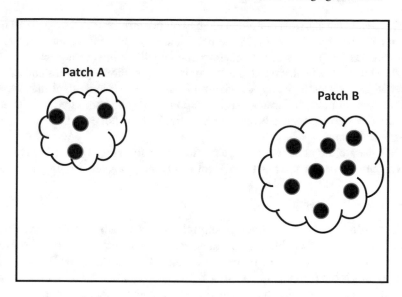

Fig. 2.2. Illustration of IFD, with patch A offering half the level of resources of patch B and consequently having four foragers vs eight foragers in patch B

world can be viewed through the lens of game theory has resulted in game theory being applied to gain understanding of the interactions of nonhuman organisms as well. A multiplicity of games exist in the natural world, including the interactions that occur in respect of mating, animal contests, communication and foraging. Foragers respond to their predators, and their prey respond to their presence. Animals may also forage competitively as part of a group. Players in a foraging game may be members of the same or different species.

In a game theory setting, the *players* are the individuals who make decisions as to what *strategy* they will play. A strategy is a complete description of how a player could play the game and can, in turn, contain multiple individual actions in response to the strategies pursued by other players. The *payoff* is the outcome that a player will receive at the end of the game. This depends on the actions of all players in the game. Players are said to be *rational* when they seek to maximise their payoff.

In some scenarios, decision makers make a single decision in isolation from each other (i.e. without knowing the strategy being used by the other player) and these scenarios are referred to as a *static game*. In other cases, the game is *dynamic*, as the players can observe the strategies being used by other players (i.e. the interaction between the players is dynamic). A dynamic game also results if a game is repeated multiple times, as players can observe the outcomes of earlier games before selecting their strategy for a subsequent interaction.

In natural settings, players learn based on experience gained in playing the game, giving rise to *evolutionary games*, where players adjust their strategy over time in response to experience. This leads to the concept of an *evolutionarily stable strategy*

(ESS) [394]. In the classical definition of an evolutionarily stable strategy, if all members of a population adopt the ESS, no mutant strategy can subsequently invade the population, as the payoffs of the mutant strategy will be lower. In population biology, the concept has been extended to study multigenerational genetics and assumes that strategies are biologically encoded and heritable. An individual may have no conscious control over its strategy, as its responses may be hardwired into its neural structure (i.e. the organism acts instinctively), and it need not even be aware that it is engaged in a game.

To illustrate the application of game theory to foraging, we outline two well-known game scenarios: the hawk–dove game and the producer–scrounger game.

2.3.1 Hawk–Dove Game

The hawk–dove game is a model of animal conflict which was first described in [394]. Two indistinguishable players must decide whether to share a resource or whether to fight for it. The resource has a value of v. In the game, hawks always chose to attack their opponent and it is assumed that when a hawk fights another hawk, it has equal (50–50) chances of (a) winning and gaining v or (b) losing and sustaining an injury with a cost of c, where $c > v$ (as is the usual situation in nature). Doves encountering a hawk flee and do not incur injury, but will attack another dove and in this case have a 50% chance of winning (again, without incurring injury). Players in the conflict must therefore chose whether to adopt a hawk strategy or a dove strategy. The general form of the game is therefore as in Table 2.1.

Table 2.1. Payoffs of Hawk–Dove strategies, with probabilities given in square brackets

	Hawk	Dove
Hawk	$(v [0.5], -c [0.5])$	$(v [1.0], 0 [0.0])$
Dove	$(0 [0.0], v [1.0])$	$(v [0.5], 0 [0.5])$

Thus, if a hawk meets a hawk, each gets an average payoff of $(v - c)/2$; if a hawk meets a dove, the hawk gets v and the dove gets 0; if a dove meets a dove, each gets an average payoff of $v/2$. However, the long-term payoff per encounter depends on the probability of meeting a hawk or a dove, that is, the relative proportion of hawks and doves in the population at the time of the encounter.

The ESS outcome is found to depend on the cost of fighting. If the cost of losing a fight is low, a hawk strategy can dominate. Conversely, if the cost of fighting is high, the hawk strategy is evolutionarily less stable, because, even though doves never win, they are also never injured. Therefore, the dove strategy can invade a population of hawks. It can be shown that when $c > v$, a Nash equilibrium arises comprising a mix of the two strategies with the proportion of hawks being v/c, and the system returns to this equilibrium point after a perturbation.

The basic hawk–dove game has been developed further to include alternative strategies such as backing off (and seeking food elsewhere) or *display, then retreat*:

bluffing aggression to intimidate the other, but backing off if the other does not back off [393]. Further strategies have been modelled (and are observed in nature), such as 'if in possession of the resource, be aggressive; if not, display, then retreat'.

2.3.2 Producer–Scrounger Game

The producer–scrounger game developed in [31] examines a common instance of *cheating* in the animal world. Cheating occurs when an individual does not cooperate with the group or only partially cooperates, but can potentially gain a benefit as a result of the cooperation of others in the group.

In foraging terms, this can occur when food finds are shared but individuals seek to scrounge on the finds of others. *Producers* spend time and energy finding food resources, while *scroungers* wait until a producer finds a resource whereupon they join in its consumption. In other words, the scrounger exploits a resource (or information about a resource) uncovered by another. Clearly, if the entire population plays a scrounger strategy, no exploration takes place, no resources are found and the entire population starves. If few individuals adopt a scrounging strategy, they can do well and even outperform producers as they do not incur the costs of finding food.

The producer–scrounger game is based on a number of assumptions. Individuals can play either a producer or a scrounger strategy, but cannot simultaneously play both strategies. It is also assumed that the payoffs received by a scrounger reduce markedly as the number of scroungers in the group increases. For specific implementations of the second assumption, the model predicts the mixed evolutionarily stable strategy of producers and scroungers that will emerge in a population.

A significant literature concerning variants of the basic model has been developed, including analysis of the case where producers gain a temporary *finder's advantage* on discovered resource patches before they are invaded by scroungers [589].

An alternative model of foraging-group-joining decisions to that of the producer–scrounger game is presented in the literature on *information sharing*. In an information-sharing group, all individuals search for food independently, and at the same time maintain watch on the behaviour of other members of the group. If another member of the group is observed to find food, all other members of the group stop searching and converge on the food resource [38].

2.4 Predator–Prey Models

Foraging takes place within the context of a food web (Sect. 2.6.2) where most individual organisms are simultaneously predators and prey. Predator species can also be in competition with one another if they consume the same prey. Consequently, the activities of an individual species can produce ripple effects within its ecology, such as the introduction of a new predator to a territory resulting in a significant reduction in the population of its prey. The interactions of predators and prey have been described as an *ecology of fear* by Brown and Kotler [88], as not only do predators impact on prey species by eating them, but they also impact on their well-being by making them afraid.

A consequence of being afraid is that a prey species will spend more time on vigilance as its risk of predation increases, leaving less time for its own foraging activities. Prey species may become less willing to enter risky locations such as thickets, or may only be willing to forage in groups or at certain times of the day, further reducing their foraging efficiency. Food chains can also be considered as *hierarchies of fear*. Counter intuitively, predators may have more impact on prey population size resulting from the fear they induce in prey (with consequent impacts as above) than they cause by their direct consumption of the prey species.

A large body of research has examined the population dynamics of species of animals when they have a predator–prey relationship. The simplest (if not most realistic) of these models assume that predators do not affect the behaviour of their prey but do affect their number due to consumption. This is termed an *N-driven predator–prey system* [88]. In contrast, in a *μ-driven predator–prey system* [88], predators impact on the population size of their prey not primarily by consumption but rather because of the behavioural effects they induce in the prey species, which in turn reduces the effectiveness of the prey's foraging (less effective foraging reduces the ability of the prey to reproduce).

Lotka–Volterra Predator–Prey Model

The oldest and best-known model of population dynamics in a predator–prey setting is the *Lotka–Volterra predator–prey model* [372, 597]. The model is based on a number of assumptions:

- the food supply for prey is unlimited;
- the prey population increases in proportion to its size (ignoring predation);
- the rate of predation is determined by the rate at which predators and prey meet;
- predators have unlimited appetite for prey.

The canonical predator–prey model consists of two nonlinear differential equations (2.1) and (2.2) which describe the dynamics of a predator–prey system, i.e. how the two populations change over time:

$$\frac{dx}{dt} = \alpha x - \beta x y \tag{2.1}$$

$$\frac{dy}{dt} = \delta x y - \gamma y. \tag{2.2}$$

Here, the term x is the number of prey items, y is the number of predators, dy/dt and dx/dt are the rates of change of the two populations over time, and α, β, δ and γ are parameters. The rate of growth of the prey species is governed by αx. The rate of predation is driven by the degree of interaction between the two species, $\beta x y$. The term $\delta x y$ is the rate of growth in the predator population and γy parameterises the loss rate of predators due to death (including starvation).

In this model, the populations of predators and prey follow a deterministic path based on the specific parameter values of the model. A common dynamic is for growth in the prey population, resulting in a growing number of predators with more food being available for them. As the predators consume greater numbers of prey, the prey population falls, leading in turn to a reduction in the population of predators due to starvation. This leads to an increase in prey items due to reduced predation and the cycle begins again.

Output from a discrete-time version of the continuous-time model outlined in (2.1) and (2.2) is illustrated in Figs. 2.3 and 2.4.

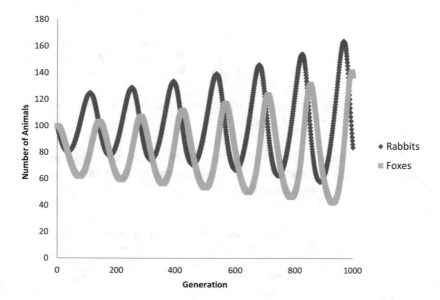

Fig. 2.3. Illustration of predator–prey dynamics in a population of rabbits and foxes. Note the cycles in both populations, with each responding to increases or decreases in the other, the fox population lagging the rabbit population

In constructing this example, we assumed an initial population of 100 rabbits and 100 foxes on an isolated island. The foxes prey on the rabbits, and the dynamics of the process is as follows:

$$R_{t+1} = R_t + GR_t - aF_tR_t, \tag{2.3}$$

$$F_{t+1} = F_t + aR_tF_t - mF_t. \tag{2.4}$$

Hence, in (2.3) the population of rabbits at each time step (R_t) is increased by a growth rate parameter (G) and reduced because of predation by foxes. Correspondingly, the population of foxes at each time step (F_t) grows depending on the population's foraging success and is reduced by a mortality rate (m). Table 2.2 defines

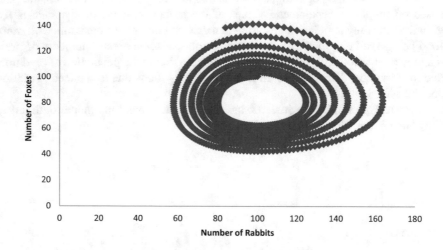

Fig. 2.4. Predator–prey dynamics, plotting each population in phase space

each of the terms in the discrete model. The values of these parameters are crucial in driving the population outcomes, and even small changes in the parameters can have a significant impact on population sizes.

Table 2.2. Terms in discrete predator–prey model

Term	Description
R_t	Rabbit population at time t
F_t	Fox population at time t
G	Growth rate of rabbit population (here, 0.4)
a	Growth rate of fox population (here, 0.0005)
m	Mortality rate of foxes due to starvation (here, 0.05)

A variety of predator–prey models have been developed under differing assumptions as to the environment and/or the interaction between the predator and prey species. A key takeaway from this literature is a reminder of the intricate nature of the ecological interactions between species, with the payoffs of the foraging strategy of an individual being determined by the strategies played by conspecifics, by competitors from other species and by prey. Predator–prey models also highlight the fact that predation risk plays an important role in the foraging decisions of animals.

2.5 Movement Ecology

Animals move in order to acquire necessary resources for survival and to avoid predators. Interest in the study of how animals move has led to the development of the multidisciplinary research field of *movement ecology*. This field is concerned with empirical and theoretical study into the movement of animals and also of other organisms such as plants and microorganisms. Areas of interest include movement phenomena surrounding foraging, seasonal migration and dispersal. Dispersal is defined here as undirected movement to new locations, and typically occurs in animals when juveniles leave the home range of their parent(s) to find a home of their own, or, in the case of plants, via seed dispersal.

In this book we restrict our attention to a subset of topics from movement ecology which are particularly relevant from a foraging perspective, specifically:

  i. random walk models of foraging movement;
 ii. free and home range behaviours; and
iii. navigation strategies.

We initially consider the modelling of foraging movements as a random walk and explore the implications of this. Animals exhibit differing levels of attachment to a territory, and this impacts on the nature of their movements when foraging. In some cases, they exhibit close attachment to an area and display a marked preference to remain within this area (home range behaviour), whereas other animals roam freely. Each of these tendencies has implications for the animal's foraging behaviour. Finally, we discuss the navigation strategies that animals employ. These impact on their ability to move efficiently from place to place when foraging, and in more extreme cases when engaging in long-distance migration.

2.5.1 Random Walk Models of Foraging Movement

In the simplest case, we can ignore cognition and sensory inputs, corresponding to a case where cognition and sensory capabilities are either nonexistent or too limited to effectively aid the search process. This allows focus to be placed on the question of how best to move when no information is available. Under these assumptions, foraging movement can be modelled as a random walk.

The best known random walk models assume Brownian motion, and it was long thought that this could be used to approximate the diffusion of biological organisms because of the central limit theorem whereby the distribution of the sum of independent identically distributed (i.i.d.) random variables with finite variance converges to a Gaussian [591]. However, an assumption of Brownian motion ignores aspects of real-world foraging, including the *directional persistence* typically exhibited by organisms. Animals rarely undertake 180° turns and revisit a just-sampled site.

2.5.2 Lévy Flight Foraging Hypothesis

A common characteristic of foraging is *area-restricted search behaviour*, whereby the behaviour of an animal changes on detection of a prey item or on detection of evidence of a prey item being in the vicinity [307]. At this point, foraging search is restricted to a local area (Fig. 2.5). When engaging in area-restricted search, the predator typically slows down and changes its direction of movement more frequently, in order to remain within the local area.

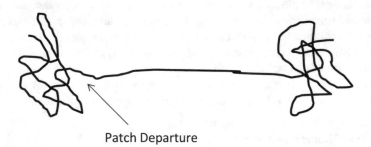

Patch Departure

Fig. 2.5. Illustration of area-restricted search behaviour, with an animal leaving one patch and recommencing area-restricted search having detected a prey item in a second location

The same behaviour can occur after consumption of a prey item, and this form of local search is sometimes termed *success-motivated search* [47]. In essence, the strategy intensifies the search for prey in the vicinity of previous prey encounters. A variant on this strategy is *near–far search* [519] where a forager searches near the location of the last food item found once the yield obtained is sufficiently high, but moves far away from that location otherwise.

Based on the above behaviours, the distribution of the movement of foraging animals might plausibly display *fat tails*, having a greater number of very short and very long jumps than would be expected under an alternative assumption that movement step sizes followed a Gaussian distribution. When this proposition was tested using empirical animal movement data, the results indicated that, particularly in cases where resources were sparsely and randomly distributed, the foraging movements of many organisms could be better described as a Lévy flight. A Lévy flight is a random walk in which the step lengths (jump sizes) l have a power-law distribution, $P(l) \sim l^{-\mu}$, where the power-law exponent is constrained to $1 < \mu \le 3$. An advantage of Lévy flights from a foraging perspective is that they avoid oversampling previously visited regions. The empirical discovery that the movement pattern of some animal species resembles a Lévy flight gave rise to the *Lévy flight foraging hypothesis* [590, 592].

However, it is recognised that neither Gaussian or Lévy distributions perfectly describe how organisms move, and there is no single universal law governing movement that applies across all species [483, 591]. It is also not clear in the case of species which appear to exhibit Lévy search behaviours whether the movement patterns are caused by innate behaviours of the foragers themselves or whether they are an emergent property caused by the prey field distribution, with predators curtailing their movement steps in response to prey encounters [242].

2.5.3 Free Range and Home Range Behaviour

Animals display differing degrees of *attachment* to a territory, which in turn impacts on the range of their movement. While some animals are free ranging, many display a tendency to remain within a *home range*. A home range is an area within which an animal or group of animals habitually lives and moves during the course of day-to-day activities [95]. The size of the home range depends on several factors including the degree of dispersal of resources required by an animal, the size of the animal (larger animals typically have larger home ranges), population density and sometimes the gender of the animal. Home ranges of animals are not necessarily exclusive and may overlap. Within the home range, some animals may be central place foragers, returning to a roost, nest or burrow after foraging. Home range behaviour results in attraction towards memorised localities and generates movement patterns in a bounded space with a bias towards previously visited locations.

Territoriality

Some species display *territoriality* and will defend all or a portion of their home range area against conspecifics or other animals. In essence, territoriality is a form of food hoarding and is often more strongly displayed towards conspecifics, as they present direct competition for the same resources.

The concept of *economic defendability* was introduced by Brown [86] to help explain when an animal might display territorial behaviour. The defence of a resource or territory has costs in terms of energy expenditure, time and risk of injury in performing defence duties as well as providing benefits in terms of priority access to the resources defended.

In economic terms, territorial behaviour should be favoured whenever the benefits of the behaviour are greater than its costs. If resources are of low density, the gains from excluding others may not be sufficient to pay for the cost of territorial defence. In this scenario, territorial behaviour is less likely and the animal may abandon territory and move elsewhere, assuming that resource patches are otherwise available. At the upper end of resource density (a very resource-rich territory), defence of a large territory is not economical as the number of invaders is likely to be high and the defender may not be able to make use of all resources in a large territory anyway [132]. It will not be efficient for an animal to defend a territory that is larger than what it needs to survive. Access to food is not the only item that could give rise to territorial behaviour, as other resources such as current or potential mates, offspring, nests or lairs, and display areas may also be defended.

2.5.4 Navigation

Another aspect influencing the movement patterns of animals is their *navigation strategies*. Navigation can be defined in a variety of ways, including 'directed movement towards a goal' [219], 'the process which enables a course or path from one place to another to be identified and maintained' [490] or 'the ability to organize behavior adaptively to move from one place to another' [619]. An example of directed movement would be the movement of central place foragers between their nest or den and a known food location [482]. Navigation strategies can be broadly considered as encompassing the methods that animals use for navigation.

Importance of Navigation

As noted by May-Britt Moser and Edvard Moser, cowinners of the 2013 Nobel Prize (with John O'Keefe) in Physiology or Medicine, the ability to figure out where we are, where we have been, and where we need to go is key to our survival [415].

Solving problems such as how to get from *here* to a known food source or from a food source back to a *den*, or knowing what locations to avoid in order to minimise the risk of predation is of obvious importance. There will be evolutionary pressure to select for more efficient navigation capabilities. Consider the case of a foraging animal returning to its den in order to feed its offspring. An inefficient navigation strategy when returning from a food source would result in longer travelling time, reducing both the quantity of food delivered to the offspring per unit of time and the time available for parental care, in turn impacting on offspring survival. Efficiency in this instance will encompass more than just route length, as factors such as the risk of predation along a specific route will also be relevant in route selection and navigation.

Common Navigation Strategies

The need to develop navigation capabilities appeared early in the evolution of life. At its simplest, navigation may be movement towards or away from a stimulus such as heat, light, sound or a chemical gradient. At a more sophisticated level, it can involve the formation of complex internal representations of the environment, and the determination by an animal of its position in that environment relative to the location of its desired destination(s) [619]. All but the simplest navigation strategies make use of memory. Three common forms of navigation strategy identified in [173] are illustrated in Fig. 2.6.

Under *beacon-based navigation*, animals memorise the perceptual signature of one or more *beacons*, defined as conspicuous objects closely associated with intended destinations. The animal navigates to desired destinations within its home range by moving from one beacon to another.

In *location-based navigation*, the animal maintains a memory of spatial relationships between a destination and surrounding nearby objects, called *landmarks*. Landmarks are environmental features which are typically located further from the destination than beacons. Beacons and landmarks are local cues as to location.

The use of landscape features to help orientate an animal in its environment gave rise to the question as to whether some animals can form *cognitive maps* of their environment. Tolman [573] was the first to propose that animals (including humans) can

Fig. 2.6. Illustration of three common navigation strategies: beacon-based, location-based and landscape-independent route-based

form a sophisticated mental representation, or cognitive map, of their environment and then use this for navigation (Fig. 2.7). These maps contain information outside the individual's current perceptual range concerning the location of various places of interest [478]. Although we now have a relatively good understanding of the neural circuits involved in the creation of mental maps in some species (particularly in rodents) [415], the question as to whether other species use cognitive maps can be hard to determine without detailed neuroscience experimentation, as when the issue is investigated observationally it is often difficult to conclusively rule out the possibility that an organism is actually using simpler navigation mechanisms [490, 619].

Under *route-based navigation*, an animal maintains a homing vector which allows it to move directly from its current position back to its starting point without having to retrace its entire outward path. The homing vector may be updated using global location cues such as *compass* information based on the position of celestial bodies (the sun or stars) or based on information gleaned from the earth's magnetic field.

An alternative method of route-based navigation is *path integration*, where an animal maintains a continually updated record of its location relative to a reference point (such as its den or a food source) using dead reckoning (Fig. 2.8). A crucial characteristic of this process is that the internal calculation of the animal's current position relative to its start point is carried out without reference to any external landmarks. Recent research in neuroscience has indicated that there are specific neural structures which monitor the direction and speed of travel, along with the angle of turns as an animal moves away from its starting point [415]. These are used to continually update the vector of travel required to return to the reference point [54].

Long-Distance Navigation

Navigation during long-distance migration poses a particular challenge and usually entails the use of multiple sensory modalities. To navigate over long distances, an

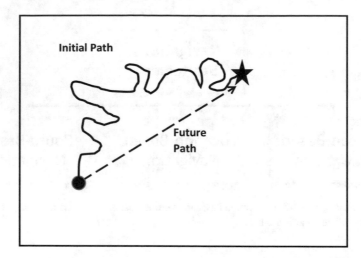

Fig. 2.7. Illustration of mental map. Initially, an indirect path was taken to the food source during its discovery. Thereafter, the animal takes a direct route, based on its mental map of the terrain

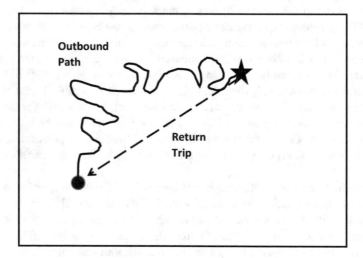

Fig. 2.8. Illustration of path integration, with an animal being able to return directly to its den having found food

animal needs to determine the position of its goal relative to its current position (i.e. the direction in which it needs to travel). The animal also needs a compass mechanism so that it can orientate to its goal.

The processes by which migrating animals can determine the appropriate direction in which they need to travel are poorly understood. In some cases, the animals may acquire a mental map via experiential learning on successive migratory trips with older conspecifics. However, there are plentiful examples of animals (e.g., homing pigeons) navigating to a desired destination when released from novel locations and in such cases it is difficult to see how a previously acquired map would be useful. It is speculated that nonvisual cues may be relevant in some cases. It is known, for example, that seabirds have an excellent sense of smell and it is thought that the emission of scents by phytoplankton may create an odour landscape or map that they can use for navigation and homing behaviours [219].

Compass information stems from a number of potential sources. Celestial cues such as sun position (or star positions at night) provide considerable information about direction. In the case of the sun, it moves across the sky from east to west during the day, and if an animal can learn the relationship between sun position and time of day, it has an accurate compass [219]. Honeybees were amongst the first organisms in which a time-compensated sun compass capability was found [601]. Some animals orientate with respect to polarised sun light (arising from the scattering of sunlight by air molecules in the atmosphere, which results in a strong band of polarisation at 90° to the sun's current position), which can be advantageous in overcast conditions when direct sight of the sun is more difficult.

In a similar fashion, knowledge of star field positions can allow nocturnal animals to determine the north or south pole point, in turn enabling them to orientate. There is evidence that some insect families have evolved neural arrangements which embed an imprinting of the sky pattern [118], implying that in these species, celestial navigation could be an evolved trait rather than arising from lifetime learning. In contrast, studies of some species of birds trained in a planetarium indicated that they could be trained to any 'north' or 'south' star (the one that moves least in the observable sky), suggesting in this case that their celestial navigation is a learnt response [118]. Use of a sun or star compass typically also requires that the animal or insect has an internal clock in order to interpret compass information correctly, as the position of the sun or the star field depends on latitude, time of day and season.

As will be discussed in Sect. 3.3.6, some animals appear to have the capability to sense magnetic fields, and this could allow them to use the earth's magnetic field to inform an internal magnetic compass, keeping them orientated during long-distance travel.

Organisms often employ a variety of navigation strategies rather than relying on any single strategy. A honeybee may initially use a sun compass on leaving its hive, and then use odour and visual cues as it gets close to its target flower patch [219]. In many cases there is redundancy in the navigation systems that animals use to orientate themselves, and this enables the cross-checking and recalibration of navigation systems against one another [490].

2.6 Networks

Essentially, *networks* (or *graphs*) model *connectivity*: the *connections* between things. The things are called the *vertices* or *nodes* of the graph and the connections are called *edges*, *arcs* or *links*. The connections capture *binary* relationships. Given vertices i and j, if there is an edge ij between these vertices, then we say that vertices i and j are *adjacent* and that the edge ij is *incident with* the vertices i and j. Of course, even nonadjacent vertices may be *(path-)connected*: if i and j are vertices then we say that there is a *path* between i and j if there is an ordered sequence of distinct vertices $i, a_1, \ldots, a_k, j$ such that each vertex is adjacent to the next in the sequence (that is, there is an edge connecting i and a_1, an edge connecting a_1 and a_2 and so on). If each edge has a *weight* (representing, for example, a length or cost), we can consider questions about least-weight paths (i.e. shortest or least-cost paths).

2.6.1 Applications of Network Models

The graph model is simple and general: it has uses ranging from analysis of road networks to social network analysis, data-clustering approaches and other applications. A multitude of problems in operational research and/or combinatorial (discrete) optimisation can be modelled as networks and thus solved. A network generally allows multiple paths between two given vertices i and j; thus, natural questions arising include:

- which path(s) are the shortest (the *shortest path problem* or SPP) by some measure such as total distance, total travel time, number of edges, or some other cost or objective function?
- how many paths are there between i and j?
- if one or more paths were removed from the network, would it still be possible to get from i to j (that is, is the network *resilient* or *fault tolerant*)?
- what is the shortest *tour* of the network that visits each vertex exactly once (this is the famous travelling salesman problem or TSP and is discussed in more detail in Sect. 9.5)?

Another standard application of network models is in the study of *flow* or *flux*: given one or more vertices (called *sources*) where flow (e.g. a fluid, or a good transported in a supply logistics network) enters the network and other vertices (called *sinks*) where flow exits the network, and a maximum throughput or *capacity* on each edge, typical questions include:

- what is the maximum flow that can be sent through the network?
- which edges are *saturated* (are carrying as much flow as they can)?
- how sensitive is the maximum flow to removal of edges?

2.6.2 Biological Networks and Algorithmic Design

Biological systems often require extensive networks to transport resources and information (along paths). These networks may be internal or external. Examples include the vasculature of animals and plants, ant trails, and the interconnected tubular networks of foraging slime moulds and fungi (Chap. 15). Transport networks are ubiquitous in both human-designed and biological systems. It is plausible that networks in both will share common features such as a short path length between source and destination nodes, and resilience to faults. Biological networks have been subjected to successive rounds of evolutionary selection balancing their cost, efficiency and resilience [567]. Resilience is likely to be a crucial factor in biological networks as even a partial failure could result in the death of an organism.

In this book, we encounter networks in three ways:

- a food web (network), which shows which species preys on which;
- a *food (transport) network* or *foraging network*, where the vertices are food items/stores and the edges are routes joining vertices; this is a foraging problem to be solved by an individual forager in real life; and
- as abstract problems (e.g. arising in combinatorial optimisation) for which we may derive solution algorithms inspired by natural foraging behaviour.

For the first of these, the pattern of foraging interactions between species can be viewed as a food web network, with nodes representing the species and links representing the interactions (who eats whom?), in other words, flows of biomass [652]. The properties of these networks can be examined, and it appears that they share some common features independent of locality, habitat and species composition. For example, the distribution of feeding links across species which characterises the relative abundance of specialist or generalist feeders is similar across food webs in that most species are specialised and a few are general [652]. It is an open question as to whether food webs have had common properties across time. Another feature of food webs is their degree of robustness to change, such as the extinction of a particular species or group of species. Obviously, the collapse of a food web is likely to be particularly detrimental to apex predators that are dependent on the existence of sizeable prey for their survival.

For the second type of network, as far back as [36], it has been observed that many animals move about their home territory or range along well-defined routes which they have cleared of obstacles, including vegetation. When these routes branch or interconnect, they end up forming a network that facilitates the flow of resources through the environment and the sharing of information among individuals [454]. Foraging networks, i.e. the networks of routes used to transport food, are particularly well documented in ants [92, 349, 371] and may be sufficiently conspicuous to be followed on the ground even without ants using them. For further discussion of foraging networks, particularly in the context of foraging slime moulds (whose interconnected tubular networks are actually internal as opposed to external networks), see Sect. 15.2.6.

Finally, several algorithms for addressing operational research and/or combinatorial optimisation problems, such as the SPP and TSP, have been inspired by natural foraging approaches: see Chap. 9, Sect. 13.3.2 and Sect. 15.4 for a discussion of some of these.

2.7 Summary

Foraging has been a canonical setting for the study of reward seeking and information processing across a range of fields, including biology, economics and computer science [308]. Unsurprisingly, a plethora of models which cast light on different aspects of foraging behaviour have been developed and applied across these and other domains. Although the models are typically simplifications of real-world settings, they can be useful in focusing attention on important theoretical issues and can lead to the generation of testable research hypotheses concerning foraging behaviour.

Optimal foraging theory points out that all foraging decisions are made under constraints and that the currency for a foraging decision is not always as simple as the net energy gain per unit of time. Crucially, this underscores the point that the foraging strategies of individual organisms are highly context specific, and therefore we should be wary about claims that an algorithm inspired by a specific strategy will necessarily perform well on a wide range of optimisation problems. The critique of OFT also points out that it is unduly simplistic to assume that elements of foraging strategies in nature such as the search stage during foraging are 'perfectly optimised', as selection processes are driven by the outcomes of a range of behaviours (including foraging) but do not specifically select for optimality in any individual behaviour.

IFD, game theory and predator–prey dynamics highlight that the payoff of an organism's foraging strategy depends on the ecology of strategies employed by other predators and prey within which it is employed. Players in the 'foraging game' will typically change their strategies adaptively. Commonly, organisms employ multiple strategies, varying depending on the circumstances they face, as uncritical use of any single strategy will leave an organism open to exploitation by others. Finally, the study of how animals move has given rise to the field of movement ecology and studies of actual movement patterns during foraging.

Apart from providing insight into aspects of foraging behaviours, the models in this chapter have directly inspired a number of optimisation algorithms and this motivates our discussion of these models. In Chap. 17 we introduce a number of algorithms which have been directly inspired by: optimal foraging theory (the optimal foraging algorithm, described in Sect. 17.1); the producer–scrounger game (the group search optimiser algorithm, described in Sect. 17.2); the predator–prey model (the predatory search algorithm, described in Sect. 17.3 and the predator–prey optimisation algorithm, described in Sect. 17.4); and animal migration, which falls within the movement ecology literature (the animal migration optimisation algorithm, described in Sect. 17.5). Patch selection theory, stemming from optimal foraging theory, provides inspiration for the foraging weed colony optimisation algorithm described in Sect. 16.5.2. As will be seen in Sect. 7.1.1, Sect. 16.5.4 and Sect. 16.5.6, concepts from the movement ecology literature have also influenced the design of several optimisation algorithms which embed a Lévy flight movement process. This list of algorithms which have been inspired by formal models of foraging is not exhaustive, and further scope exists to draw on these models for the purposes of algorithmic design.

3

Sensory Modalities

In the last chapter we introduced a number of theoretical models which can be used to gain insight into various aspects of foraging. A shortcoming of these models is that they ignore the sensory and cognitive processes of organisms. While this level of abstraction helps make the models more general across animal taxa, it is a significant limitation [524]. A comprehensive understanding of foraging activity needs to embed consideration of sensory ecology which is concerned with the ways organisms acquire, process and act on state information about themselves and the external world [3]. Processes of perception, selective attention, learning and memory are vital for these tasks.

Over the next two chapters, we examine selected aspects of sensory ecology. In this chapter we initially introduce a stylised model of the sensory and cognitive processes surrounding foraging in higher animals. A critical input to this model is information about the environment which is captured via sensory modalities. In this chapter, the concept of a perceptual world and a number of common sensory modalities are introduced. We also consider the 'cost' of sensory modalities as it is evident that different organisms 'invest' varying amounts in these capabilities leading to the obvious question as to why this is the case.

We follow this in Chap. 4 with a discussion of the role of learning, both individual and social, in a foraging setting.

3.1 An Internal Model of Foraging

Taking a high-level perspective, a stylised internal model of the foraging process is presented in Fig. 3.1. The key steps include the acquisition of information by an animal via its sensory modalities concerning the internal and external environment. These information flows are integrated and processed in order to form a *perception* of the environment. As the volume of sensory information being detected by an animal at any point in time is typically vast, it cannot all be processed. To overcome this problem, *selective attention* is paid to incoming information, with not all being equally weighted. Perception is also influenced by past experience (memory) and the

© Springer Nature Switzerland AG 2018

A. Brabazon, S. McGarraghy, *Foraging-Inspired Optimisation Algorithms*,
Natural Computing Series, https://doi.org/10.1007/978-3-319-59156-8_3

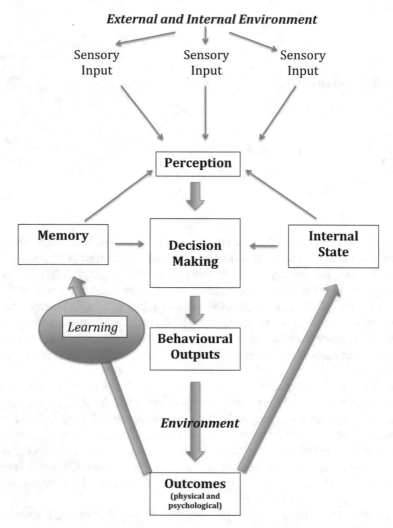

Fig. 3.1. Stylised model of sensory and cognitive processes concerning foraging. The model is simplified as not all information flows and feedback loops are illustrated

internal state, with expectations as to the external environment influencing what is perceived. As noted in [322], perception can be considered as a process of inference, in which bottom-up sensory inputs and top-down expectations are integrated.

Perception is the nexus between sensory inputs and cognition and plays a key role in the decision making of an animal [298]. In turn, these decisions produce outcomes which alter both the external environment and the internal state of the organism. Feedback with respect to decisions can also result in changes in memory via learning.

Taking the case of the internal state, one of the better-studied aspects of this in the context of foraging concerns the effect of starvation on both foraging and predation-avoidance behaviours. Most animals experience fluctuations in food availability over time and alter their behaviour in response, for example by moving to a new habitat or increasing the time they spend on foraging activities [309]. Starvation usually decreases the sensitivity of foraging animals to predation risk and, consequently, they may forage in places or at times that they would not otherwise. Animals become more likely to attack prey and may consume less favoured prey items when they are starving [507].

In summary, sensory inputs provide a critical input into an animal's perception of its environment. However, the same sensory inputs will not necessarily result in an identical perception of the environment over time or between different animals, as attention, past experience and internal state will vary between settings.

While we separate our discussion of sensory and learning capabilities over the next two chapters in order to provide a clearer discussion of each, in reality these capabilities, along with the physical characteristics of an organism such as its locomotion capability, interact heavily in order to determine an organism's foraging success and survival. The foraging success of an organism is determined by how well these capabilities are coadapted, in the context of the ecological niche which it inhabits.

As we will see in later chapters, most foraging-inspired optimisation algorithms can be captured in the general framework presented in Fig. 3.1. Typically, in these algorithms, individual agents collect information on their surrounding environment. This information is then used in conjunction with information they have previously gained in the search process (personal learning) and/or information transmitted by another agent (public information) in order to decide where next to move to in the search space. Deepening our understanding of real-world processes for sensory perception, learning and memory can assist in strengthening our capability to critique existing foraging-inspired optimisation algorithms and to design novel algorithms.

3.2 The Perceptual World

All life forms are capable of sensing and using information concerning their local environment in order to improve their foraging efficiency and effectiveness. In more complex organisms, such as higher-order animals, specialised tissue has evolved to create a central nervous system (brain and spinal cord) which can capture information from *sensory neurons*, process this information, and generate physical responses in muscles or glands via *motor neurons*. However, even organisms without any nervous system, such as plants and microscopic life, are also capable of adapting their foraging activity based on external and internal state information.

Sensory systems work within a larger framework, typically including, in higher animals, a nervous system and a brain (readers who are interested in a comprehensive description of this framework in humans are referred to the reference text of Kandel et al. [298]). All sensory systems require receptors which detect stimulus energy from the external environment. Sensory receptors in human beings can broadly be classified into three types, namely:

 i. photoreceptive cells, including the visual receptor cells found in the eyes;
 ii. mechanoreceptive cells, including the auditory and cutaneous receptors found in the ears and the skin (these cells can detect stimuli, including pressure and temperature); and
 iii. chemoreceptive cells, including the taste and olfactory sensory receptors found in the tongue and in the nose.

These receptors are very sensitive, with as little as a single photon of light or a single molecule of a chemical (odour) being sufficient to trigger a receptor neuron. External stimuli are converted into a stream of electrical signals (this process is known as *transduction*), which are in turn passed along afferent nerves ('afferent' meaning towards the central nervous system; in contrast, 'efferent' nerves transmit signals from the central nervous system to effectors in the body such as muscles and glands).

Most sensory pathways arriving in the brain (those devoted to olfaction being a notable exception) initially pass through the thalamus, located at the base of the brain on top of the brain stem, and from there are directed to distinct sensory processing areas in the cerebral cortex. These processing areas are interlinked and are also linked to areas of the brain which are associated with memory.

Sensory processing is a complex and incompletely understood phenomenon encompassing all the mechanisms by which the brain processes incoming information, assesses the significance of this information, and converts it into appropriate behavioural and motor responses. Aspects of this include the coherent integration of information from multiple sensory modalities, resulting in the creation of an internal perception in the brain of the external environment and the state of the body. In turn, the purpose of this complex information processing is to allow an organism to interact with its environment via motor system outputs.

3.3 Sensory Modes

Information about the environment is acquired via a range of sensory modalities, and these sensory systems sit at the interface between the environment and an organism's behaviour [431]. All organisms operate on the basis of a model or *representation* of the physical world [133, 431]. Every organism has a *unique* perceptual world, or *Umwelt* [580], based on the subset of environmental factors detected by it. This can be contrasted with the *Umgebung* [580] or the larger reality of the environment, which will usually transcend the organism's *Umwelt*.

The sensory world of an organism may be (and often is) completely alien to that of a human observer, with some animals being able to form 'images' of their environment using sound, magnetic fields, electrical fields or ultraviolet light. These animals can perceive and process environmental information that humans cannot. Often the human observer is unaware that this information is being processed and influencing the animal's behaviour.

Even within individual sensory modalities, the perceptual world of animals can vary markedly depending on the way that the modality is operationalised in a species. Sensory systems encode both static qualities of a stimulus (e.g. the shape of an object) and its kinetics (e.g. speed and direction) [562]. The limits with which stimulus kinetics can be resolved for each sense (i.e. its temporal resolution) can vary significantly between species.

Sensory Capabilities

The study of the sensory capabilities of organisms forms the subject matter of the field of *sensory biology*, a subset of sensory ecology. Six of the better-known sensory modalities are illustrated in Fig. 3.2, and these are discussed in the following sections.

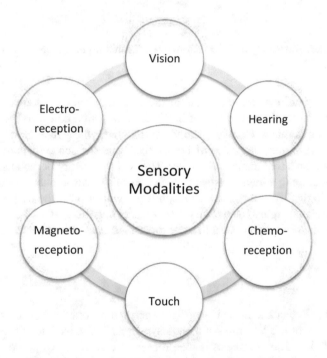

Fig. 3.2. Some of the better-known sensory modalities

There are also a number of other lesser-known sensory capabilities which make up the *somatosensory nervous system* (Fig. 3.3). These can be split into three main groupings. *Interoception* is the sensing of internal states from within the body such as hunger, stretching sensations from lungs, swallowing and so on. *Proprioceptor organs* are embedded in muscles, tendons, ligaments and joints. Their function is

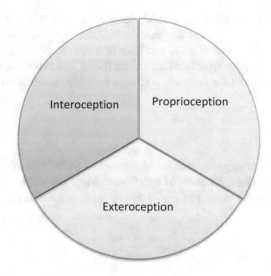

Fig. 3.3. Taxonomy of the somatosensory capabilities of higher-order animals

to detect position, tension and stress in the musculoskeletal system and allow the brain to coordinate the movement of limbs. *Exteroception* concerns the direct interaction of the external world with the body and includes the senses of touch, pressure, motion and vibration, the sense of hot or cold, and the sense of pain in response to external events that damage the body. The somatosensory nervous system along with the *autonomic nervous system* (which controls unconscious responses to regulate physiological functions such as respiration and those of the heart, bladder and endocrine-related organs) makes up the *peripheral nervous system*. Information from the somatosensory nervous system receptors is fed back to the central nervous system.

3.3.1 Vision

The ability to detect and make use of light energy developed very early in the evolutionary history of life on earth, with photosynthesis dating back to at least 2.8 billion years ago [146]. Photosynthesis is the process of conversion of photon energy into stored chemical bond energy in the form of ATP (adenosine triphosphate), which can then be used to fuel cellular metabolism [78]. The evolution of an ability to capture energy from sunlight was a major advance for biological organisms as it provided access to an energy resource orders of magnitude larger than that previously available from oxidation–reduction reactions associated with weathering and hydrothermal activity [146]. In addition to the energy gains from photosynthesis, the development of a sensing mechanism for light provided access to a range of adaptive capabilities, including an ability to measure time of day and season.

Starting from a basic capability to differentiate light from shade, it is thought that the emergence of visually guided predators, initially in aquatic environments, during the Cambrian epoch some 540 million years ago [118], led to a rapid arms race that forced prey to respond by evolving visual systems of their own. Eyes are believed to have evolved independently at least 40 times, indicating the significant adaptive impact of visual systems [346].

What we commonly refer to as vision is more correctly described as the detection of energy in a narrow band of electromagnetic radiation. This radiation travels as a wave, with the distance between one peak and the next being referred to as its wavelength. Electromagnetic radiation ranges in wavelength from gamma rays (short wavelengths below 10^{-9} cm) through ultraviolet (wavelengths circa 10^{-6} cm) to infrared (wavelengths circa 10^{-3} cm), microwaves and radio waves (Fig. 3.4).

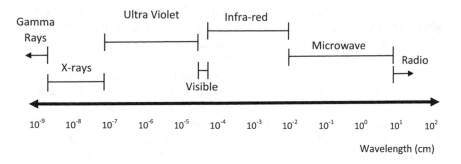

Fig. 3.4. Electromagnetic spectrum

The Visual System in Humans and Others

In the human eye, light initially passes through the cornea, which, in conjunction with the lens, focuses the incoming light onto the retina at the back of the eye. The retina is a photosensitive tissue containing photoreceptor cells called rods and cones. The retina in each eye has some 126 million sensory receptors, comprising roughly 120 million rods and 6 million cones. Rods can work at low light levels, and only register images in black and white, whereas cones require higher light levels and allow us to distinguish colour.

When light energy enters the outer segment of a rod or cone cell, it interacts with molecules which can absorb energy at specific wavelengths of light. This initiates a cascade of biochemical reactions which activate an adjoining bipolar neuron, which in turn synapses with ganglion neurons. These neurons comprise the optic nerve and send visual sensory information to the brain at a rate of about 10 MB per second [548]. The optic nerve from each retina is not a single nerve fibre but rather is a bundle of about 1 million individual nerve fibres (there are more sensory receptors than individual optic nerve fibres, with several receptors feeding into each nerve fibre). After leaving the eyes, the nerve fibres from each eye partly cross over at the optic chiasm; hence signals from the left visual field in each eye are processed by the right visual cortex and vice versa.

As with all senses, the key processing of information about the external environment takes place in the brain, involving multiple distinct regions. Much of the processing of features such as size, shape, colour and movement takes place in the visual cortex, located in the occipital lobe at the back of the skull. Not all information detected by the visual sensory receptors is processed, with the brain paying selective attention to the incoming information.

Although visual systems are common across animal taxa, the mechanisms for vision, and the ways that organisms process and interpret visual sensory information, vary considerably between species. One obvious reason for differences in visual systems is that species inhabit nonhomogenous light environments. Consider the light environment of the deep ocean relative to that of a tropical forest, or the differing light environments faced by diurnal and nocturnal animals. As would be expected, the nature of the visual mechanism evolved for an organism will be adapted to the challenges of the ecological niche which it inhabits.

The human eye can detect electromagnetic radiation with wavelengths of approximately 380–760 nanometres (a nanometre is one billionth of a metre), and this range of wavelengths is referred to as *visible light*. Other animals can see different parts of the electromagnetic spectrum. For example, some birds can see in the ultraviolet spectrum, and some insects and some reptiles such as snakes can see in the infrared spectrum [631]. In simpler creatures, a wide variety of visual mechanisms exist, including the compound eyes of many insects. These consist of thousands of individual photoreceptor units with the perceived image being a combination of inputs from all of these. These eyes can have a very large viewing angle and are very suited for the detection of movement. In addition to having a varying sensitivity to different wavelengths of light, some species, including 90% of invertebrates, are sensitive to the polarisation of light. This changes as the sun moves and therefore can be used as a navigation mechanism by means of a sun compass (Sect. 2.5.4).

The consequence of differences in mechanisms for vision is that organisms see the world in fundamentally different ways.

Uses of Visual Systems

Visual systems have multiple uses, including entrainment of internal body processes to a circadian rhythm, orientation, navigation through an environment, and identification of predators, prey and conspecifics. As noted in [118] some organisms have general-purpose visual systems and use them for all of these tasks, whereas other organisms have special-purpose visual systems which concentrate on performing one primary purpose. For example, the visual systems of jellyfish and worms are focused on detection of shadows (which could signal the approach of a predator), and their eyes and related nervous systems are coadapted for this task [118].

3.3.2 Hearing

Sound can be defined as the propagation of a perturbation in local pressure away from an initial location [78]. As sound propagates through a medium, physics and the nature of the medium play a critical role in the signal propagation characteristics. Sound travels more effectively underwater than in air (sound travels 4.5 times faster and with much less attenuation than in air [382]), particularly at low frequencies. Many aquatic animals, including many species of fish and amphibians, can hear quite well.

Hearing in Humans

Any form of hearing must detect sound wave energy. Typically, this energy is converted into a vibration of an organ that in turn stimulates mechanoreceptor sensory neurons. In the case of humans, sound waves excite the eardrum and the motion of this is detected by ciliary hairs in the cochlea, which move in response to the sound energy. This movement produces electrical signals in the auditory nerves. The frequency to which each hair cell responds depends on its position in the cochlea. Auditory signals pass through intermediate brain regions to the primary auditory cortex in the temporal lobe. Direct connections facilitate quick reaction to loud sounds, with other pathways facilitating processing of auditory signals en route, including the coordination of signals from each ear. Distinct structures in the brain process different aspects of sound energy, including Wernicke's area, which is essential for language comprehension.

Other Hearing Mechanisms

From an evolutionary perspective, cochlea-based hearing mechanisms are a relatively recent development and many species use other mechanisms to sense sound energy. The earliest vertebrates including fish did not have a cochlea but were capable of sensing sound energy using a set of organs containing *otoliths* (these are also known as *statoliths* in invertebrates) [572]. Otoliths are small crystals of calcium carbonate. When exposed to sound energy, the otoliths vibrate and displace sensory hairs which in turn trigger nerve signals via mechanoreceptors.

These forms of hearing mechanisms share a common evolutionary history with sensory mechanisms for the detection of orientation [572]. Statolith- and otolith-based sensory systems can be used to determine the orientation of an animal. If the crystals are contained in a sealed compartment containing sensory hairs, they will be pulled downwards in the compartment by gravity as the animal changes orientation, generating a signal which informs the animal as to which direction is 'down'. This mechanism is also found in plants and helps ensure that roots can determine which direction is down (so that the roots grow down into the ground) and shoots can determine which direction is up. Humans have otolith organs in the ear and these organs are part of the vestibular system, which provides sensory information about motion, equilibrium and spatial orientation.

In addition to gravity, statoliths and otoliths can also be affected by sound energy, so the same basic system that originally evolved to sense gravity and orientation can be adapted to sense sound.

Sound as an Active Sensory System

While humans usually think of hearing as being a passive sense, some animals are capable of emitting sound energy and using sound as an active sense. In *echolocation*, an animal emits a pulse of sound and uses the reflection (echo) of these pulses from its surroundings to help detect, localise and classify objects, including obstacles and prey.

The best-known echolocating animals are bats. Unlike daytime foragers, which can rely primarily on vision to identify prey, nocturnal hunters, including many species of bats, need either to develop highly acute visual systems which can function in a low-light environment, or alternatively to develop nonvisual senses to allow them to hunt effectively. It is speculated that echolocation in bats arose as a result of an evolutionary adaptation to hunt at night rather than compete for food during the day [178, 290]. During echolocation, the acoustic information contained in returning echoes is processed by the animal's brain to form images of its surroundings by comparing the outgoing pulse with the returning echo. An echolocating animal obtains a snapshot of environmental information from each sound pulse and, therefore, echolocation is a discrete-time sensory system. The computational complexity of processing echolocation information is underscored by the fact that it may involve both a moving source (bat) and moving target. There may be a complex echo pattern arising from the many differing objects within the detection range [382].

As echolocation is an expensive sensory modality because of the need to expend energy to emit sound, it is found in a relatively limited number of species. Apart from bats, echolocation capabilities have been discovered in a number of small nocturnal animals including rats [287] and the Vietnamese pygmy dormouse (*Typhlomys chapensis*) [448], and some marine species, including dolphins and whales [402]. A number of cave-dwelling birds use simplified versions of echolocation.

An interesting variant on the use of echolocation is provided by the aye-aye (*Daubentonia madagascariensis*), a lemur that is native to Madagascar. It is the world's largest nocturnal primate, weighing about 2–3 kg and being about 30–37 cm long. The aye-aye uses a foraging method known as *percussive foraging*, where it taps on trees using a highly specialised finger and listens to the echo in order to locate hollow chambers which may contain an insect. On hearing a suitable echo, it digs a hole in the bark of the tree to access the insect or grub [171].

Although humans do not echolocate, there is some evidence that both blind and sighted humans can learn to use echo information to gain information about their environment [323].

3.3.3 Chemoreception

Chemoreception refers to the ability to detect specific chemicals in the environment. This sense dates from the earliest days of the development of life, as even unicellular organisms need the ability to detect and selectively absorb the various chemicals needed for cellular metabolism. The ability to detect chemicals can therefore be considered as the 'first sense'.

In animals, chemoreception usually occurs via the olfactory reception of airborne chemicals but chemoreception can also occur through direct contact with a chemical. An analogue in human terms for each of these would be a sense of smell and a sense of taste. When a smell enters the human nose, it comes into the proximity of some 400 types of odour receptors, each of which is designed to attach to a specific set of molecules [96]. When a molecule attaches to a receptor, a chemical cascade is initiated, generating an electrical signal which is transmitted directly to the olfactory cortex in the brain. The incoming signal is then further processed in the piriform cortex. Smell and taste can initiate anticipatory physiological changes to facilitate the digestion of food and, in many species of animal, chemoreception can trigger innate behavioural responses.

As animals live in an odour landscape, individuals will be under strong selection to be able to localise odours from potential predators, mates and food. Mosquitoes provide a good example of the use of chemoreception to detect and home in on potential prey. While both male and female mosquitoes feed primarily on nectar and other plant products, females of many species of mosquito need to obtain nutrients and proteins from blood in order to produce eggs. Female mosquitoes hunt for prey by detecting the carbon dioxide (CO_2) contained in exhaled breath and a number of other attractant chemicals such as octenol ($C_8H_{16}O$), a chemical found in exhaled breath and sweat. The mosquito has some 27 odour receptors on its antennae to assist in sensing odours from potential prey. At close range, mosquitoes also use body heat and vision to sense the exact location of their prey.

Pheromones and Allelochemicals

Two special cases of chemical emission by organisms are commonly distinguished, namely, *pheromones* and *allelochemicals*. Pheromones are chemicals emitted and detected by organisms of the same species (i.e. they carry information between conspecifics) and are intended to impact on the behaviour of the receiving individual. Examples of uses include the marking of territory or a trail, attracting a mate, or as an alarm substance. Allelochemicals are chemicals emitted by an organism which are intended to influence the behaviour of members of another species. An example would be the emission of a chemical signal for defensive purposes.

Chemical Trail Marking

Chemicals are used extensively by organisms to mark trails to resources, with one of the best-known examples of this being the laying down of chemical trails by certain species of ants during foraging. The ants deposit trail-marking chemicals when returning to their ant colony with food. These trails provide a chemical roadway which can be subsequently followed by other foraging ants back to the food resource.

A similar process is used by Norway rats, which deposit scent trails from food finds when returning to their nests, which in turn bias subsequent foraging path choice by conspecifics [200]. Therefore, both ants and Norway rats can use socially acquired information from conspecifics to increase the efficiency with which they find food resources. Even simple organisms such as bacteria are capable of detecting and following a chemical gradient to find a food resource.

Advantages and Drawbacks of Chemical Signals

The advantages of chemoreception include the ability to receive sensory information day or night and the potential to detect information at a considerable distance (consider a wolf detecting a wind-borne scent several kilometres from its source). The drawbacks of chemical signals include the problem that the diffusion of odourant molecules follows an irregular path and hence there is weak directional information in airborne scents. Another drawback is that the speed of propagation of odours is usually slow. The irregular path of odourant chemicals also implies that unlike sound, which maintains any temporal pattern in its emission, this pattern can be lost in cases of olfactory signals [78].

Although olfactory signals do not contain unambiguous directional information, intensity gradient information can be obtained either by *tropotaxis* (simultaneous sampling by two or more paired receptors separated on an organism's surface with this information then being used to estimate the spatial gradient) or by *klinotaxis* (sequential sampling by a single receptor as it moves from one place to another, with a memory, therefore allowing the determination of a temporal gradient) [78]. Generally speaking, animals that use chemical information to orientate and find objects act relatively slowly, often travelling up chemical gradients that can have very low concentrations [546].

3.3.4 Touch

Nearly all organisms can detect direct contact between themselves and other objects [78]. Unlike other senses such as vision or hearing, there is no central sense organ that detects touch; rather, touch receptors are dispersed over the body. Although there are some specialised receptors in humans for touch such as Merkel's discs and Meissner's corpuscles which detect light touch, many of the receptors are simply free nerve endings (i.e. axon terminals). All these receptors connect back to the central nervous system via the somatosensory nervous system. Mechanical forces on the receptors open ion channels, which in turn generate electrical signals for onward transmission.

Hydrodynamic Sensing

A common adaptation of this sense is a *hydrodynamic sensory capability*, in which stimuli generated by displacements of a fluid medium such as water or air can be detected by an organism. All animals moving in a fluid medium create currents, eddies and vortices and the ability to sense these can be valuable in the context of detecting nearby predators or prey [78].

The best-known example of hydrodynamic sensing is provided by the mechano-sensory *lateral line* found in most species of fish and in some amphibians. The lateral line is a linear or curvilinear array of hydrodynamic sensors. The most conspicuous lateral line canal in fishes, the trunk canal, runs along the side of the body from near the gills to the tail fin. This provides a very sensitive detection mechanism for water movements in the surroundings of the fish.

The structure consists of a row of perforated scales opening into a tube lying lengthways under the skin. Sensory cells in the tube have a hair-like structure which responds to even slight pressure changes in the water in the tube, such as those caused by another fish swimming nearby in the water. The lateral line mechanism plays an important role in the schooling, predation and orientation behaviours of fish. Some mammals, including seals, walruses and sea lions, also have hydrodynamic sensors (whiskers) [78]. Many insects, including cockroaches and crickets, make use of similar mechanisms, such as small hairs, which are very sensitive to changes in air currents around them.

Speed of Response

An interesting aspect of mechanoreception mechanisms such as touch or hydrodynamic sensing is that these typically trigger sensory nerve responses which are 10–100 times faster than either photoreception or chemoreception, as touch stimuli produce receptor depolarisations directly, whereas vision and chemoreception require a chemical cascade to link the initial stimulus to eventual nerve excitation [78]. Touch or hydrodynamic sensing receptors are sometimes wired as *last chance evasion* sensors, protecting the organism from approaching predators. When these are triggered in flying insects by nearby air turbulence, indicating an imminent attack by a predator, the response is typically the immediate firing of motor neurons which results in a sudden change in flight direction. In many cases, this neural circuit completes at the spinal cord, removing the delay in sending the sensory information to the brain for processing.

3.3.5 Electroreception

Electroreception refers to a sensory capability to detect electrical fields. This ability has given rise to the evolution of an *electrosensory system*, whereby some organisms can sense electrical fields in their environment and can use this information to build up an 'image' of their surroundings. As all living organisms produce electrical fields, this image can be used for prey localisation and predator avoidance. Electroreception is an ancient sensory modality, having first evolved more than 500 million years ago, and has been lost and subsequently reevolved a number of times [115].

As air has a high resistance to electric fields, electroreception is more commonly found in aquatic environments, with multiple species of fish including lampreys, paddlefish, catfish and elasmobranchs (this branch of species includes sharks, skates and rays) having this capability [78]. Electroreception will be most useful in environments where organisms cannot easily depend on other sensory modalities. Vision is of limited use in dark conditions or in murky water. Chemical signalling is also challenging in aquatic environments, as water currents can obscure the source of signals.

A small number of mammals, including monotremes (mammals such as the platypus and four species of echidnas, or spiny anteaters, that lay eggs instead of giving birth to live young) and the Guiana dolphin [126], are capable of electroreception. So too are some species of birds, such as wading birds. Insects, including bees and cockroaches, are also sensitive to electrical fields but their perception of these fields is more typically based on mechanoreception. This often occurs via an induced movement resulting from exposure to an electrical field in body hairs or antennae, for example, in the Johnston's organs in the antennae of bees [225].

3.3.6 Magnetoreception

Multiple species of mammals, birds, fish and insects are sensitive to magnetic fields, as are many microorganisms. The widespread distribution of organisms that are magnetoreceptive suggests that the origin of this sensory capability is quite old [319].

The impact of magnetic fields on animals has attracted a good deal of study over the years. The results obtained are curious and not always easily explained. Some large mammals, including deer and cattle, display a preference for adopting a magnetic north–south resting position with their heads pointing north [45]. Why this is so is not known.

Magnetoreception and Bird Migration

Perhaps the best-studied cases of magnetoreception concern the ability of some species of birds to migrate over vast distances and the ability of homing pigeons to find their way home (Sect. 2.5.4). While it was suggested as early as the 19th century that birds could use the earth's geomagnetic field as a directional guide [603], it was not until the 1960s that the use of a magnetic compass by migrating birds was experimentally demonstrated [111]. Placing small magnets on birds was found to significantly interfere with their ability to navigate (this did not happen in control groups on which equivalent brass weights were placed), supporting a hypothesis that the birds were indeed using information from the earth's magnetic field for this purpose. The degree of inclination of the earth's magnetic field varies from 0° at the equator to ±90° at the magnetic poles and the intensity of the magnetic field at a particular location provides a second reference point, potentially allowing the combination of these items to assist migratory or homing species. In the case of homing pigeons, it is thought that the birds learn the strength and gradient (both direction and slope) of both the total and the vertical magnetic field at their home location (i.e. the magnetic coordinates of this location) during early local flights, and this subsequently allows them to return to their home even if released in an unfamiliar area [221]. Similar magnetic coordinate imprinting is thought to occur in other species, with, for example, loggerhead sea turtles being able to return to their natal beach to lay eggs many years after their birth [84].

Mechanisms for Magnetoreception

A practical problem concerning studies of magnetoreception is that while animals appear to be responding to magnetic fields, it is unclear what detection mechanism underpins the sense. There is no obvious sensory organ for magnetoreception; however, as magnetic fields can pass through the body, the relevant sensory organs could plausibly be internal and therefore less evident. Two detection mechanisms are discussed in the literature, namely biogenic magnetite (an iron mineral that is magnetic) and radical-pair biochemical reactions [320].

In the first case, it is proposed that crystals of magnetite which have their own permanent magnetic field interact with the earth's magnetic field, creating a pressure to align the two. This pressure creates small forces in receptor cells connected to the nervous system, and the resulting sensory signal is used by the organism for orientation [220]. Magnetite has been found in many organisms which are responsive to magnetic fields, including bacteria, honeybees, and various species of fish, amphibians, reptiles, birds and mammals.

Under the second mechanism, it is proposed that magnetic fields generate an effect on photochemically generated radical pairs in the eye [320]. In the case of birds, the state of a molecule called cryptochrome which is found in the retina can be influenced by magnetic fields, leading to a different chemical outcome when light hits the retina. It is speculated that birds may 'see' magnetic fields as an overlay of patches of light or colour on their vision [103]. However, the exact nature of the interaction of the two sensory systems (vision and magnetoreception) in this case is still poorly understood [118].

In summary, while there is evidence for the existence of a magnetoreception sense in some animals, research is still under way in order to conclusively identify the receptor mechanisms which underpin it.

3.3.7 Multisensory Capabilities

Because of the inherent limitations of any individual sensory modality, most organisms draw information from multiple modes. Taking the case of honeybees, a combination of visual, tactile, olfactory and electric stimuli (Sect. 10.3.1) is used when foraging.

Sometimes there will be overlap in the information being captured by various sensory modes, with similar information being received on multiple sensory channels. However, given that all sensory channels are subject to noise, confirmation of signals by more than one sense can be useful. Does a nearby rustle of leaves indicate a hidden predator or is it merely due to the wind?

There is a broad relationship between body size and the sensing modes available to an organism, with larger animals typically having a greater number of sensory modes. Figure 3.5, drawing on [388], illustrates this, outlining the general hierarchical relationship between body size and sensory modes for marine animals. Chemoreception is almost ubiquitous across living organisms. Mechanosensing is an important sense in marine environments, and even small organisms such as zooplankton

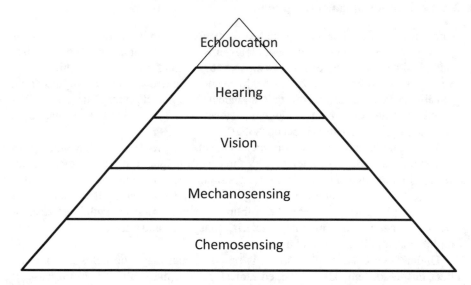

Fig. 3.5. Hierarchy of senses in marine animals with larger animals possessing additional senses

will typically possess this sensing capability. As marine animals get larger, vision and hearing sensory mechanisms become common, with echolocation being found in the largest, such as toothed whales. Other specialist sensory mechanisms also exist in some marine animals, such as electroreception and magnetoreception.

3.4 Cost of Sensory Capabilities

As the development of enhanced sensory capabilities allows the capture and processing of more information about internal state and the external environment, an obvious question is, why has evolution not driven all species to develop highly competent sensory mechanisms?

Sensory systems differ in the volume of data they collect and the speed with which this data is transmitted within an organism. The resulting data must then be processed in order to extract useful signals. Typically, olfactory sensors react slowly and transmit data at a low rate. In contrast, visual and hearing sensory systems are faster, require greater amounts of data processing, and are therefore more energetically expensive [548]. The energy 'cost' of each sensory capability varies and will also depend on the degree of acuity of each sense.

Enhancement of all capabilities, be they locomotive, sensory or cognitive, comes at a price in terms of the daily energy intake required in order to support these capabilities. In addition, as the structural complexity of an organism is increased, it will usually require a longer maturation time before adulthood, thereby requiring greater investment from its parents. Longer maturation times also delay the onset of reproductive maturity. Therefore an implicit cost–benefit test is applied by evolution to determine which adaptations can pay for themselves. Enhanced sensory and other capabilities are only useful if their benefit exceeds their cost given an organism's lifespan and the characteristics of the ecological niche which it inhabits.

Temporal Resolution of Sensory Information

Although it is widely recognised that organisms will have varying acuity in their senses due to physical differences in their design, the rate or *temporal resolution* at which information from the environment can be perceived and processed has received far less attention [249]. The level of temporal resolution of visual, auditory and olfactory systems can vary markedly between organisms [562].

The capability to capture and process environmental information at a high temporal resolution is likely to incur higher energy costs but may enable quicker identification of predators and prey in the environment, thereby providing a benefit. For example, predators of faster-moving prey may require higher temporal resolution than predators of slower-moving prey.

It is also suggested that there is a relationship between sensory temporal resolution and both metabolic rate and physical form (as measured by body mass). As an organism becomes larger, it finds it more difficult to respond quickly to physical stimuli. Therefore, high-temporal-resolution information from senses is of less benefit. Small organisms such as flying insects will typically have higher visual temporal resolution than large vertebrates [249].

The Expensive-Tissue Hypothesis

The case of neural tissue provides an interesting example of a cost–benefit trade-off in a foraging context. Brain tissue is comparatively expensive in terms of its energy requirements, with an adult human brain making up some 2–3% of body mass but consuming about 20% of an adult body's resting energy requirements (this consumption is up to 60% in the case of a child) [4, 209]. In contrast, the smaller brains of apes consume only some 8% of resting energy requirements [239]. The development of the human brain has come at a notable cost in terms of its impact on energy requirements, raising the question as to how this 'cost' has been 'paid for'.

The question as to why the human brain has increased substantially in size over evolutionary timescales has attracted a lot of study. Unsurprisingly, a number of the explanations have revolved around improvements in foraging capabilities arising from the enhanced cognition capabilities of larger brains. One hypothesis which seeks to explain the increase in brain size is the *expensive-tissue hypothesis* of Aiello and Wheeler [4]. Digestive system tissue is also expensive in terms of its energy requirements, and Aiello and Wheeler [4] hypothesised that an evolutionary trade-off occurred between the length of the human intestinal tract and brain size, with the former shrinking and the latter expanding over time. The hypothesis suggests that enhanced human cognition, facilitated in part by an expansion in brain tissue, gradually resulted in the development of more complex foraging strategies and the development of technologies such as cooking which allowed a switch to the consumption of higher-energy animal products in human diets. Cooking is a form of external digestion, as cooked food is easier to digest (modern humans spend less than an hour a day chewing food; in contrast, chimpanzees spend more than five hours a day foraging and eating [209]). By changing the chemistry of food, cooking turns foods such as wheat, potatoes or rice that humans cannot digest in their natural forms into useful food products. Cooking also changes the biology of raw food, as it can kill pathogens thereby making food safe to eat [239].

The expensive-tissue hypothesis suggests that the introduction over time of more animal products into human diets, and the development of cooking, facilitated the gradual substitution of brain tissue for digestive system tissue as higher-energy cooked food became available and therefore a shorter digestive system became viable.

Other hypotheses advanced to explain the increase in human brain size include the *social brain hypothesis*. Under this view, the need to create functionally cohesive social groups as a means of solving ecological problems, resulting in increasingly complex social settings (including the requirement to anticipate and manipulate the behaviour of other individuals), led to the evolution of a large human brain [85, 164, 165]. Whilst this hypothesis has attracted a considerable research literature, [136] notes that studies investigating the relationship between brain size and various measures of complexity of social setting have produced conflicting results, suggesting instead that brain evolution was primarily driven by selection on foraging efficiency.

A significant literature has developed concerning competing theories for primate (including human) brain evolution. As commented by Dunbar and Shultz (2017) [165] in a good review of this area, most studies on this topic are based on correlational evidence and therefore, it is difficult to distinguish between the merits of competing hypotheses. The debate concerning the drivers of the evolutionary increase in human brain size continues.

3.5 Summary

A key enabler for all foraging behaviours is the acquisition of environmental information. Organisms have evolved an incredibly diverse array of sensory capabilities and employ those capabilities in quite differing ways. Each sensory modality has its own strengths, weaknesses and costs, with (for example) visual signals being easily obstructed by obstacles and chemical signals suffering from relatively slow rates of information transmission [546].

Of course, sensory capabilities are only one component in determining an organism's foraging behaviours. Sensory mechanisms are deployed in conjunction with cognitive and locomotive capabilities in tackling foraging and other problems [650]. All these systems are coevolved and adapted to the ecological niche faced by the organism's predecessors.

In the design of foraging-inspired optimisation algorithms, agents have to have some means of perceiving their environment. Typically, in these algorithms, the implemented sensory mechanism is significantly simpler than real-world sensory capabilities, frequently being restricted to a single simulated sensory modality. In Chap. 5 we return to this issue and provide a taxonomy of the main families of foraging algorithms based on the primary sensory modality employed by the algorithms.

4

Individual and Social Learning

Foraging can be an individual or a group activity, with the payoffs being divided amongst the participants in the latter case. In each setting, learning plays a key role. In this chapter we introduce some aspects of learning which are relevant in a foraging context, in particular social learning, where an animal learns from others. Examples of social learning range from an animal being taught how to forage by a parent to the case where an animal gleans information from observing the actions of another animal and uses this to assist in its foraging activities.

Following a brief introduction to learning and predictive modelling, the chapter provides a discussion of social foraging and considers why cooperative foraging may emerge and persist in a population. A key enabler for social learning is communication, and this is also discussed.

4.1 Learning

Learning is a term that is difficult to define comprehensively (for good discussions of this issue, see [32, 138]) and, indeed, a variety of definitions are suggested in the multiplicity of literatures concerning various aspects of learning. These literatures include psychology, neuroscience, education, economics, anthropology and computer science, with each domain having its own focal point of interest and its own 'language' in describing learning.

Learning is multilayered and occurs over multiple timescales. In the broadest sense, it can encompass genetic adaptation as genes encoding novel capabilities in organisms spread through a population over successive generations, personal lifetime learning, and cultural learning such as occurs with humans, where ideals and beliefs are passed from one generation to the next. Another taxonomy of learning is to divide it into *asocial learning*, where an organism learns through its own experience via trial and error, and *social learning* where an organism learns through interactions with others.

Arbilly et al. [14, p. 573] define learning as 'a process of acquiring information, storing it in memory, and using it to modify future behaviours'. This definition can

© Springer Nature Switzerland AG 2018
A. Brabazon, S. McGarraghy, *Foraging-Inspired Optimisation Algorithms*,
Natural Computing Series, https://doi.org/10.1007/978-3-319-59156-8_4

encompass both asocial and social learning. The definition identifies two important aspects of learning, namely *memory* and the role of learning in enabling an organism to make better *predictions* about future events in its environment, leading to enhanced decision making.

4.1.1 Memory

Memory, like learning, has many definitions. In this chapter, we adopt the perspective of Squire [540] and consider memory to be 'the persistence of learning in a state that can be accessed at a later time'. Hence, the process of memory covers the encoding, storage and later retrieval of knowledge. Memory and learning can be considered as two sides of the same coin, as learning cannot occur without the existence of some structure to maintain the acquired knowledge.

A substantial literature in both psychology and neurobiology considers the processes of learning and memory formation in humans and animals. In neurobiology, the focus of research is on how memories are recorded and maintained in the physical structure of the brain.

The basic structural units of the brain are individual neurons. The connection structure among these neurons is *plastic*, being altered in the process of learning. The concept of plasticity was first suggested over a century ago by William James [277], and the *synaptic plasticity hypothesis* lies at the centre of most research on memory storage [498].

While there are many different types of neurons, the canonical model of information flow at a neuron (the *neuron doctrine*) is that the cell body of a neuron integrates electrical signals which enter the cell through dendritic nerve fibres (Fig. 4.1). A neuron typically has a dense web of input dendrites and these connect, via synapses, to axon terminals (nerve fibres) of other neurons at small structures known as dendritic spines. These spines can grow or shrink and are constantly extending out of and retracting back into each dendrite, changing the connection structure between neurons. If the total input signal into a neuron in a time period exceeds a threshold level, the cell *fires* and sends an output electrical signal along its axon. The firing of one neuron can result in a cascade firing effect in other neurons.

Another important feature of neuron behaviour is *spike time dependent plasticity*, which governs changes in the strength of synaptic connections between neurons. If the input from one neuron to another spikes, on average, just before the second neuron fires (spikes), then the synaptic connection between the two neurons is strengthened. On the other hand, if an input spike occurs immediately after an output spike from the second neuron, the connection is weakened. Therefore, strong connections are built over time between neurons that are working together to pass on signals, with other input signals being deemphasised.

As learning takes place, the physical network of connections between neurons adapts. Changes result from the growth or retraction of dendritic spines, and changes also take place at synaptic junctions, enhancing or reducing the ease with which electrical signals can cross the synaptic gap. Learning therefore alters the physical structure of the brain and this encodes our memory.

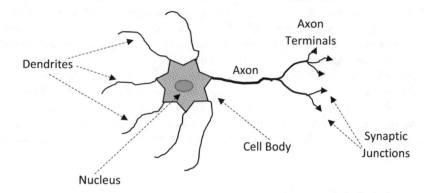

Fig. 4.1. Stylised representation of a neuron. Input signals from other nerve cells flow into the cell body along the dendritic nerve fibres. If the cell body fires, a signal is sent out along the axon. The axonic nerve fibres in turn connect to dendritic nerve fibres of other neurons via synaptic junctions. Neurons with differing functions will vary in the complexity of their dendritic trees, the length of the axon and the number of synaptic terminals. For example, axons can vary considerably in length, with some extending beyond one metre in the human body [298]

Although learning and higher order decision making involve complex neurological processes, not all actions of animals involve processing by the brain. An example of a basic neural circuit is a *reflex arc*, a neural pathway that controls an action reflex. A reflex arc can be termed *monosynaptic* when it consists of a sensory neuron linked to a motor neuron, i.e. there is a single synapse on the arc. In more complex cases (termed *polysynaptic reflex pathways*), one or more *interneurons* connect the afferent (sensory) and efferent (motor) signals. Even in higher animals, most sensory neurons synapse in the spinal cord rather than the brain, allowing reflex actions such as pulling one's hand away from a fire or 'last chance evasion' in the case of flying insects (Sect. 3.3.4) to occur quickly without requiring the intervention of the brain.

As we will discuss later in Chaps. 14, 15 and 16, the importance of learning and memory is not restricted to neuronal organisms (organisms with neural tissue) and considerable investigation of these processes in nonneuronal organisms such as bacteria, slime moulds and plants is now being undertaken by researchers.

Developmental Processes and Learning

The degree of plasticity of a neural system, and therefore its capacity for learning, is linked to the developmental process of that organism. Humans have a long developmental period after birth in which perceptual, cognitive and motor abilities are immature. The structure of neural circuitry during this developmental phase is very plastic and is strongly influenced by the sensory experience of the child, thereby helping to tailor the neural circuitry to the child's environment. One advantage of this delayed maturation process is that it allows the design of the neural system to be strongly driven both by genetically determined connectivity and by early childhood environmental factors. In other animals, the postbirth development phase is typically much shorter and the neural circuitry of the adult animal is predominately fixed by genetic factors at birth. Hence, much of their behaviour and problem-solving capability are *instinctual*, based on genetically stored learning. This does not necessarily preclude the ability of these animals to learn during their lifetime but the overall degree of neural plasticity is less than for humans. Consequently, we would expect such animals to perform less well if they faced an environment that differed from that of their ancestors.

4.1.2 Predictive Modelling

Past experience is particularly useful if it provides some insight into the future conditions which may be faced by an organism, i.e. if it enables a prediction of future events, or of the outcome of a specific action. For previous learning to be relevant, there must be some degree of commonality between past and future environments. Of course, a prediction may be incorrect, and we can consider either asocial or social learning within a Bayesian framework, where a current learning event is combined with a prior mental model of the environment in order to produce an updated predictive model.

The degree of cognitive complexity varies significantly between organisms, with, for example, humans being able to use memories of past events to guide decisions which may have a temporally distant payoff. In contrast, it has long been thought that other animals do not use memory in this way but, rather, exist in a constant stream of present needs unable to plan for the future [555]. While some recent studies challenge this perspective [293], it is likely that the 'predictive horizon' will vary among organisms of differing cognitive complexity. However, even simple creatures have demonstrated functional learning capabilities.

Associative and Nonassociative Learning

One of the best known forms of predictive modelling arises with *associative learning*, where an organism learns to associate stimuli or behaviours that have previously occurred concurrently with significant events. An organism may learn how sensory inputs (for example, the scent of a predator) are linked with important events such as the nearby presence of a predator [631] or learn that a particular action, such as the pressing of a lever in an experimental setting, is associated with a food reward (operant conditioning).

In contrast, *nonassociative learning* occurs when an organism changes its behaviour towards a stimulus in the absence of any associated output stimulus or event such as a reward or a punishment. The two most common forms of nonassociative learning are *habituation* and *sensitisation* which can be considered as mirror images of each other. Under habituation, an organism learns to ignore stimuli which have proven to be irrelevant in the past. Becoming habituated to irrelevant stimuli allows greater concentration to be placed on more important stimuli from the environment. The basic neural circuit underlying this behavioural response involves a sensory neuron which is connected to a postsynaptic motor neuron via a synapse, with the strength of the synapse decreasing over time in response to a repeated stimulus.

Sensitisation refers to an increase in the likelihood that a behaviour occurs in response to a repeated stimulus. As for habituation, a basic neural circuit can be described in which the strength of the synapse connecting the presynaptic and postsynaptic cells is increased over time in response to the stimulus.

Both habituation and sensitisation can substitute for associative learning, as they provide a method for responding to relevant environmental stimuli, without having to make 'predictions' of future states of the environment.

More complex learning arises when organisms become capable of generalising from past experiences to help deal with novel future environments.

Predictive Modelling and Algorithmic Design

The issue of predictive modelling also arises in the design of optimisation algorithms. The key mechanism in all optimisation algorithms concerns the choice of the next move by the searching agents, with this choice being determined by a strategy which typically blends feedback obtained by individual agents on the success of their recent moves and, in populational algorithms, feedback on the success of other members of the population. The strategy (or move rule) implicitly creates a prediction as to where a better location might be, based on this information.

4.2 Social Learning

Whether an animal engages in solitary or group foraging, its behaviour will usually be influenced by social learning. Social learning can be defined in many ways, with one definition from Hoppitt and Laland [268] (which is in turn derived from [72, 257]) being 'learning that is facilitated by observation of, or interaction with, another individual (or its products)'. Social learning is common in animals, as much of the behaviour of individuals is observable by others and is therefore public information. In the context of foraging, social learning could provide information on where to find food, what to eat and how to eat it.

A related term, originated by Galef [198], is *social transmission*, which refers to 'cases of social learning that result in increased homogeneity of behaviour of interacts that extends beyond the period of interaction' [199]. In essence, social transmission refers to the dissemination of information in a population, as a result of social learning, which results in an increasing similarity of a behaviour. Many foraging algorithms implicitly use social transmission as information on good regions in the search space is disseminated in the population, resulting in the movement of agents to that region, thereby intensifying search there.

Characterising Social Learning

Wynne and Udell [631] characterise social learning as follows:

- it must be a learned behaviour;
- it must be acquired through some form of social transmission, often by watching or interacting with another individual; and
- it must persist even in the absence of the demonstrator.

Wynne and Udell draw a distinction between social learning and a wider set of social influences. *Social facilitation* arises when individuals do something (for example, flee from their current location) just because they see others doing so. In this case it is not apparent that the individual learns something new from their behaviour. Although this is a socially influenced behaviour, it is not considered to be an example of social learning.

Social Learning Mechanisms

Social learning concerning foraging can occur in many forms (Fig. 4.2), including

Fig. 4.2. Some of the mechanisms for social learning

imitation of the actions of another animal. Imitation may result from a visual sensory input, with the observed behaviour being reproduced exactly by the observing animal. However, other sensory inputs can also lead to imitative behaviour such as the case where a bird mimics a song or a call.

A closely related form of social learning is *emulation*, where an animal observes another interacting with objects in its environment. The observer then attempts to bring about a similar effect on those objects, but does not necessarily use the same method. In *stimulus enhancement*, the observer learns the relationship between a stimulus and an outcome, but does not directly copy the behaviour of the experienced individual. *Local enhancement* occurs when a demonstrator attracts an observing animal's attention to a particular object or to a particular part of the environment [569]. An example of local enhancement would be the case where the attention of birds searching for food is drawn to a carcass as a result of other birds foraging there. There are multiple forms of social learning and not all entail direct imitation. Indeed, authors including De Waal [142] suggest that there is a drive towards *conformism* in many species which in turn underpins imitative-type behaviours. Under this thesis, social learning is not necessarily driven by a focus on rewards or incentives but rather by a desire for social connection with De Waal [141] referring to this process as *bonding and identification-based observational learning* (BIOL). In structured social groups, individuals are usually most likely to learn from social models such as knowledgeable, older, high ranking members of the same group [141].

Social learning need not result from an intentional process. An animal whose behaviour is being observed will often be sharing information unintentionally. Contrast the case of a lioness teaching her cubs to hunt and handle prey (intentional information sharing) with the unintentional information sharing that occurs when one animal observes another hiding food in a cache.

Teaching

Perhaps the most sophisticated form of social learning is directed instruction, where one animal explicitly teaches another a useful skill or useful knowledge.

Teaching is defined by Caro and Hauser [101] as follows: 'an individual (the tutor) modifies its behaviour in the presence of a naïve observer (the pupil), at some cost (or at least with no immediate payoff) to itself ... as a result, the pupil acquires knowledge or learns a skill earlier in life or more rapidly or efficiently than it might otherwise do so, or would not learn at all'.

For a long time it was assumed that teaching occurred only in humans and perhaps in some other primates. It is now known that teaching behaviours are much more widespread, and the existence of teaching across a range of organisms, with varying degrees of cognitive ability, challenges the long-standing assumption that teaching requires complex cognitive abilities [267].

One of the first examples of nonhuman teaching under the above definition was found in ants of the species *Temnothorax albipennis* [187, 189]. These ants engage in tandem running, whereby an ant that has found a food source leads a naïve ant from the nest to the resource in order to teach it the route, with signals between the two ants controlling the speed and course of the run. The behaviour embeds bidirectional feedback between the teacher and pupil, as the run only continues when the pupil periodically taps on the legs or body of the teaching ant (indicating that the pupil is following). Tandem running imposes a cost on the teaching ant as it could otherwise proceed faster to the food resource. As knowledge of the location of the resource is transferred from one ant to the next, pupil ants become teachers in turn.

There is good evidence that teaching has also arisen in other species, including meerkats and pied babblers [267]. It is expected that other examples of teaching will be uncovered as the range of observational studies investigating these behaviours increases [568].

4.2.1 Is Social Learning Always Useful?

Social learning is potentially a low-cost, low-risk way of acquiring valuable information as the learner does not incur the costs and risks of personal trial and error learning. Consequently, it is sometimes thought that the use of social learning will invariably lead to more effective foraging outcomes at an individual or group level. This is not always the case [24].

For individuals, learning from others comes with its own risks as the acquired information may be outdated, misleading or inappropriate. Another practical problem that can emerge is that there may be a conflict between information gathered from social learning and information obtained through their own experience [629], leaving the animal with a decision as to which information to use.

At the group level, the producer–scrounger dilemma (see Sect. 2.3.2) illustrates another limitation of social learning. In the context of foraging, producers can be considered to be asocial learners who uncover (produce) new information. In contrast, pure social learners are scroungers as they only exploit information discovered by others. Game theory models of producer–scrounger interactions highlight that scroungers only outperform producers when they are a small portion of the total population.

Rogers' Paradox

The question of how best to learn in a changing environment was examined by Rogers [491] within a producer–scrounger framework. The key finding was that social learning does not in itself increase mean populational fitness, as its utility is frequency dependent. While copying others (i.e. social learning) is advantageous at low frequency because social learners acquire their information (and avoid the costs of asocial learning) primarily from asocial learners who have directly sampled the environment, copying becomes disadvantageous as it increases in frequency, as indiscriminate social learners find themselves copying other copiers. The problem is intensified in a dynamic environment, as information becomes outdated owing to environmental change.

This gives rise to *Rogers' paradox*, wherein at equilibrium both pure social learners and pure asocial learners persist with the same average fitness, i.e. at equilibrium, the payoffs of each strategy (asocial and social learning) are equal. Therefore, it is not correct to claim that social learning *always* enhances the performance of a population.

A more nuanced learning strategy for an individual is to selectively use *both* asocial and social learning, directly sampling the environment some of the time and combining the information gained from this with information gained from social learning [344]. The range of potential foraging strategies which combine both social and asocial learning is infinite.

The above discussion has clear implications for the design of optimisation algorithms, as it points out that a suitable balance between exploration of the environment by individual agents needs to be maintained with exploitation via social learning of information concerning the environment which has been gained by other agents. The relative balance between exploration and exploitation is also critically dependent on the rate of change of the landscape, with exploitation being less useful as information becomes outdated owing to environmental change.

4.2.2 Social Learning Strategies

Although the use of social learning is widespread, being observed in organisms as diverse as primates, birds, fruit flies and crickets [486], appreciating when and how individuals learn from others is a significant challenge.

Understanding how to take advantage of social information, while managing the risks associated with its use, has become a focus for research on social learning strategies. A *social learning strategy* can be defined as 'a decision rule as to how to employ social learning in different contexts' [268]. Each of these strategies needs to specify [268, 344, 486]:

  i. when an individual should copy;
 ii. from whom they should learn; and
iii. what information should be copied.

Organisms will typically not employ a single invariant learning strategy but will choose from a repertoire of learning strategies. Sample choices which can be chained together to produce a learning strategy are illustrated in Fig. 4.3. Evidence from empirical studies suggests that even simple animals are capable of flexibly employing multiple strategies as environmental conditions change [486].

The question of *when to copy* covers the decision as to when to seek social information. A simple strategy is to copy another animal if dissatisfied with one's current payoff. Under this strategy, an animal need not assess the relative costs and benefits of alternatives but will continue with its current strategy until the payoff drops below a threshold level, at which point a copying behaviour is motivated. This approach has been uncovered in honeybees [232] and Norway rats [200, 201], amongst other organisms. Imitation may also be triggered in cases where an animal is uncertain as to what strategy to employ (i.e. their private knowledge is limited or imperfect) or where asocial learning would be costly or dangerous.

Whom to copy may depend on factors such as the social structure of the population and the ability of the individual to recognise whether other individuals are

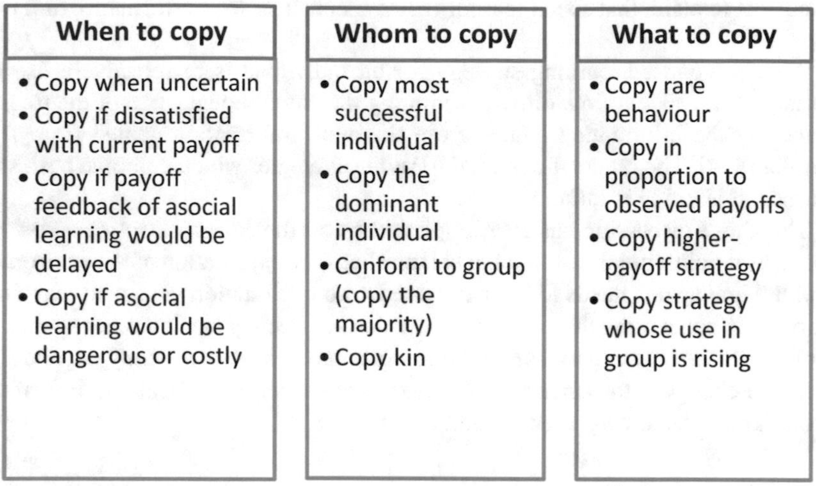

When to copy	Whom to copy	What to copy
• Copy when uncertain • Copy if dissatisfied with current payoff • Copy if payoff feedback of asocial learning would be delayed • Copy if asocial learning would be dangerous or costly	• Copy most successful individual • Copy the dominant individual • Conform to group (copy the majority) • Copy kin	• Copy rare behaviour • Copy in proportion to observed payoffs • Copy higher-payoff strategy • Copy strategy whose use in group is rising

Fig. 4.3. Some elements of a social learning strategy

obtaining higher payoffs. Possibilities include the copying of the most successful individual, copying of kin or adherence to a social norm by copying the majority.

What to copy considers which strategies to copy. Choices could include the copying of high-performing strategies or the copying of novel strategies.

4.2.3 Social Learning Strategies and Optimisation Algorithms

Rendell et al. [486] note that computer scientists are now starting to use concepts of strategic social learning, and its interaction with individual learning, to develop optimisation algorithms. In a foraging context most animals employ more than one learning strategy. They can alter whom they choose to copy and the relative weight they place on social vs asocial learning, depending on their internal state and the context of the current environment [268]. Notably, many foraging-inspired algorithms omit this capability and instead embed a single, fixed, social learning strategy. Any algorithm which uses a fixed search strategy will have limitations particularly when attempting to optimise in dynamic environments.

What Does Social Learning Mean in Foraging Inspired Optimisation Algorithms?

While the embedding of social learning mechanisms in algorithmic designs is an inherently attractive approach in order to encourage the dissemination of useful information across a population, it is important to note that there are critical distinctions between the meaning of social learning in the multiple literatures which touch on aspects of this topic and the way that social learning mechanisms are typically implemented in foraging-inspired optimisation algorithms.

Although the term 'social learning' appears quite intuitive, it can mean varying things to researchers, and the term is rarely defined when it is used in the foraging-inspired optimisation literature, or indeed in the metaheuristic literature more generally. In metaheuristics practice, the term social learning is most commonly used in a very broad sense to include any information transfer from one searching agent to another. This can encompass both intentional and/or unintentional transmission of information among agents, as well as information transmission mediated by signals left in the environment (e.g. chemical signals altering the actions of conspecifics or other organisms). We follow this convention of using the term *social learning* in a very broad sense when discussing the various algorithms later in this book, but note that this does not accord with its more tightly prescribed definitions in the literature on learning.

Another feature of the use of the term social learning in the foraging-inspired optimisation literature is that it is often technically incorrectly applied to diversity-generating operators which operate on solution representations. The issue here is that we can only meaningfully speak of social learning at a phenotypic level, i.e. as occurring between organisms at a behavioural level. Again, we will follow common usage when discussing algorithms later in this book, and consider learning to be social if it involves any information transfer from one searching agent (or representation thereof) to another during the algorithm.

The usual justification for this approach is that a utilitarian perspective is being adopted in which inspiration may be drawn from concepts in the social learning literature for the design of optimisation algorithms, even if the resulting algorithms apply those ideas in a biologically implausible way.

4.3 Optimal Level of Learning

Although it appears reasonable at first glance to assume that more sophisticated learning systems would enhance survival, it is clearly the case in nature that the learning capabilities of organisms vary widely in their complexity. As discussed in Sect. 3.4, sensory systems and cognition (including learning and memory) are not cost free (Fig. 4.4).

Cost of Learning

In the case of learning capabilities, costs include the energy required to encode, maintain and retrieve memories in neural or other forms [173], as well as the energy required to develop and maintain the underlying biological infrastructure to support learning capabilities. Several studies have examined the energy costs of learning [399], with [282] finding that in the case of individual honeybees, this cost is significant, and that honeybees suffer learning and retention deficits when energetically

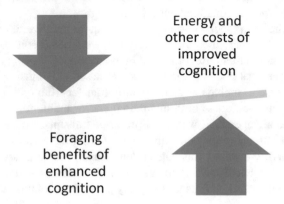

Fig. 4.4. Tradeoff between foraging benefits of enhanced cognitive capabilities and the increased energy cost of those abilities

stressed. Additional costs of learning include the attention diverted from other cognitive tasks such as scanning for predators while exploiting existing knowledge, and the cost of learning errors which arise when an incorrect association is made between a stimulus and a subsequent event [14, 163].

Sleep plays a key role in the process of memory consolidation, with experiences being replayed in the hippocampus during sleep. It is now known that memory reactivation during sleep not only is found in humans but also occurs in other organisms, including rodents and some invertebrates such as the honeybee [667]. A requirement for sleep in order to facilitate the consolidation of long-term memories in the brain [74] imposes costs including lost foraging time and, potentially, an increased risk of predation whilst sleeping.

The Tradeoff

As the complexity of learning and memory structures in an organism increases, the associated costs will also rise. From an evolutionary point of view, enhanced learning capabilities will only evolve in a population when the costs of these capabilities are outweighed by the benefits that they can provide when employed in a specific ecological niche. As the demands posed by the environment vary between species, the optimal level of learning for individual organisms will also differ, and more complex learning capabilities will not always be adaptive. An organism using simple learning mechanisms can be just as successful as, or even more successful than, a more cognitively complex organism in a given environment.

This cautions us against a general claim that foraging strategies, or algorithms based on these, which embed sophisticated personal and social learning mechanisms (and which will entail energy costs, or computational overhead in the case of algorithms) are invariably better than simpler strategies. In both biological organisms and computational algorithms, a key concern is how best to allocate energy or computational resources in order to maximise fitness or performance.

4.4 Social Foraging

Both solitary and social foraging behaviours are widespread in nature and this raises the question as to what circumstances are likely to give rise to each behaviour, or, perhaps more concisely, why social foraging may arise at all.

A common phenomenon in social animals is *food calling*, where an animal emits a vocal (or other) signal to alert conspecifics to the discovery of a food item [295]. While the benefit to the recipient of this information is evident, it is much less obvious why an animal would wish to transmit this information. The sender of a food call incurs a cost in that the food resource will be shared if others respond to the call, and also incurs a risk that a predator will use the signal to locate and attack the sender.

4.4.1 Why Would Social Foraging Arise?

For social foraging, or indeed any cooperative behaviour, to arise and persist there must be some benefit to the participants. Three scenarios in which cooperative behaviour may emerge are generally recognised [532], namely:

i. indirect selection;
ii. reciprocity; and
iii. mutualism.

Under indirect selection, an individual may not benefit directly from cooperative behaviour but if the benefits to its kin, weighted by the degree of relatedness of the kin, outweigh the cost to the individual the behaviour can still be advantageous to it. Under reciprocity, the individual incurs an initial cost but the recipient may return the favour later. Mutualism requires that both individuals benefit immediately from the interaction.

Social Recognition

The nature of the social behaviour that arises in a species is closely linked to the degree of *social recognition* displayed by that species. Recognition can occur at numerous levels, including an ability to differentiate conspecifics from members of other species and an ability to recognise offspring, kin or a mate. The most general form of social recognition is when members of a species can learn to identify individual members of the population (Fig. 4.5). While own species recognition is nearly ubiquitous across animal taxa [61], the degree of social recognition of kin and individuals varies substantially between species.

Animals typically exhibit the type of social recognition needed for their life strategy, so offspring recognition is unlikely if an animal does not care for its young. Even

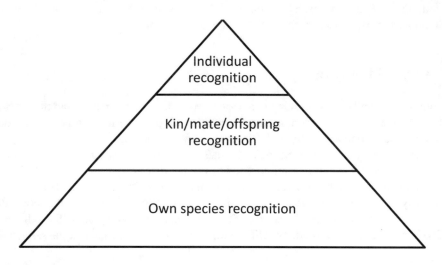

Fig. 4.5. Hierarchy of social recognition capabilities

within an individual animal taxon, there can be differences in social recognition. For example, some solitary nesting birds cannot recognise their own offspring from other nestlings (they simply remember the location of their nest), whereas congener species that nest colonially can recognise their young [61].

Indirect selection and reciprocity require social recognition (kin or individual). Differing social foraging strategies could be expected in these scenarios compared with the case where social recognition is limited to the species level. In the latter case, social foraging behaviours would rely on mutualism and therefore require immediate payoffs to all individuals concerned.

4.4.2 Aggregation and Dispersion Economies

Biologists draw a distinction between *aggregation* and *dispersion* economies. In dispersion economies any increase in group size decreases each member's fitness, so maximal benefits are obtained when individuals are dispersed and solitary.

In contrast, aggregation economies emphasise how group membership can increase the fitness of each individual. In the case of foraging, aggregation benefits can arise from [211]:

  i. improved capabilities in searching for food;
 ii. enhanced ability to capture and kill larger or more dangerous prey;
iii. enhanced ability to defend a food discovery from other groups;
 iv. improved feeding of offspring; and
  v. increased protection from predators as vigilance and defence duties can be shared.

Having more conspecifics at a feeding location means that vigilance and predator defence duties can be shared. It also dilutes the risk to an individual if a predator attacks. There may also be other payoffs in that food-sharing behaviour may attract potential mates to the food site and may help strengthen social bonds within the foraging group.

In cases where prey items are large in size or dangerous, animals may send calls in order to recruit a group of conspecifics in order to attack the prey item.

The spatial distribution of food resources also plays an important role in determining whether a species will tend to group together and forage socially. Species that feed on resources such as fruit or nuts, which are typically dispersed in large, ephemeral, patches often live in groups as, in this case, the key limitation in foraging tends to be finding the location of good food sites [132].

The decision as to whether or not it makes sense to forage socially, and/or to alert conspecifics to food resources when found, can be viewed through an economic lens of costs and benefits. Social foraging is more likely to occur when aggregation benefits exist. The idea that increases in group size could produce increasing returns to scale in terms of individual fitness is termed the *Allee effect* [7, 8].

Aggregation benefits will usually be subject to eventual diminishing returns. Although larger groups may be better able to uncover food resources, the degree of in-group competition for resources increases as the group size increases, and a larger group may attract more predators, parasites or pathogens. Accordingly, for any specific setting there will be an *optimal foraging group size* beyond which the marginal costs of additional group members begin to outweigh their marginal benefit. We can expect to see group foraging emerging in aggregation economies, but there will be a limit on group size within which aggregation benefits will be obtained.

4.4.3 Influence of Social Setting on Individual Behaviour

A significant question in the study of collective behaviour is whether individuals alter their personalities (defined here as 'consistent behavioural variation between individuals across time and contexts' [475]) and, consequently, their behavioural patterns when they are in a group setting. A study by McDonald et al. [395] considered this question by examining the foraging behaviours of three-spined stickleback fish. These small freshwater fish can be found living both as individuals and in social groupings in the wild. While individual fish exhibit consistent personalities in terms of their attitude to risk taking (or boldness) when foraging in isolation, these personalities can alter when the same fish are placed in a social foraging setting with conspecifics. Normally timid fish will rapidly follow bolder fish during foraging expeditions. Similarly, when normally bold fish are placed in a social foraging setting, they lead the foraging group but wait for other members of the group to follow them during foraging expeditions. The bolder fish become less bold in this context as the rate at which they explore the environment reduces.

The results of the study indicate that individual personalities are suppressed when decisions are made by a group, with the authors suggesting that this occurs owing to the requirement for conformity in order for the group to reach a decision. These changes in personality were not permanent, as when fish were later returned to solitary foraging, their original personalities were found to reemerge, implying that the effect of being in a group did not lead to lasting personality changes in individual fish.

While general conclusions cannot be drawn based on a single research study, the findings underscore the points that personalities and foraging behaviour can be plastic with respect to a social context, and that aspects of the foraging behaviour of an individual animal such as its risk-taking tendency may not provide an accurate proxy for the behaviour of the same animal in a social setting. Both individuals and groups can have distinct personalities and it is not necessarily trivial to determine the latter, even where there is good understanding of the individual personalities of the animals making up the group. The idea that individuals can have differing personalities depending on whether they are foraging alone or as part of a group has not yet been explored in algorithmic design.

4.5 Communicating Information About Resources

A key aspect of social foraging is the transmission of information about food finds such as quantity, quality and location. All forms of communication are closely tied to the sensory capabilities of an organism, and communication mechanisms can encompass tactile, olfactory, visual, vocal, chemical or electrical signalling [532]. Communication may be multimodal, encompassing, for example, both visual and sound signals.

Intentional and Unintentional Communication

Communication among animals need not be intentional. Transmission of information can be broadly classified as arising from:

- *signals* deliberately produced by an animal; and
- *cues* emitted as a by-product of an animal's behaviour.

Signals can be defined as 'stimuli produced by a sender and monitored by a receiver to the average net benefit of both parties' [78], whereas cues are 'assessable properties that are at least partly correlated with a condition of interest' [78], such as the intentions of a nearby conspecific or the likelihood that a prey item is nearby.

Communication via signals is intentional, for example, the emission of a food call to deliberately alert other animals to a food find. In the case of cues, communication to third parties occurs unintentionally as a result of the behaviour of an animal (for example, local enhancement effects as a result of an animal feeding at a food find).

Information may transfer between animals as a result of one animal observing the activities of another or by eavesdropping on a cue produced by another animal.

Where communication is intentional, the intended outcome is usually to alter the pattern of behaviour or the physiology of the receiving organism [624]. This can give rise to a conflict of interest between senders and receivers of communication signals with the senders attempting to manipulate the receivers [132]. A signal is considered *honest* when both the sender and the receiver of the signal benefit from the interaction but is considered *deceptive* if the signaller benefits at the expense of the receiver that responds to the signal. Successful deceptive communication is more likely when the frequency of encounter with the signal is low or when a signal exploits a sensory bias in the receiver. Hence, an important issue faced by an animal when it receives a signal is what level of trust to place in it.

How Is Information Communicated to Fellow Foragers?

Although a huge array of natural mechanisms for communication of information about food finds exist, these have been broadly categorised into four main groupings in the social foraging literature. The economic balance of benefits and costs of the four options varies. The forager that finds a food resource may [78]:

i. take up position at the food resource and broadcast a localisable signal to fellow foragers;
ii. generate a chemical or visible trail between the food resource and a central location, and then induce fellow foragers to follow this trail to the food;
iii. return to a central location and provide directions to fellow foragers on how to find the food resource; or
iv. return to a central location, recruit fellow foragers and lead them back to the food site.

Broadcasting an advertisement for a food resource from its location is relatively easy, and the related signal can be optimised for maximum range and localisation. The major cost (risk) is that of eavesdropping, as both intraspecific and interspecific competitors could use the signals to locate the advertised food resource. Predators could also use the signal to find the animal advertising the food resource.

The use of chemically mediated trails is common in species of ants, termites, and stingless bees. These trails can create large aggregations of foragers at a food find within a short period. One risk of such trails is their interception by eavesdroppers and predators. Some insects such as stingless bees reduce this risk by creating a broken (as distinct from continuous) trail, where trail pheromone is laid down only every few metres.

Perhaps the best-known example of recruitment at a central location and providing directions to fellow foragers is the honeybee dance [515, 602]. The private nature of this dance (it is performed within the hive) minimises the risk of eavesdropping by colony nonmembers. The system is costly in that it imposes a significant cognitive burden on both senders and receivers (to produce and to process the information in the dance), and the waggle dances are also an energetically expensive display.

The final mechanism of 'recruitment at a central location and subsequently leading followers back to the food site' creates fewer eavesdropping risks than broadcast signals from the food site. However, a potential leader requires a mechanism for locating likely recruits, the ability to find the food again, and sufficient compensation for the extra time and energy that leading others to the food source imposes on it.

As will be seen in later chapters, foraging-inspired optimisation algorithms make use of all of the above communication mechanisms.

4.6 Summary

This chapter contributes to our understanding of a range of foraging issues, including:

i. Why is learning useful for foragers?
ii. What are the costs of learning?
iii. How should an animal learn from other animals (i.e. what social learning strategies should it use)?
iv. When will social foraging be a sensible group behaviour?
v. How can animals communicate information about food finds to other members of their group when engaging in social foraging?

These issues have direct relevance to our ability to design and critically evaluate foraging-inspired algorithms.

An important aspect of learning is how it can be encoded and maintained in memory. The implementation of memory is crucial in the design of optimisation algorithms, as these algorithms typically bias the choice of the next potential solution (or the next state or location in the search space) to be examined using information captured in the memory of an agent from its individual search history (arising from asocial learning) and/or using information uncovered by another member of the population of agents (arising from social learning). Memory can be operationalised in many ways, with differing mechanisms giving rise to quite distinct algorithms.

Another issue that arises in the design of algorithms is the maintenance of a suitable balance between asocial and social learning. Rogers' paradox highlights the fact that strategies which rely solely on social learning are not necessarily better than asocial learning strategies. This underscores the need for agents in a population to use both asocial and social learning. In the context of algorithmic design, this has parallels with the exploration–exploitation balance of an algorithm, with undue emphasis on social learning tilting an algorithm towards overweighting exploitation at the expense of exploration.

Foraging-inspired optimisation algorithms which include social learning implicitly embed design choices as to how social information is communicated between individuals in the population. A wide variety of strategies for animal communication is found in nature, and variants on these can serve as inspiration for agent communication mechanisms in algorithm design.

Part II

Foraging-Inspired Algorithms for Optimisation

Having already introduced elements from various literatures which are relevant to the design of foraging-inspired search algorithms, Chapter 5 provides an introduction to the literature on foraging-inspired optimisation algorithms. In order to make this literature more accessible, a number of taxonomies of the various algorithms are presented. This chapter also outlines a metaframework within which most foraging-inspired algorithms can be conceptualised, as well as setting up the following parts of the book, which discuss the specifics of a wide range of foraging-inspired algorithms.

5

Introduction to Foraging-Inspired Algorithms

The observation that the foraging processes of animals often entail a search for resources whose location is not known with certainty in advance bears a clear parallel with 'a search for good locations on a partially observable terrain'. This has led to the development of a significant literature which takes metaphorical inspiration from the foraging strategies of various organisms in order to design powerful search algorithms. In the last three chapters we have introduced elements from various literatures which are relevant to the design of these algorithms.

In this chapter we aim to set up the following sections of the book, which discuss the specifics of a wide range of foraging-inspired algorithms. As noted in Chap. 1, we restrict our attention to optimisation algorithms, so we begin in this chapter by providing a brief introduction to typical optimisation problem settings.

A drawback of many discussions of foraging-inspired algorithms is that they are usually presented in isolation, with no attempt being made to place the algorithms in a coherent taxonomy. A problem with this approach is that it makes it difficult for a reader to fully appreciate the breadth of algorithmic approaches available, or to critically compare and contrast algorithms which may on the surface have very different metaphorical inspirations. In this chapter we illustrate a number of taxonomies which can be used to group the various foraging algorithms. This leads naturally into a discussion of algorithmic design issues, and we demonstrate that most foraging-inspired algorithms can be described in a relatively compact metaframework. Finally, we discuss the phenomenon of swarm intelligence in the context of foraging-inspired algorithms.

5.1 Characterising an Optimisation Problem

In this book we characterise optimisation as being the search in a *solution space* for the best solution to a problem of interest, where the quality of a solution can be determined according to some predefined criteria. Below we describe a number of common optimisation settings.

© Springer Nature Switzerland AG 2018
A. Brabazon, S. McGarraghy, *Foraging-Inspired Optimisation Algorithms*,
Natural Computing Series, https://doi.org/10.1007/978-3-319-59156-8_5

Continuous Optimisation

Perhaps the best-known example of an optimisation problem is where we seek to find the input values which maximise or minimise the output value of a real-valued function. Given a function $f : X \rightarrow \mathbb{R}$, we wish to determine the vector $x_0 \in X$ such that $f(x_0) \leq f(x)$ for all $x \in X$ (for a minimisation problem) or, alternatively, $f(x_0) \geq f(x)$ for all $x \in X$ (for a maximisation problem).

Constrained Optimisation

Frequently, optimisation problems will be subject to constraints and not all solutions will be considered feasible. A constrained optimisation problem (assuming that the objective function is to be maximised) can be stated as follows. Find a vector $x = (x_1, x_2, \ldots, x_d)^T$, $x \in \mathbb{R}^d$, so as to:

$$\text{Maximise: } f(x) \tag{5.1}$$

subject to

$$\text{inequality constraints:} \qquad g_i(x) \leq 0, \quad i = 1, \ldots, j, \tag{5.2}$$

$$\text{equality constraints:} \qquad h_i(x) = 0, \quad i = 1, \ldots, k, \tag{5.3}$$

$$\text{boundary constraints:} \qquad x_i^{\min} \leq x_i \leq x_i^{\max}, \quad i = 1, \ldots, d, \tag{5.4}$$

where j and k are the numbers of inequality and equality constraints, respectively. Not all constrained optimisation problems will necessarily have all three of the above categories of constraints.

Multiobjective Optimisation

Another common scenario in optimisation problems is that they may be multiobjective. In this case the aim is to optimise over several, possibly conflicting, objectives. The solution space may also be constrained in that all solutions may not be feasible. A multiobjective problem can be generally formulated as follows.

Assume that there are n objectives $f_1, \ldots, f_n$ and d decision variables $x_1, \ldots, x_d$ with $x = (x_1, \ldots, x_d)$, and that the decision maker is seeking to minimise the multiobjective function $y = f(x) = (f_1(x), \ldots, f_n(x))$. The problem, therefore, is to find the set (region) $R \subseteq \mathbb{R}^d$ of vectors $x = (x_1, x_2, \ldots, x_d)^T$, where $x \in \mathbb{R}^d$, in order to:

$$\text{Minimise: } y = f(x) = (f_1(x_1, \ldots, x_d), \ldots, f_n(x_1, \ldots, x_d)) \tag{5.5}$$

subject to

$$\text{inequality constraints:} \qquad g_i(x) \leq 0, \quad i = 1, \ldots, j \tag{5.6}$$

$$\text{equality constraints:} \qquad h_i(x) = 0, \quad i = 1, \ldots, k \tag{5.7}$$

$$\text{boundary constraints:} \qquad x_i^{\min} \leq x_i \leq x_i^{\max}, \quad i = 1, \ldots, d. \tag{5.8}$$

Any specific member x of R (corresponding to a set of decision variable values) will produce a unique y (vector of objective values in objective space), and the above formulation aims to determine the members of R which are *nondominated* by any other member of R. This set is termed a *Pareto set*. A solution is termed *Pareto optimal* if an improvement in any one component of the objective function implies a disimprovement in another component.

Combinatorial Optimisation

Optimisation problems need not be real-valued. Other common instances include discrete optimisation, where some or all of the solution variables are restricted to assuming a discrete set of values. Closely related to discrete optimisation is the field of combinatorial optimisation which includes well-known graph-based problems (Sect. 2.6 provides a short introduction to networks and related terminology) such as the travelling salesman problem (TSP).

Dynamic Optimisation

Many of the most challenging optimisation problems are of a dynamic nature. That is, the environment in which the solution exists, and consequently the optimal solution itself, changes over time. Examples of dynamic problems include trading in financial markets, routing in telecommunication networks and, of course, foraging.

A particular issue in dynamic environments, particularly those of a biological nature, is that *robust* solutions—i.e. solutions or strategies that produce good results in a variety of environments—will be important, as a nonrobust solution can lead to death.

Scope Limitation

Given the large number of foraging-inspired algorithms which have been developed, no single text can hope to cover all of them comprehensively. In writing this book, we have chosen to focus on breadth, selecting algorithms from differing metaphorical inspirations. As would be expected, in most well-developed families of foraging-inspired algorithms a multiplicity of algorithmic variants have been developed to cater for each type of optimisation application. In the following chapters we focus attention in most cases on the canonical version of the relevant algorithm, which is typically designed for application to either real- or discrete-valued unconstrained optimisation problems.

5.2 Categorising Foraging-Inspired Algorithms

Because of the multiplicity of foraging-inspired algorithms, it is not an easy task to create a taxonomy which encompasses all of them satisfactorily. In the subsections

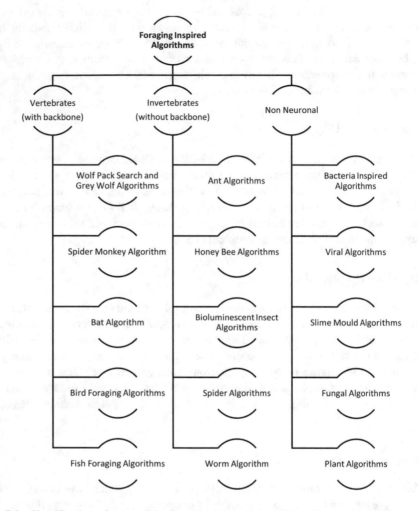

Fig. 5.1. Classification of some of the foraging-inspired algorithms discussed in this book

below we present a number of exemplar taxonomies which can be used to segment the population of extant algorithms. This is not an exhaustive listing of potential taxonomies and many others could be proposed. By definition, any choice of taxonomy focuses attention on a particular facet of the algorithms and therefore each taxonomy will have its own strengths and weaknesses.

5.2.1 Tree of Life

In Fig. 5.1 we divide the algorithms into three main classes, initially distinguishing between those inspired by the foraging activities of vertebrates and those arising from the foraging activities of invertebrates. In both of these classes, the organisms possess a central nervous system. The third class of algorithms correspond to those inspired by the foraging activities of nonneuronal organisms, in other words, organisms without a central nervous system or brain. In each of the three classes, we further subdivide the algorithms, based on their domain or kingdom from the tree of life. For the purposes of this exercise, we include viral inspired algorithms as a separate category within the nonneuronal algorithms grouping although, as discussed in Sect. 14.6, there is a long-standing debate as to whether viruses can be considered as 'being alive'.

Classifying the algorithms using this taxonomy helps illustrate the wide variety of life forms whose foraging activities have inspired the design of computational algorithms. It also helps highlight some gaps, namely, those life forms whose foraging activities have not yet been explored as a possible source of inspiration for algorithmic design. The drawback of a taxonomy based on the tree of life is that it does not highlight design commonalities and overlaps between algorithms, as it does not provide granular detail on the search mechanisms embedded in each algorithm. This makes it difficult to critique new algorithms in terms of their degree of similarity or differentiation from existing algorithms.

5.2.2 Foraging Capabilities

In earlier chapters, foraging capabilities were discussed along several dimensions, including:

 i. sensory mechanisms;
 ii. memory mechanisms; and
iii. learning and communication mechanisms.

These dimensions can be used in isolation or in conjunction with each other to create a variety of taxonomies for distinguishing between algorithms.

5.2.3 Sensory Mechanisms

In Table 5.1 we provide an illustration, using three exemplars of foraging algorithms which are discussed in detail in later chapters, of a taxonomy based on the primary sensory modality from which metaphorical inspiration is drawn for that algorithm: chemical sensing in the case of the canonical ant colony (ACO) and bacterial foraging algorithms (BFOA); and visual (observation of a waggle dance) in the case of the canonical honeybee algorithm (HBOA).

While there are distinct narratives concerning the sensory modalities from which inspiration is drawn in these algorithms, it is debateable to whether a taxonomy based on these provides detailed insight into the workings of the various algorithms. There are important differences between sensory systems in the real world and the implementation of sensory mechanisms in foraging algorithms. As discussed in Chap. 3, organisms have evolved a complex array of sensory modes and means of processing sensory information in order to extract useful information from their environment.

Table 5.1. Stylised mapping of sensory modalities to ant colony optimisation algorithms (ACO), bacterial foraging optimisation algorithms (BFOA), and honeybee optimisation algorithms (HBOA)

Sensory modality	ACO	BFOA	HBOA
Vision	✗	✗	✔
Sound	✗	✗	✗
Chemical	✔	✔	✗
Touch	✗	✗	✗
Magnetism	✗	✗	✗
Electric	✗	✗	✗

Sensory systems in nature are designed to capture and integrate a variety of signals which, for most sensory channels, are propagated by various carriers (such as photons of light or sound waves).

These concepts have little direct analogue in an optimisation setting, with foraging algorithms typically lacking clear mechanisms for multimode sensing, sensory integration, perception or selective attention (Sect. 3.1). As implemented in foraging algorithms, sensory capabilities are often very simple, sometimes being little more than an ability to assess the quality of the agent's current location and, perhaps, receive signals from other agents in the population as to the qualities of their locations.

5.2.4 Memory Mechanisms

The way that memories are created and maintained during the search process varies among algorithmic families. Table 5.2 illustrates the use of memory mechanisms as a taxonomy for classifying foraging algorithms using three sample algorithms. In general, memory of good locations found in the past may be maintained by individuals in the population based on their own personal experience, or by a subset of the population such as dominant animals, which decide where the entire population or pack will forage next. Memory can also be maintained externally in the environment itself, where, for example, a trail to food is marked by early foragers, allowing subsequent foragers to travel directly to the food source. In some algorithmic implementations, an exogenous memory structure is created (which may not be biologically plausible) whereby the location of good resources, once found, is maintained in an external store and the contents of this store are then known to subsequent foragers. The implementation of g^{best} in the particle swarm optimisation algorithm (PSO) [315], which stores the best location ever found by the population, is an example of an external storage mechanism.

In Table 5.2, memory is created in ACO via the simulated deposit of chemical trails in the (simulated) environment, which, in turn, influence the behaviour of subsequent foraging agents. In the case of BFOA, each simulated bacterium has a short term memory which allows it to assess whether the gradient of chemical trails (corresponding to attractant food sources) is increasing or decreasing, enabling the bacterium to travel up a chemical gradient to find the food source. In the HBOA, each simulated honeybee can remember the location of a good food source which it has found.

Table 5.2. Stylised mapping of memory mechanisms of ant colony optimisation algorithms (ACO), bacterial foraging optimisation algorithms (BFOA), and honeybee optimisation algorithms (HBOA)

Memory mechanism	ACO	BFOA	HBOA
Personal memory maintained by individual	✗	✓	✓
Populational memory maintained by dominant individual	✗	✗	✗
Memory maintained in the environment	✓	✗	✗
Memory maintained in external structure in algorithm	✗	✗	✗

An additional dimension of creating a memory mechanism is how many past locations should be recorded, i.e. how far back should the memory extend? Having a more comprehensive memory mechanism could potentially be useful when attempting to adapt to a moving optimum location in a cyclic environment.

5.2.5 Learning and Communication Mechanisms

Foraging algorithms can be divided between those where each agent in the population forages alone, with no interaction with other individuals, and algorithms where there is social interaction, facilitating social learning, between foragers. The majority of algorithms embed social learning mechanisms in the form of a predefined social learning strategy. As discussed in Sect. 4.2.2, social learning strategies include specific choices of when to copy, whom to copy and what to copy. A taxonomy based on these choices could be used to compare and contrast algorithms embedding social learning.

A key enabling factor in using a social learning strategy is the availability of social information. This is directly impacted by the design of communication mechanisms within that population.

Communication Mechanisms

Algorithms can be categorised based on their mechanism(s) for social communication of information about food finds. As described in Sect. 4.5, four scenarios can be distinguished. An agent may:

  i. take up position at a food resource and broadcast a localisable signal to fellow foragers;

ii. generate a chemical or visible trail between a food resource and a central location, and then induce fellow foragers to follow this trail to the food;

iii. return to a central location and provide directions to fellow foragers on how to find the food resource; or

iv. return to a central location, recruit fellow foragers and lead them back to the food site.

In the BFOA an important mechanism is the recruitment of conspecifics to good resource locations via the emission of a simulated *attractant* chemical into the environment by a bacterium, akin to broadcasting a food signal (communication mechanism i). As increasing numbers of bacteria congregate in this region, the strength of the attracting signal increases further. In ACO the communication is via a chemical trail (mechanism ii), whereas in the HBOA the emphasis is placed on provision of information as to resource location to conspecifics in the hive (mechanism iii). Table 5.3 illustrates the use of communication mechanisms as a taxonomy for classifying foraging algorithms.

Table 5.3. Mapping of communication mechanisms to ant colony optimisation algorithms (ACO), bacterial foraging optimisation algorithms (BFOA), and honeybee optimisation algorithms (HBOA)

Communication mechanism	ACO	BFOA	HBOA
Broadcast signal from resource	✗	✔	✗
Generate chemical or visible signal en route to/from resource	✔	✗	✗
Return to central location and provide information on location	✗	✗	✔
Return to central location and lead followers to resource	✗	✗	✗

Communication Network Topology

A closely related issue is the specification of how information disseminates across a population, in other words, who transmits information in a group of foragers and who listens? In each case, we can consider the topology of the communication network as creating a *neighbourhood* of foragers whose activities can influence one another.

Differing network topologies will alter the speed with which a population converges to exploit new information. At one extreme, the neighbourhood could include all foragers in the population (i.e. a *complete* or fully connected network), where each forager can learn of the relative successes of all members of the population (Fig. 5.2). Alternatively, a neighbourhood could be defined using a distance metric such as Euclidean distance. In this case a scenario could arise where there is no other foraging organism within this neighbourhood, leading to zero opportunity for social learning where an individual has no neighbours.

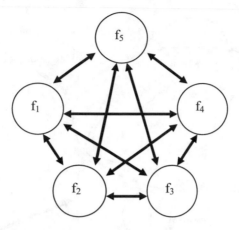

Fig. 5.2. Illustration of a *complete network* communication topology where all foragers are aware of each of the others' success. Each forager is given a unique identifier (here, index number 1,...,5) which remains unchanged during the course of the run of the algorithm

A wide variety of topological structures for interaction can be implemented including fixed structures where the network is defined at the start of the algorithm and subsequently remains unchanged (Fig. 5.3 illustrates one exemplar), to topological structures which alter dynamically during the algorithm. The nature of the topology of the implemented communication network could be used to generate a taxonomy of foraging algorithms.

5.2.6 Stochastic Component of Foraging Process

In virtually all implementations of foraging algorithms, a stochastic component is incorporated. This parallels real-world foraging in that organisms rarely employ a completely deterministic approach to foraging. From a design perspective, embedding stochastic mechanisms in search algorithms is generally essential, in order to achieve good performance across a range of problem settings. The objective in doing this is to ensure that the same inputs will not invariably produce the same next step on the search landscape. This can be achieved in multiple ways, with differing foraging algorithms implementing alternative approaches. Below, we highlight four examples of stochastic mechanisms, with Table 5.4 illustrating how the choice of stochastic mechanism can be used as a taxonomy for classifying foraging algorithms.

Stochastic Weight Selection

In foraging algorithms, several factors are combined in order to determine the next step for an individual simulated forager. Take the case of how much reliance a forager should place on private versus public information. The weight applied to each

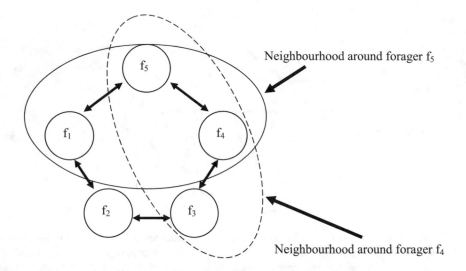

Fig. 5.3. Illustration of fixed communication topology defined using overlapping neighbour-hoods consisting of three foragers. Each forager communicates with the same two other for-agers regardless of the Euclidean distance between them

Table 5.4. Stylised mapping of stochastic mechanisms to ant colony optimisation algorithms (ACO), bacterial foraging optimisation algorithms (BFOA), and honeybee optimisation algorithms (HBOA)

Mechanism	ACO	BFOA	HBOA
Stochastic weight selection	✗	✗	✗
Stochastic decision making within an algorithm	✔	✔	✗
Stochastic forager placement around good locations	✗	✗	✔
Stochastic reinitialisation of forager location	✗	✔	✗

could be varied stochastically during the algorithm so that, in some instances, private information will play a dominant role in determining the next move of the forager, while, in other cases, social information may be the dominating factor. In essence, this approach ensures that the simulated forager employs more than one strategy whilst foraging, as the precise strategy used in each iteration of the algorithm is determined stochastically. An example of this approach of using stochastic weights is provided by the predator–prey optimisation algorithm discussed in Sect. 17.4.

Stochastic Decision-Making Within an Algorithm

Typically, a simulated forager must make decisions, such as in what direction to move next or what step size to take during the search process. In making these decisions, the forager is tasked with selection of an option or a parameter from multiple possibilities. Frequently, these choices are implemented stochastically. Consider the case provided by the canonical ant colony optimisation algorithm (i.e. Ant System in Sect. 9.5). This discrete optimisation algorithm is frequently applied in a network setting, where, for each forager (ant), the choice of which arc to select next is a probabilistic function of the arc's perceived quality, as measured using a heuristic, and its degree of prior use by the entire ant population. Arcs which are short and which have been previously heavily used by the population are more likely to be chosen. However, as the choice is stochastic, it is not guaranteed that such arcs will be selected, allowing a degree of exploration of alternative options. Another example of a stochastic choice is provided in the canonical bacterial foraging optimisation algorithm (BFOA) (Sect. 14.3), where, periodically during the algorithm, a bacterium (forager) stochastically decides in which direction to move next during a *tumble* movement.

Stochastic Forager Placement Around Good Locations

An alternative approach for incorporation of a stochastic element into a foraging algorithm is to randomly select a new trial foraging location for a simulated forager, based on already-discovered high-quality locations. An exemplar of this is provided in the canonical bees algorithm (Sect. 10.2.1), where, in each iteration of the algorithm, a subset of the better locations are selected, and foragers are placed at a number of random locations within a hypersphere around each of these, in an attempt to find even better locations.

Stochastic Reinitialisation of Forager Location

A *reinitialisation* mechanism is applied in some algorithms, in order to reduce the chance of the search process stagnating. Under this approach, the location of a forager may be reinitialised to a random location if it has not improved the quality of its location for some time. More generally, an algorithm could periodically reinitialise the location of some or all of the foraging population, regardless of their recent search performance. A version of this is applied in the canonical bacterial foraging optimisation algorithm (BFOA) (Sect. 14.3), where an *elimination–dispersal* mechanism is implemented, in which each bacterium (simulated forager) is stochastically subject to a death event, being followed by the generation of a new bacterium which is randomly placed in the search space.

Exploration and Exploitation

In algorithmic design, a critical consideration is the balance between exploration and exploitation. Too much exploration tends towards random search, as little attention is paid to prior knowledge gained during the search process. On the other hand, too much exploitation of existing information risks insufficient exploration of the environment leading to the premature convergence of the algorithm to a local optimum. Algorithmic mechanisms, such as memory and transmission of social information concerning good locations in the environment, focus on exploitation of existing knowledge. These need to be balanced by mechanisms that promote exploration in order to ensure good algorithmic performance.

Stochastic mechanisms can assist in promoting exploration but, depending on the way they are implemented, can promote *either* exploitation or exploration. Stochastic mechanisms need to be designed in conjunction with the other mechanisms in an algorithm, in order to ensure a suitable overall exploration–exploitation balance. Of course, the problem environment determines where this balance should lie, with dynamic environments generally requiring a higher degree of exploration, in order to maintain the populational diversity required for robust search performance over time.

Comparing and Contrasting Algorithms

In the paragraphs above we have described several taxonomies which can be used to compare and contrast differing foraging-inspired algorithms. The above exemplars are not intended to be exhaustive, but rather to illustrate some of the dimensions along which algorithms can be compared and contrasted.

5.3 A Metaframework for Foraging-Inspired Algorithm Design

In spite of the multiplicity of foraging-inspired algorithms in the literature, a relatively compact metaframework can be outlined which encapsulates most of the algorithms. If we consider the stylised model of foraging presented in Sect. 3.1 and combine this with the discussion of social learning from Sect. 4.2, four key elements (Fig. 5.4) can be abstracted in order to inspire the design of foraging algorithms.

Each individual (a simulated agent in the case of an algorithm) is typically capable of capturing information about its immediate environment via its sensory capabilities. These capabilities can range in sophistication from a basic ability to assess the worth of the individual's current location to an ability to determine (sense) a local gradient towards a preferred resource, and often encompass an ability to extract information from signals transmitted by or emanating from other organisms or objects in the environment.

Individuals may have a memory of previously successful and unsuccessful food-foraging locations, which they can refer to in deciding where to forage next. This is likely to be particularly useful when food locations are persistent for a period of time or where food locations regenerate cyclically.

Information about food finds can disseminate across a population based on the social learning strategy (or strategies) in use. This process is critically underpinned by the communication (and of course, the sensory) mechanisms which exist in the population.

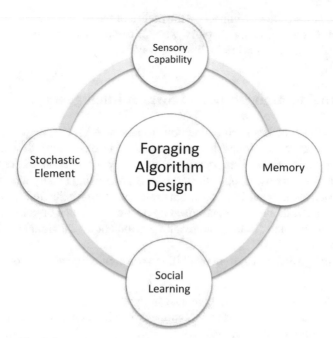

Fig. 5.4. Four key elements in foraging algorithm design

In addition to the previous three influences, there is also a stochastic element to animal behaviour when searching for new food resources, in other words a degree of randomness as to where an animal will forage next.

These components are combined to create a predictive model (Sect. 4.1.2) wherein an individual forager attempts to determine the next location at which it should forage, given the information gained so far during the search process. Most foraging-inspired algorithms embed each of these four components to some degree, and therefore the algorithms have notable high-level similarities.

While the metaframework is quite compact, it can give rise to an unlimited number of specific algorithms, depending on how its elements are operationalised. Decisions facing the algorithmic designer include:

- How do individuals sense the environment?
- What memory mechanisms are implemented?
- Who transmits information in a group of foragers?
- How do they transmit information and who listens?
- How do individuals weight/combine private and social information?
- How is 'randomness' incorporated into the algorithm?
- Can individuals employ multiple foraging / search strategies or do they employ a fixed strategy?

These decisions critically impact on the nature of the search process in the resulting algorithm. In the following chapters, we will explore a variety of algorithms which operationalise the framework in differing ways.

5.4 Foraging Algorithms and Swarm Intelligence

Swarm intelligence has been defined by Bonabeau et al. [67] as 'the emergent collective intelligence of groups of simple agents'. More generally, swarm intelligence can be considered as a system- or population-level phenomenon, where the collective behaviour of relatively unsophisticated agents, interacting locally with their environment or each other, results in a coherent problem-solving behaviour at a populational level. Critically, there is no centralised control or *top-down* organisation of the activities of individuals. Rather, the resulting population behaviours are *bottom-up* or *emergent*.

Four properties of a system exhibiting swarm intelligence are commonly identified [67]:

i. It is composed of many individuals.
ii. The individuals are relatively homogeneous (i.e. either they are all identical or they belong to a few typologies).
iii. The interactions among the individuals are based on simple behavioural rules that exploit only local information. This information is exchanged directly between individuals or via the environment.
iv. The overall behaviour of the system results from the interactions of individuals with each other and with their environment. The group behaviour self-organises.

Examples of swarm intelligence include the remarkable ability of social insects such as ants and bees to coordinate their activities in a bottom-up manner for purposes including foraging, cooperative transport, nest building and division of labour [67].

Swarm Foraging

On examination, it is apparent that a number of families of foraging algorithms, including ant algorithms (Chap. 9), honeybee algorithms (Chap. 10) and most of the nonneuronal families of algorithms draw on a swarm intelligence metaphor, with some individual algorithmic variants in other families of foraging algorithms doing likewise (such as the fish school search algorithm (Sect. 8.2)).

However, not all foraging activity can be encompassed in this framework, with many organisms engaging in solitary foraging rather than foraging as a group. Even in the case of socially foraging organisms, many species have higher-order individual cognitive capabilities, involving the use of both private and public information during foraging, producing a repertoire of complex behaviours which transcend a swarm metaphor. Curiously, as we more closely examine the behaviours of cognitively simpler organisms such as insects, we are finding growing evidence that individuals in these species are not completely homogeneous but can differ in terms of their behaviours, based on genetic factors or based on their personal lifetime experience [607], giving rise to differing personalities (see Sect. 4.4.3). It is also speculated that insect colonies (for example, of honeybees) may have distinctive 'cultures' as they can display differences in their propensity for group-level activity in foraging, defence and nest search [607].

While swarm phenomena are undoubtedly important in contributing to the foraging behaviours of many species, we should bear in mind that an assumption that all agents are homogeneous, responding only to local cues, is likely to be a simplification of real-world behaviour. More plausibly, even simple organisms forage based on a subtle combination of individual and swarm influences.

5.5 Summary

In this chapter we have characterised a number of typical optimisation problem settings, provided several taxonomies which can be used to classify the algorithms discussed in this book, illustrated that most foraging-inspired algorithms can be described in a relatively compact metaframework, and introduced the phenomenon of swarm intelligence which arises in several foraging algorithms.

In the following chapters we describe a wide range of foraging-inspired optimisation algorithms. As noted in Sect. 5.1, we intentionally focus on a limited number of variants of algorithms from each family in order to provide broad coverage of a diverse range of foraging-inspired algorithms.

Part III

Vertebrates

The subphylum Vertebrata comprises a grouping of some 66,180 species, or about 5% of all species if vertebrates and invertebrates are combined [628]. Major groupings within the classification of vertebrates include mammals (totalling approximately 5510 species), amphibians (approximately 7300 species), reptiles (approximately 10,040 species), birds (approximately 10,425 species) and fish (approximately 32,900 species) [628]. The key distinguishing feature of vertebrates is that they possess a central nervous system, in which a spinal cord runs from the brain of the animal through a hollow backbone with the peripheral nervous system branching out from the spinal cord.

Most vertebrates have well-developed sensory capabilities and are generally capable of quite sophisticated cognition. Accordingly, they display a diverse array of foraging behaviours. In the next three chapters we describe a number of algorithms which have been inspired by the foraging activities of various mammals (Chap. 6), birds (Chap. 7) and fish (Chap. 8).

6

Mammal Foraging Algorithms

The earliest traces of mammals in evolutionary history date to the late Triassic period, approximately 210 million years ago [90]. During the subsequent Jurassic and Cretaceous periods, mammals remained a relatively minor grouping, with reptiles and dinosaurs dominating. The end-Cretaceous extinction event of 66 million years ago, arising from an asteroid impact [9] (most likely at the Chicxulub crater in the Gulf of Mexico) exacerbated by volcanic eruptions in what are now the Deccan Flats in India, wiped out approximately 75% of all species [273], particularly larger herbivores, resulting in the extinction of the dinosaurs. The extinction event also impacted on mammals, with the placentals (mammals which give birth to live young) becoming the dominant grouping of mammals [90]. Modern mammals share multiple common features including a neocortex, a brain region unique to mammals. They are also endothermic and require significant energy from food in order to maintain body temperature.

Many of the best-known animals in our environment are mammals, and it is evident that these species harvest an array of food resources using a variety of foraging behaviours. In this chapter we introduce two families of algorithms based on the foraging behaviours of mammals, one drawn from pack hunters such as wolves, humpback whales and spider monkeys, and one drawn from the foraging behaviour of bats.

Pack hunting arises when a number of animals come together to hunt as a unit. An example of this is provided by wolves, and two algorithms inspired by this behaviour, the *wolf pack search* and the *grey wolf optimiser* algorithms, are discussed in this chapter. Also discussed is the *whale optimisation algorithm*, another pack-hunting algorithm, which derives inspiration from the bubble net feeding behaviour of humpback whales.

Another social grouping behaviour exhibited by some species is *fission–fusion* in which groupings of varying size may form based on the coalescence of smaller groups or the splitting of larger groups. In these societies the size of hunting groups may also vary over time. One species that exhibits fission–fusion while foraging is spider monkeys, and this behaviour has inspired the *spider monkey optimisation algorithm*.

© Springer Nature Switzerland AG 2018
A. Brabazon, S. McGarraghy, *Foraging-Inspired Optimisation Algorithms*,
Natural Computing Series, https://doi.org/10.1007/978-3-319-59156-8_6

The final family of algorithms introduced in this chapter is inspired by the foraging activities of bats. An interesting aspect of bat foraging is that many bat species rely on echolocation, an *active* sensing mechanism, whereby bats emit sound energy and interpret the resulting echoes in order to determine the location of prey items. This mechanism has inspired the development of the *bat algorithm*.

6.1 Pack Hunting

Although most predatory carnivores are solitary and hunt alone, a number of these species hunt their prey as a pack. Some of the best-known carnivorous pack-hunting animals include grey wolves, lions and hyenas. Pack hunting is also observed in other mammals including chimpanzees, dolphins and whales. While pack-hunting behaviour was long thought to be restricted to higher animals, it has now been identified in many other lineages including fish, birds and insects [370]. The complexity of the tactics employed by pack hunters varies considerably, with some animals primarily relying on weight of numbers to subdue prey and others appearing to use specialist teamwork to achieve this.

Pack hunting is more likely to evolve when foraging in groups yields higher payoffs than solitary foraging. Pack hunting can provide a number of potential advantages to predators, including:

- enabling the search of a larger area thereby increasing the chance of encountering prey;
- enhancing the probability of a successful kill on encountering prey; and
- subsequently protecting the kill from scavengers.

A pack of animals can disperse during the search phase of hunting and use social communication to call the attention of other pack members to a prey item once encountered. The chances of a successful kill can also be enhanced by pack foraging. Frequently, during the attack phase of a hunt predators need to isolate a single prey item from a herd, and this is easier with a pack. The physical risks of attacking a prey item are also lessened when a pack rather than a single predator are involved. After making a kill, a pack can more easily defend a kill from scavengers. Not all of the above advantages need be present simultaneously to promote the development of pack-hunting behaviours.

6.1.1 Pack Behaviour of Grey Wolves

The pack behaviours of grey wolves (also known as timber wolves in North America and white wolves in the Arctic) have been extensively studied. Wolves are the largest members of the dog family, having a body approximately 1.0–1.6 m in length, standing about 0.8 m high at the shoulder and weighing on average 40–43 kg (male) and 36–39 kg (female). They are carnivores, typically eating large hoofed animals such as deer, moose, elk or bison but also attacking smaller animals such as hares, rodents or birds.

Wolves typically group together in a pack of 7–10 related animals. At the apex of the pack's dominance hierarchy is a breeding pair, the *alpha* male and female. The pack also contains a group of nonbreeding adults and younger immature wolves and pups. The alpha wolves are considered to be the key decision makers, determining such things as when the group should go out to hunt or when it should move from one place to another. The alpha wolves are dominant at food kills and are the first to eat. The next level of hierarchy in the pack is the *beta*, the male or female wolf most likely to replace the respective current alpha. The lowest-ranking wolf in the pack is termed the *omega*, and all other wolves are termed *subordinate*, equivalently referred to as *gamma* or *delta* (Fig. 6.1). Although the idea of a dominance hierarchy in wolf packs is a long-standing one, this view has been critiqued by some authors as most packs consist of a mated pair and their offspring. Therefore, a dominance hierarchy could alternatively be viewed as a reflection of the age and reproductive structure within the group.

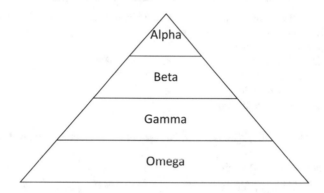

Fig. 6.1. Dominance hierarchy ranging from alphas to omegas

The hunting of prey by wolves is frequently a group activity. The wolves commence their hunt by moving through their territory until they come across the trace of a prey animal via scent, hearing or vision. The wolves then stalk the prey, attempting to close in from a downwind direction without alerting the prey to their presence. Wolves require the stimulus of fleeing prey to trigger an attack and chase. On an encounter with a prey item, the wolves attempt to goad the prey into flight. At this stage, if they are chasing a herd of prey, the wolves will try to isolate an individual animal. During the final attack, wolves usually attack the side or rear of the animal in order to lessen the potential for injury that could occur from attacking the prey animal head on. When the prey animal is finally killed, the pack feed on it in order of dominance hierarchy.

A number of optimisation algorithms have been developed based on inspiration taken from various hunting behaviours of wolves. In this chapter we introduce two of these algorithms, namely *wolf pack search* [634] and the *grey wolf optimiser* [403, 407].

6.1.2 Wolf Pack Search

One of the earliest studies introducing an algorithm inspired by wolf pack behaviour was that of Yang et al. [634], who described the *wolf pack search* (WPS) algorithm. The algorithm is loosely based on the idea that a wolf pack searches randomly until the scent of a prey animal is found and then seeks to follow the scent in the direction of its strongest gradient towards the prey. The wolf closest to a prey animal attracts other wolves to its position simulating a communication mechanism. Pseudocode for WPS is outlined in Algorithm 6.1.

Algorithm 6.1: Wolf pack search algorithm [634]

Generate an initial population of n wolves and locate each one's position vector x_i,
  $i = 1,\ldots,n$, randomly in the search space;
Initialise the step parameter s;
Choose value for t_{max}, the maximum number of iterations;
Set iteration counter $t := 1$;
for *each wolf* $i = 1$ **to** n **do**
  | Calculate fitness f_i of wolf i;
end
Determine which wolf $g^{best} \in \{1,\ldots,n\}$ has best fitness and let $x_{g^{best}}$ be its location;
repeat
  | **for** *each wolf* $i = 1$ **to** n **do**
  | | $x_i := x_i + \frac{s}{\|x_{g^{best}} - x_i\|}(x_{g^{best}} - x_i)$;
  | | Calculate fitness f_i of wolf i;
  | | **if** $f_i > f_{g^{best}}$ **then**   // if i has higher fitness than g^{best}
  | | | Update g^{best};
  | | **end**
  | **end**
  | $t = t + 1$;
until $t = t_{max}$;
Output the best solution found;

In WPS, each wolf is initially assigned to a random location in the search space and the fitness of that location is determined, as is the location $x_{g^{best}}$ of g^{best}, the wolf which has highest fitness. Next, each wolf moves from its current location by taking a step in the direction of the location $x_{g^{best}}$ of g^{best}, with the size of this step being determined by the value of a parameter s and how close the wolf currently is to $x_{g^{best}}$. After all wolves in the population have changed location, the fitness of the new location for each wolf is determined. If one of these locations produces a higher fitness than that of g^{best}, then g^{best} is updated. The process iterates until a terminating condition is reached.

In essence, WPS implements a local search process and was used for this purpose in [634], implementing a local search step in a honeybee algorithm. The canonical WPS algorithm would require modification in order to increase its exploration capabilities if it was to be employed as a stand-alone optimisation algorithm.

6.1.3 Grey Wolf Optimiser

The *Grey Wolf Optimiser* (GWO) algorithm, developed by Mirjalili et al. [403, 407], implements a more complex suite of wolf pack behaviours for the purposes of developing an optimisation algorithm. It embeds a pack leadership hierarchy distinguishing among alpha (α), beta (β), delta (δ) and omega (ω) wolves. In the following, the notation of [407] is largely followed, except that vector arrowheads are omitted for consistency with the rest of this book, and the notation . for a componentwise product of vectors is changed to $\odot$ to avoid confusion with the more usual use of . for the dot or scalar product of vectors: if $x = (x_1, \ldots, x_n)$ and $y = (y_1, \ldots, y_n)$ are vectors, then the componentwise product is $x \odot y := (x_1 y_1, \ldots, x_n y_n)$. Note that for $D = (d_1, \ldots, d_n)$, $y = (y_1, \ldots, y_n)$ and $x = (x_1, \ldots, x_n)$, the notation $D = |y - x|$ is a shorthand for the vector whose ith component is $d_i = |y_i - x_i|$, as occurs, for example, in (6.1) and (6.5).

Algorithm

In the algorithm, each wolf corresponds to an agent attempting to uncover the optimal solution in a search space. The best location found so far, corresponding to the fittest solution α, is designated by x_α and the locations of the second and third best solutions found so far, β and δ, are designated by x_β and x_δ respectively. The locations of α, β or δ are assumed to provide a guide to the location of the prey item (corresponding to the global optimum).

As the algorithm iterates, conceptually the wolves update their position around the prey item as follows:

$$D_i = |C \odot x_{\text{p}}(t) - x_i(t)|, \tag{6.1}$$

$$x_i(t+1) = x_{\text{p}}(t) - A \odot D, \tag{6.2}$$

where t is the current iteration, A and C are coefficient vectors, x_{p} is the position of the prey and x_i is the position vector of the ith grey wolf. D_i is a distance vector and is determined by the location of a wolf i relative to the position of the prey, modified by the value of C. The vectors A and C are calculated as follows:

$$A = 2a \odot r_1 - a \tag{6.3}$$

$$C = 2r_2 \tag{6.4}$$

where the components of a are decreased linearly from 2 to 0 during the algorithm and r_1 and r_2 are random vectors with each component drawn from $[0, 1]$. Using the above equations, a grey wolf updates its position in each iteration of the algorithm and the search process produces a stochastic search process based on a hypersphere around the location of the prey item.

Operationalising the Algorithm

Equations (6.1) and (6.2) require the determination of a prey item location in order to update the location for each wolf. In order to operationalise the algorithm it is assumed that the α, β and δ have the best current knowledge about the potential location of prey and the information contained in the positions of these wolves is stored and used to update the position of each other wolf i using

$$D_\alpha = |C_1 \odot x_\alpha - x_i|, \quad D_\beta = |C_2 \odot x_\beta - x_i|, \quad D_\delta = |C_3 \odot x_\delta - x_i| \tag{6.5}$$

where $x_\alpha, x_\beta, x_\delta$ are the positions of the α, β, δ wolves, respectively, C_1, C_2, C_3 are random vectors, and x_i is the location of the current solution.

Equation (6.5) calculates the approximate distances between the current solution i and α, β, δ, respectively. After determining the distances, the final position of the current solution is calculated as follows:

$$x'_\alpha = x_\alpha - A_1 \odot D_\alpha, \quad x'_\beta = x_\beta - A_2 \odot D_\beta, \quad x'_\delta = x_\delta - A_3 \odot D_\delta. \tag{6.6}$$

Equation (6.5) determines the step sizes of wolf i towards α, β, δ respectively, and the final updated position for wolf i is determined using the average:

$$x_i(t+1) = \frac{1}{3}(x'_\alpha + x'_\beta + x'_\delta). \tag{6.7}$$

Hence, the locations of α, β and δ are used to estimate the position of the prey, and each of the other wolves updates its position using this information to move to a random location in a hypersphere around the prey item (Fig. 6.2).

The vectors A and C adaptively control the degree of exploration and exploitation in the algorithm. In order to encourage convergence of the wolves during the algorithm, the value of a is decreased (that is, each component of a is decreased), in turn reducing the value of (each component of) A. Each component A_j of A is a random value in the interval $[-a_j, a_j]$, where a_j is reduced from 2 to 0 during the algorithm.

When the random values of A are in $[-1, 1]$, the next position of a wolf can be anywhere between its current position and the location of the prey; therefore $|A| < 1$ (that is, each $|A_j| < 1$) implies a simulated attack (or exploitation of the information in the prey's current estimated location). When the random values of $|A| > 1$ (that is, each $|A_j| > 1$), the wolf moves further away from the estimated position of the prey item, producing exploration in the search process.

The parameter vector C also encourages exploration. This vector contains random values in the range $[0, 2]$. When $C > 1$, the effect of prey location in defining the distance in (6.5) is emphasised (and when $C < 1$ the effect of prey location in defining the distance in (6.5) is deemphasised). This process helps to ensure that the algorithm never converges completely.

With this notation and these equations in place, the GWO algorithm may be given as in Algorithm 6.2.

The performance of the GWO algorithm on 29 test functions was benchmarked against that of several well-known metaheuristics including Particle Swarm Optimisation (PSO) [314, 315], Differential Evolution (DE) [550, 551] and Evolution Strategies (ES) [476, 477, 512, 513], and found to produce competitive results [407].

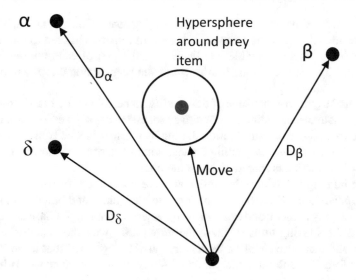

Fig. 6.2. Position update in the GWO algorithm. The individual is moved into a random location in a hypersphere around the prey item. The prey item is estimated using the current positions of α, β and δ

Algorithm 6.2: Grey Wolf Optimisation Algorithm [407]

Generate an initial population of n wolves and locate each one's position vector x_i, $i = 1, \ldots, n$, randomly in the search space;
Initialise parameter vectors a, A and C;
Set iteration counter $t = 1$;
Choose value for t_{max}, the maximum number of iterations;
Calculate fitness of each wolf;
Determine the locations x_α, x_β and x_δ of the three highest-fitness wolves;
repeat
    for *each wolf $i = 1$* **to** n **do**
      | Update position of wolf i using (6.7);
    end
    Update a, A and C;
    Calculate fitnesses of all wolves;
    Update x_α, x_β and x_δ;
    Set $t = t + 1$;
until $t = t_{max}$;
Output the best solution found;

6.1.4 Pack Hunting: Emergence or Complex Cognition?

Prima facie, from the observation of pack foraging behaviours in real life, it is notable that they often appear to be quite coordinated. In turn, this has led to the suggestion that this indicates considerable intelligence on the part of each hunting animal as it dynamically responds to the movement of the prey item and to the movement of other members of its pack.

The underlying mechanisms of this intelligent behaviour are often attributed to a number of sources, including the cognitive processing capacity of individual animals, the benefits arising from information sharing by means of social communication, and the role of *social hierarchy* within a pack, which determines who makes decisions and how information is shared amongst members of the pack.

Two hunting behaviours in particular were considered to display intelligence: *ambush behaviour*, where one member of the pack hides and other members of the pack chase prey to them; and *relay running*, where a continuous chase is created with pack members taking turns to chase the prey [420] (with the object of exhausting the prey and therefore enabling its capture). It was suggested that these behaviours required foresight and the anticipation of future events: in other words intelligent, planned hunting.

An alternative perspective is that complex (and successful) hunting behaviours could actually emerge as a result of individual wolves following quite simple rules, rather than necessarily resulting from complex cognition and communication mechanisms (see Sect. 5.4 for a discussion of swarm intelligence).

This hypothesis was examined in a multiagent simulation study by Muro et al. [420, 421] which adopted two simple, decentralised rules to govern the movement of each simulated wolf. The two rules were 'move towards the prey until a minimum safe distance to the prey is reached' and 'when close enough to the prey, move away from the other wolves that are close to the safe distance to the prey'. Crucially, in the simulation, no dominance hierarchy and no explicit interagent communication mechanism was implemented. The only information available to individual wolves was the location of the prey item and the position of other pack members. In the simulation, varying pack sizes and different prey movement patterns were also considered.

The simulation results showed that the two rules, when combined with the prey's own dynamic movement, were sufficient to ensure that the simulated pack behaved like a real pack, tracking prey, carrying out a pursuit, closing in and circling a prey item, and seemingly engaging in ambush behaviour.

This suggests that a hypothesis that wolf pack-hunting strategies can be explained as emergent phenomena is plausible, and this offers a simpler explanation for observed hunting behaviours than assumptions of high levels of wolf intelligence and sophisticated mental processes.

Boids Simulation

Examples of emergent behaviour arising from simple rules have been observed in simulations of bird flocking and fish schools. In the famous *boids* simulation of Craig

Reynolds [484] (the term *boid* refers to a simulated agent and comes from a fusion of the word *bird* and a science fiction term for a robot, *droid*), three simple movement rules were implemented for each individual boid, namely:

i. attempt to match velocity to the average velocity of local boids;
ii. fly towards the centre of mass of local boids; and
iii. avoid collisions by moving away from local boids that are too near.

These were found to be sufficient to produce emergent *birdlike* flocking behaviour in a population of simple agents whose movements were based solely on a local perception of their environment [485]. The results of these simulation studies demonstrate that it can be difficult to reverse engineer the precise mechanisms which underlie complex behaviours such as hunting purely by observation. The results also remind us that we need to guard against a tendency to attribute human-like behaviours, cognition and intentionality to animals [542].

Of course, there are plausible reasons as to why real-world animals may follow similar simple rules resulting in *behaviour matching*. If animals get too close during movement, they risk collision and potential injury. Conversely, if an animal becomes isolated from a group it may lose grouping benefits, and be at higher risk of predation. As noted by [116], hard wired rules that govern animal interactions are subject to natural selection, and evolution can generate systems that produce highly coordinated movements without a need for sophisticated cognition on the part of individual animals.

6.1.5 Other Pack-Hunting Approaches Observed in Nature

Other cooperative hunting approaches have been observed in nature. Here, we mention two such approaches observed in aquatic mammals, namely, pelagic dolphins and humpback whales.

Benoit-Bird and Au [49] examined cooperative prey herding by pelagic dolphins, specifically spinner dolphins. They identified that spinner dolphins have four clear stages of foraging:

i. in the first, *wide line* search stage, the dolphins swim in a line perpendicular to the shoreline;
ii. in the second, *tight line* stage, the spacing between the dolphins in the line decreases while the animals continue to swim forward, bulldozing the prey;
iii. in the third, *circling* stage, the animals in this tight line then close into a circle from offshore, concentrating the prey and increasing its density;
iv. in the fourth and final stage of foraging, called *inside circle*, pairs of dolphins take it in turns to actually feed.

These stages followed a very strict three-dimensional geometry, with tight timing and orderly turn taking. There is a clear progression from exploration to exploitation in this approach. [49] note that there is 'evidence of a prey density threshold for feeding[, which] suggests that feedback from the environment may be enough to favor the evolution of cooperation'.

Another cooperative feeding behaviour, observed among humpback whales, is subsurface bubble net feeding for zooplankton or small fish [286, 215]. In this, a group or a single submerged whale will swim in a tightening spiral under a group of prey, releasing large bubbles or clouds of bubbles in patterns, ranging from lines to complete or partial circles. The prey move to avoid the bubbles and so become corralled within a shrinking cylinder or net. The density of prey within this cylinder rises as the diameter shrinks. Other whales may dive deeper to drive prey towards the surface, while yet others may vocalise to herd prey into the net. Finally, one or more whales will swim vertically through the concentrated school of prey, with mouths open to feed. The spiral can start at approximately 30 m in diameter, and as many as ten to twelve whales may cooperate.

In each of these two cases of cooperative cetacean feeding behaviour:

i. herding exploits the prey's own avoidance behaviour to achieve food densities that would not be achieved otherwise; and
ii. an individual animal has more access to prey by working cooperatively than it would have if feeding on its own, with the resulting energy benefit outweighing the extra cost of cooperative manoeuvres.

Each of these aquatic mammalian feeding approaches is analogous to a landscape adjustment or deformation, in the sense that a flat landscape is being changed to a high-walled valley (akin to a potential well in physics). Such foraging behaviours in relatively food-sparse environments would appear to have potential to be developed into algorithms that adjust the search space geometry to enhance search and optimisation.

This topic appears to be in its infancy, but some algorithms have appeared recently which draw on these ideas. The dolphin echolocation algorithm [311, 312] mentioned in Sect. 6.3.2 has a cooperative feeding element. Also, an algorithm related to the GWO algorithm has been developed based on humpback whales' bubble net feeding [405], and this is the subject of the next section.

6.1.6 Whale Optimisation Algorithm

The *Whale Optimisation Algorithm* (WOA) algorithm, developed by Mirjalili and Lewis (2016) [405, 406], takes inspiration from the subsurface bubble net feeding behaviour observed among humpback whales. According to the authors [405], it is related to the GWO algorithm but differs in that:

i. a random or the best search agent chases the prey; and
ii. a mathematical spiral is used to simulate the whales' bubble net attack method of swimming in a tightening spiral under the group of prey.

The WOA uses both exploration and exploitation and so performs a global search.

In the following, as was done for the GWO algorithm, the notation of [405] is largely followed, except that vector arrowheads are omitted for consistency with the rest of this book, and the notation $\cdot$ for a componentwise product of vectors is changed to $\odot$ to avoid confusion with the use of $\cdot$ for the dot product of vectors:

if $x = (x_1, \ldots, x_n)$ and $y = (y_1, \ldots, y_n)$ are vectors, then the componentwise product is $x \odot y := (x_1 y_1, \ldots, x_n y_n)$. As for the GWO algorithm, for vectors $D = (d_1, \ldots, d_n)$, $y = (y_1, \ldots, y_n)$ and $x = (x_1, \ldots, x_n)$, the notation $D = |y - x|$ is a shorthand for the vector whose ith component is $d_i = |y_i - x_i|$.

Algorithm

In the algorithm, each whale corresponds to a search agent attempting to uncover the optimal solution in a search space. The best location found so far, corresponding to the fittest solution, is designated by x^*. The location x^* is assumed to provide a guide to the location of the target prey (corresponding to the global optimum). The coefficient vectors A and C are calculated as follows:

$$A = 2a \odot r_1 - a \qquad (6.8)$$

$$C = 2r_2 \qquad (6.9)$$

where the components of a are linearly decreased from 2 to 0 during the algorithm run and r_1 and r_2 are random vectors with each component drawn from $[0, 1]$. The vectors A and C adaptively control the degree of exploration and exploitation in the whale optimisation algorithm, in precisely the same way as in the GWO algorithm.

The WOA starts with a set of random solutions (corresponding to locations of whales). As the algorithm iterates, conceptually, each whale updates its position with respect to either a randomly chosen search agent (exploration, when $\|A\| \geq 1$) or the best solution obtained so far (exploitation of information, when $\|A\| < 1$).

In a bubble net attack—which is the exploitative phase—each whale i updates its position around the prey item (the best solution found so far) $x^*(t)$ as follows:

$$D_i = |C \odot x^*(t) - x_i(t)|, \qquad (6.10)$$

$$x_i(t+1) = x^*(t) - A \odot D, \qquad (6.11)$$

where t is the current iteration, A and C are the coefficient vectors described above and x_i is the position vector of the ith whale. D_i is a direction vector and is determined by the location x_i of whale i relative to the position of the best solution found so far, modified by the value of C.

To simulate the bubble net hunting behaviour of humpback whales, two approaches are used:

i. Shrinking encircling mechanism: This behaviour (akin to that of the GWO algorithm) is simulated by decreasing the value of a in (6.11). The fluctuation range of A is also decreased by a, in the sense that component A_j of A is a random value in the interval $[-a_j, a_j]$. By setting random values for components of A in $[-1, 1]$, the new position of a whale can be placed anywhere between its original position and the position x^* of the current best whale. Using (6.10) and (6.11), a whale can update its position in each iteration of the algorithm and this gives a stochastic search process based on a hypersphere around the location of the best solution found so far. Conceptually, for $|A| < 1$ this simulates whales encircling the prey. This is an exploitative process (but the algorithm also incorporates exploration, for $|A| \geq 1$, as explained later).

ii. Spiral updating position: this mechanism first calculates the distance between the whale located at x_i and the prey (best solution obtained so far) located at x^*. A logarithmic spiral equation is set up, using the positions of whale i and the prey item to model the tightening helical upwards movement of humpback whales in bubble net feeding:

$$x_i(t+1) = e^{bl}\cos(2\pi l)D' + x^*(t), \qquad (6.12)$$

where the vector $D' = |x^*(t) - x_i(t)|$ indicates the direction and distance of whale i to the prey, b is a constant defining the shape of the spiral and l is a random number drawn from $[-1, 1]$.

The algorithm design assumes that the two mechanisms—the shrinking encircling mechanism and the spiral model—for updating the positions of whales are equally likely, so a random number p drawn from $[0, 1]$ is used to choose between them according as $p < 0.5$ or $p \geq 0.5$. The rule adopted is to calculate $x_i(t+1)$ by using (6.10–6.11) if $p < 0.5$ (provided $\|A\| < 1$; when $\|A\| \geq 1$ the whale explores as described next), and by using (6.12) if $p \geq 0.5$.

The algorithm also allows for search for prey (exploration): humpback whales also search randomly according to each other's positions. The same approach of varying the vector A can be used to search for prey: $\|A\| \geq 1$ will force a search agent to move away from a reference whale. In this exploration phase—as against the exploitation phase—the position of whale i is updated relative to a randomly chosen reference whale instead of the best whale found so far. The approach used is:

$$D = |C \odot x_{\text{rand}} - x_i| \qquad (6.13)$$

where x_i is the location of the current solution, x_{rand} is the position of a randomly selected reference whale and C is a random vector as in (6.9). Then the location of whale i is updated by:

$$x_i(t+1) = x_{\text{rand}} - A \odot D. \qquad (6.14)$$

With these notations and equations in place, the WOA can be described by Algorithm 6.3.

The performance of the WOA on 29 test functions and six structural optimisation problems was benchmarked against that of several well-known metaheuristics including Particle Swarm Optimisation (PSO) [314, 315], Differential Evolution (DE) [550, 551] and Evolution Strategies (ES) [476, 477, 512, 513], and found to produce competitive results [405].

6.2 Spider Monkey Foraging

Spider monkeys comprise a group of seven distinct species of New World monkey which live in the rainforests of Central and South America. The monkeys are social

---

Algorithm 6.3: Whale Optimisation Algorithm [405]

Generate an initial population of n whales and locate each one's position vector x_i,
  $i = 1,\ldots,n$, randomly in the search space;
Initialise parameter vectors a, A and C using (6.8–6.9);
Set iteration counter $t := 1$;
Choose value for t_{max}, the maximum number of iterations;
Calculate fitness of each whale's position;
Determine the location x^* of the highest fitness whale;
while $t < t_{max}$ **do**
  for *each whale i = 1 to n* **do**
    Draw a random number p from $[0,1]$;
    if $p < 0.5$ **then**
      if $\|A\| < 1$ **then**
        | Update position x_i of whale i using (6.10) and (6.11);
      else
        | Select a random whale and let x_{rand} be its location;
        | Update position x_i of whale i using (6.14);
      end
    else
      | Update position x_i of whale i using (6.12)
    end
  end
  Update a, A and C;
  if *any whale has gone outside the search space* **then**
    | Amend its position to bring it back within the search space;
  end
  Calculate fitnesses of all whales;
  Update x^* if there is a better solution;
  | $t = t + 1$;
end
Output the best solution found;

---

and typically live in bands of 15–20 animals, although occasionally they can band in larger groups of up to 40 animals [105]. Spider monkeys are amongst the most intelligent of New World monkeys [134]. It is speculated that their communication capabilities (mediated by a range of vocalisations, physical postures and stances), the opportunities for social and lifetime learning resulting from their long lifespans (20–30 years typically), their close social bonds, and a relatively lengthy period of maternal care all contribute to this.

The monkeys forage in the high canopy and are primarily frugivores with fruit comprising some 70–80% of their diet. They will also consume other foods including nuts, seeds, leaves, insects and bird eggs, if fruit is scarce. Each group of monkeys has a home range and is territorial. If monkeys from groups in adjoining territories meet, the males in each group display antagonistic and territorial behaviours such as calling or barking. The groups usually interact at a distance, and direct physical conflict is rare.

Foraging Behaviour

Feeding occurs most intensively early in the day from dawn until mid morning and only occasionally thereafter. A particular characteristic of spider monkeys is that they exhibit a *fission–fusion social system* in that the group splits into smaller subgroups for foraging purposes, recombining to associate at a nocturnal sleeping site. Fission–fusion group dynamics is relatively rare in mammals, although the behaviour is displayed in some species including chimpanzees, orang-utans, humans, elephants [503] and dolphins [142].

The foraging subgroups typically consist of between two and eight individuals but are fluid in membership. They can change in composition during the day [105, 106]. The groups are led by a dominant female and move and feed within the home range of the bigger group. If the female cannot find sufficient food for the group, the group splits into smaller groups that forage separately.

Inspiration has been taken from the fission–fusion dynamics of spider monkeys to design the *spider monkey optimisation algorithm* [27]. In the next section we provide a short introduction to fission–fusion social systems, concentrating on why they might arise in nature. We then outline the detail of the optimisation algorithm.

Fission–Fusion Social Systems

Although the concept of a fission–fusion social system is straightforward, an obvious question is why some species display this method of social organisation and others do not. In other words, what adaptive benefits does this method of organisation provide?

As discussed in Sect. 4.4.2, larger social groups offer benefits such as better mating opportunities and enhanced protection from predation, at the expense of greater competition for food from other members of the group [106]. Trading these factors off against one another suggests that the optimal size of a foraging group will vary depending on the nature of the specific ecological environment faced by a species.

Spider monkeys feed primarily on fleshy fruits that are located in dispersed patches [105]. If all members of the group fed in the same tree, there would be considerable competition for food resources. It is plausible that this environment favours the evolution of fission–fusion social behaviours whereby a large group of spider monkeys will break into smaller groups for foraging purposes but remain part of a larger social unit otherwise.

Three dimensions of fission–fusion dynamics were identified in [472], namely, subgroup size, dispersion and composition. The ability to vary the size, dispersion and composition of foraging groups is advantageous for flexibly exploiting food resources such as seasonal fruits which are distributed heterogeneously in time and space [472]. A number of studies have examined whether subgroup size in spider monkeys varies in response to changes in environmental conditions, such as the relative scarcity of food resources. The findings support a claim that fission–fusion dynamics allows individuals to minimise food competition, leading to more efficient foraging. Subgroups are found to be generally larger in periods of food abundance and smaller in periods of scarcity. Subgroups are also found to be larger when food patches in the environment are larger [21].

An interesting aspect of foraging group size in some species of spider monkeys is that their food-calling behaviour is conditional on the degree of food abundance and the dominance rank of the calling monkey (dominant monkeys call more as they are less threatened by new arrivals) [106]. This suggests that spider monkeys can alter the information they broadcast about food finds in order to manipulate their subgroup size, thereby relating feeding competition to the abundance of available resources.

6.2.1 Spider Monkey Optimisation Algorithm

The propensity of spider monkey groups to flexibly split into groups of smaller size for foraging purposes, depending on the degree of availability of food resources, inspired Bansal et al. [27] to develop the *spider monkey optimisation* (SMO) *algorithm*. Four features of spider monkey foraging underlie the algorithm [27]:

i. Monkeys live in groups of up to 40 animals but divide into smaller subgroups when foraging.
ii. A female (denoted in the algorithm as the *global leader*) leads the group and is responsible for searching for food resources. If insufficient food is found, the group is divided into smaller subgroups which forage independently.
iii. Subgroups are also led by a female (denoted in the algorithm as the *local leader*) who attempts to plan an efficient foraging route each day.
iv. The members of the subgroups can communicate within and without their subgroups depending on the availability of food resources.

The algorithm comprises six elements, namely, the local leader phase, global leader phase, local leader learning phase, global leader learning phase, local leader decision phase and global leader decision phase. Each of these is discussed below, following the description provided in [27].

In essence, the group commences foraging and evaluates the quality of the locations that the group and its local leader find. Based on this information, subgroup members stochastically update their positions and evaluate the quality of their new location. Next, the location of each individual is stochastically influenced by the location of the global leader. Information from both local leaders and the global leader influences the movement of all members of the population. The positions of the local and global leaders can change as a result of these updates, if better positions are found.

If the local leader's position does not change for a set number of iterations (this parameter is denoted by $\lim^{local}$), indicating that the search process may be stagnating, all members of the local group start individual searches in different directions, thereby promoting explorative search. Similarly, if no new global best is found for some time (i.e. within $\lim^{global}$ iterations), the population splits into smaller subgroups and the search process continues.

The algorithm embeds elements of both local (exploitation) and global (exploration) search. It also embeds self-organisation and division of labour.

Algorithm

Initially, a population $\{M_i : i = 1, 2, \ldots, N\}$ of N simulated spider monkeys is created, where each monkey is represented by a D-dimensional vector corresponding to a potential solution to the problem of interest. M_i is the ith spider monkey in the population. At the start of the algorithm, each is initialised to a random location in the search space using

$$M_{ij} = M_j^{\min} + U(0,1) \cdot (M_j^{\max} - M_j^{\min}), \quad j = 1, 2, \ldots, D, \tag{6.15}$$

where M_{ij} is coordinate j of monkey i, $M_j^{\min}$ and $M_j^{\max}$ are the lower and upper bounds on the jth variable for the problem of interest, and $U(0,1)$ is a uniformly distributed random number in $[0,1]$.

Local Leader Phase

In this process, selected spider monkeys alter their current position based on information from the search experience of their local leader and based on information from other members of the local group. If the fitness value of the new position is higher than that of the monkey's current position, the monkey moves to the new position. The position update for M_i in the kth local group is

$$M_{ij}^{\text{new}} = M_{ij} + U(0,1) \cdot (LL_{kj} - M_{ij}) + U(-1,1) \cdot (M_{lj} - M_{ij}), \tag{6.16}$$

where M_{ij} is the jth dimension of M_i and LL_{kj} is the jth dimension of the local group leader for group k. The term M_{lj} is the jth dimension of M_l, where l is randomly selected from within the kth group ($l \neq i$). $U(-1,1)$ is a uniformly distributed random number in the range $[-1, 1]$.

The update process in the local leader phase is outlined in Algorithm 6.4. In this, and subsequently, $G_{\max}$ is the maximum number of groups in the population, G is the current number of groups in the population and r_{pert} is the perturbation rate, which controls the degree of perturbation in the current position ($r_{\text{pert}} \in [0.1, 0.9]$). The value of r_{pert} is a critical factor, as higher values of r_{pert} imply that this update step will be applied to fewer members of each subgroup.

Global Leader Phase

The global leader phase starts after the local leader phase is completed. In this step, the position of each monkey is updated based on the search experience of the global leader and also based on information from other members of the local group. The update equation for monkey i is

$$M_{ij}^{\text{new}} = M_{ij} + U(0,1) \cdot (GL_j - M_{ij}) + U(-1,1) \cdot (M_{lj} - M_{ij}). \tag{6.17}$$

Algorithm 6.4: Position update process in local leader phase

---

for $k = 1,\ldots,G$ do
    for *each member* $M_i \in$ kth *group* do
        for $j = 1,\ldots,D$ do
            if $U(0,1) \geq r_{\text{pert}}$ then
                $M_{ij}^{\text{new}} = M_{ij} + U(0,1) \cdot (LL_{kj} - M_{ij}) + U(-1,1) \cdot (M_{lj} - M_{ij})$;
            else
                $M_{ij}^{\text{new}} = M_{ij}$;
            end
        end
    end
end

The parameters have the same meaning as in (6.16), with GL_j being the jth dimension of the global leader's position and $j \in \{1, 2, \ldots, D\}$.

The position of each spider monkey (for example, M_i) is updated based on a probability value P_i which is calculated by reference to relative fitness. Better-quality individuals have a higher chance of selection for improvement. The value P_i is calculated using

$$P_i = 0.9 \cdot \frac{f_i}{f_{\max}} + 0.1, \tag{6.18}$$

where f_i is the fitness of M_i and $f_{\max}$ is the maximum fitness in the group. If the fitness of a newly generated position is better than the fitness of the current location occupied by a spider monkey, the monkey moves to the new position. The position update process is outlined in Algorithm 6.5.

Global Leader Learning Phase

In this step, the position of the global leader is updated to the location of the spider monkey in the population which has the highest fitness. If the position of the global leader is not changed, a counter parameter c^{global} is incremented by one.

Local Leader Learning Phase

The position of the local leader in each subgroup is next updated to the location of the spider monkey in that subgroup which has the highest fitness. The parameter c_k^{local} is the trial counter for the local best solution in the kth group. If the position of the local leader of group k is not changed, the parameter c_k^{local} is incremented by one.

Local Leader Decision Phase

If the position of the local leader for any group is not updated during $\lim^{\text{local}}$ iterations, then all members of that group update their positions either randomly, or using

---

Algorithm 6.5: Position update process in global leader phase

---

for $k = 1,\dots,G$ **do**
  $g_k = k$th group size;
  $c = 1;$ $// \ c$ is a counter variable
  while $c < g_k$ **do**
    for $i = 1,\dots,g_k$ **do**
      if $U(0,1) < P_i$ **then**
        Randomly select $j \in \{1,\dots,D\};$
        Randomly select M_l from kth group such that $l \neq i;$
        $M_{ij}^{\text{new}} = M_{ij} + U(0,1) \cdot (GL_j - M_{ij}) + U(-1,1) \cdot (M_{lj} - M_{ij});$
        $c := c + 1;$
      end
    end
    if $i = g_k$ **then**
      $i = 1;$
    end
  end
end

---

information from both the global leader and the local leader as follows:

$$M_{ij}^{\text{new}} = M_{ij} + U(0,1) \cdot (GL_j - M_{ij}) + U(0,1) \cdot (M_{ij} - LL_{kj}). \tag{6.19}$$

M_i is attracted towards the global leader and is repelled from the local leader. This mechanism simulates a case where the local leader does not find resources of good quality and members of the subgroup therefore engage in their own search.

Algorithm 6.6 outlines the update process for the local leader decision phase for the kth group.

Global Leader Decision Phase

The position of the global leader is monitored and, if it is not updated within a predetermined number of iterations ($\text{lim}^{\text{global}}$), then the global leader divides the population into smaller groups incrementally, starting from $G = 1$ (a single group), then splitting into two, three etc. subgroups up to a maximum number of subgroups (denoted by G_{max}). Every time the global leader decision process is initiated, a new local leader is determined for each newly established subgroup. Figures 6.3 and 6.4 illustrate the cases where a population is divided into one and two subgroups, respectively.

When the maximum number of subgroups have already been formed, and the position of the global leader remains unchanged for $\text{lim}^{\text{global}}$ iterations, the population is recombined into a single group. This mechanism simulates the fission–fusion process in spider monkey social groups and the working of this mechanism is as outlined in Algorithm 6.7.

Algorithm 6.6: Local leader decision phase

---

for $k = 1, \ldots, G$ **do**

    if $c_k^{\text{local}} > \text{lim}^{\text{local}}$ **then**

        $c_k^{\text{local}} = 0$;

        $g_k = k$th group size;

    end

    for $i = 1, \ldots, g_k$ **do**

        for $j = 1, \ldots, D$ **do**

            if $U(0,1) \geq r_{\text{pert}}$ **then**

                $M_{ij}^{\text{new}} = M_j^{\text{min}} + U(0,1) \cdot (M_j^{\text{max}} - M_j^{\text{min}})$;

            else

                $M_{ij}^{\text{new}} = M_{ij} + U(0,1) \cdot (GL_j - M_{ij}) + U(0,1) \cdot (M_{ij} - LL_{kj})$;

            end

        end

    end

end

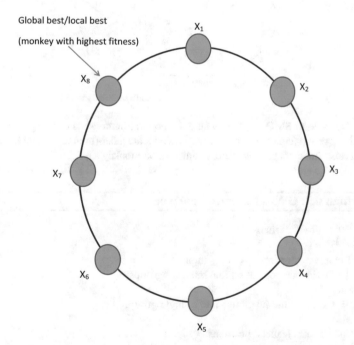

Fig. 6.3. Example of SMO topology where the population consists of a single group. In this case, the global leader and the local leader are the same monkey (i.e. monkey 8)

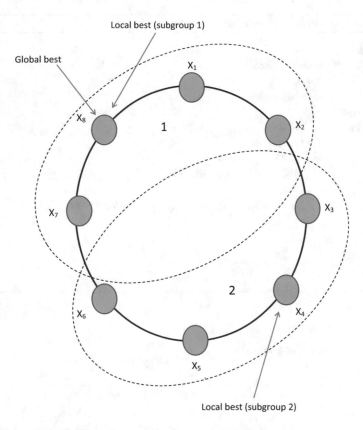

Fig. 6.4. Example of SMO topology where the population consists of two subgroups. In this case, each subgroup (denoted 1 and 2) has its own local leader (monkeys 4 and 8) and the fitter of these is also the global leader (here assumed to be monkey 8)

Algorithm 6.7: Global leader decision phase

---

if $c^{\text{global}} > \lim^{\text{global}}$ **then**

    $c^{\text{global}} = 0$;

    if *number of groups* $G < G_{\max}$ **then**

        Divide the population into more subgroups;

    else

        Combine the subgroups into a single group;

    end

    Update local leaders' positions;

end

Overall Algorithm

Algorithm 6.8 brings together the individual elements of the SMO as described above.

Algorithm 6.8: Spider monkey optimisation algorithm [27]

Initialise population locations, $\lim^{local}$, $\lim^{global}$ and r_{pert};
Calculate fitness of each individual;
Select global and local leaders;
while *termination criteria not satisfied* **do**
 Generate new locations for all monkeys stochastically, using their current location
  and location of local leader using Algorithm 6.4;
 if *new location has higher fitness than existing location for M_i* **then**
  | Move monkey to new location;
 end
 Calculate P_i for all group members using (6.18);
 Generate the new locations for M_i selected, using Algorithm 6.5;
 for *each individual* **do**
  if *new location is better than current* **then**
   | Move individual;
  end
 end
 Update the location of local and global leaders (if necessary) by reference to the
  highest-fitness location in the subgroup(s) and overall population (implement
  local leader learning and global leader learning phases);
 if *any local group leader has not updated its position in last* $\lim^{local}$ *iterations*
 then
  | All members of that subgroup engage in personal search using Algorithm 6.6;
 end
 if *global leader has not changed position in last* $\lim^{global}$ *iterations* **then**
  | Divide population into smaller groups as per Algorithm 6.7;
 end
end
Output the best solution found;

Parameters

The SMO algorithm has a number of key parameters, including $\lim^{local}$, $\lim^{global}$, G_{max} and r_{pert}. Suggested settings for these parameters according to [27] are:

- $G_{max} = N/10$ so that the minimum number of spider monkeys in a subgroup is at least 10 (N is the population size),
- $\lim^{global} \in [N/2, 2 \cdot N]$,

- $\lim^{\text{local}} = D \cdot N$, and
- $r_{\text{pert}} \in [0.1, 0.9]$.

The results of sensitivity analyses in [27] indicate that the performance of SMO is sensitive to the choice of r_{pert} and it is suggested that a useful approach may be to linearly increase the value of r_{pert}, $r_{\text{pert}} \in [0.1, 0.4]$, as follows:

$$r_{\text{pert}}(t+1) = r_{\text{pert}}(t) + \frac{(0.4 - 0.1)}{t_{\max}}, \tag{6.20}$$

where t is an iteration counter (with $r_{\text{pert}}(1) = 0.1$) and $t_{\max}$ is the maximum number of iterations. It is also noted that the results from SMO are sensitive to the population size, with suggestions that a size of 40 may give acceptable results.

Variants on the canonical SMO algorithm of [27] have been developed which implement slightly different search processes. For example, in [341] the local leader and global leader phases (outlined in Algorithms 6.4 and 6.5) are modified to include golden section search [317], a method for finding the maximum or minimum values of unimodal functions (or of a more complex function if the region in which the global maximum or minimum lies can be identified a priori) by successively narrowing the range of values inside which the maximum or minimum values exist. Noting that the canonical SMO algorithm can have slow convergence due to the selection of a random monkey in the local leader phase and global leader phase processes (i.e. the selected monkey may not be in a good location), [522] developed a number of variant algorithms which seek to address this issue.

Comparison with Other Heuristics

On examination of the canonical SMO algorithm, it bears some parallels with the PSO algorithm, in that both employ a blending of local and global information in determining the movements of individuals that are searching the environment.

In canonical PSO, information on the search history of an individual is captured in the p^{best} (personal best, i.e. best location ever found by that individual member of the population) of each individual and this information along with the location of the g^{best} (global best, i.e. best position ever found by a member of the population) in the swarm impacts on the position update of each particle as the algorithm iterates. Some variants of PSO use an l^{best} (local best) concept, whereby each particle can only 'see' a subset of the entire swarm (i.e. each individual is connected to a subset of the particles in the swarm and information as to the best location found by these particles thus far is available to all members of this subset). In this case, the position update of each individual is governed by its current location relative to its p^{best} and l^{best} (the location of the best individual to which it is connected). Differing search processes are obtained depending on the nature of the l^{best} topology defined at the start of the algorithm.

In SMO, the topology of the links between individuals in the population, and the associated information flows, vary as the algorithm iterates. The location of the global leader monkey can influence the search process but the size of the subgroup in which a monkey finds itself (and the associated local leader) will vary. Hence, the SMO does not bear simple comparison with either the g^{best} or l^{best} version of PSO. Another distinction from PSO is that the individual monkeys in SMO do not maintain a memory of their personal search history (i.e. there is no memory or p^{best} mechanism).

The fission–fusion dynamic is a key element of the SMO algorithm. While no variants of PSO explicitly use this approach, there are some commonalities between it and multiswarm versions of PSO [59, 60, 253, 582]. In multiswarm PSO, several swarms search independently of each other. Occasionally, particles and/or information migrate between the swarms. The use of multiple swarms can help encourage populational diversity, which can provide obvious benefits in environments which are dynamic, in environments which have many local optima or in multiobjective problems where the aim is to uncover a diverse set of potential solutions (for example, a Pareto front). In designing multiple-swarm systems, the key decisions include the number of swarms that will be used and how information is to be passed between the individual swarms.

In summary, the canonical SMO algorithm is quite complex and, as above, there are some parallels between elements of it and PSO. Further work is required to more clearly delineate the distinctions between SMO and the multiple variants of PSO.

6.3 Bat Foraging

Bats along with birds share the accolade of being the only vertebrates that are capable of powered flight [290]. Over 1300 species of bat have been identified, comprising some 20% of all mammalian species, ranging in size from the small bumblebee bat (weighing 2 g) to the Indian flying fox bat, which weighs some 1500 g [16, 535]. Most species feed on small insects, with other species feeding on a variety of items including fruit, nectar and small vertebrates (including fish). A notable aspect of bat behaviour is that many species of bat live in colonies and display complex social behaviours [137]. Some species of bat can live for over 30 years, providing ample time for extensive individual and social learning to occur [178].

Echolocation

A particular feature of most species of bats is that they use echolocation, or *active sensing*, in which pulses of acoustic energy are emitted and the resulting echo is resolved into an image of their surrounding environment [226, 227]. This is used to detect objects and to locate food resources such as flying insects, or—in the case of the greater bulldog bat—ripples in water made by fish just below the surface, which the bat then catches by trawling the water surface and scooping them up in its claws.

Echolocation calls of bats are sometimes noted as being an example of *autocommunication*. Autocommunication occurs when the sender and receiver are the same individual and the sender emits a signal that is altered by the immediate environment during propagation. The individual then retrieves the altered signal and uses this to gain information about the environment [78].

Echolocation calls are species specific and they differ based on the sex, age and size of a bat. They also differ between colonies and even individuals for some species [289, 181, 596]. Therefore, echolocation calls contain a considerable amount of information which can potentially be exploited by other bats. In addition to generating echolocation calls, bats are also capable of generating *social calls*, which they can use for communication purposes. This communication can be quite nuanced, containing multilayered information concerning the identity of the emitter of a call, its context (i.e. why the call is being made) and the intended individual addressee [463]. Some species of bats also engage in bird-like 'singing' behaviours [535].

The nature of the echolocation calls produced by bats varies, with some being broadband signals (encompassing a wide range of frequencies with bandwidths of up to 100 kHz, typically of short duration), and others being narrowband signals (consisting of a narrow range of frequencies, circa 5 kHz, and of relatively long duration) [16, 287]. Narrowband signals are good for ranging distant items (or prey) through neuronal processing of the time elapsed between the emission of the pulse and the return of the echo, with broadband signals being well adapted for the fine grained localisation of items (note that the shorter the wavelength of an echolocating call, the greater the degree of acuity of the resulting image). This leads to a phenomenon where as insectivorous bats home in on their aerial (insect) prey, they switch from narrowband to broadband signals, which are emitted at an increasingly rapid rate as the bat approaches the prey, resulting in what is known as the *feeding buzz*.

The acuity of the echolocating signal can be quite remarkable. As a signal reflects off the moving wings of a prey item such as a flying insect, the returning echo will contain distinctive peaks in frequency and intensity, termed 'acoustic glints' [181]. In addition to providing information on target velocity and direction, these glints can allow a bat to identify different types of prey, depending on the pattern of sequential glints as the insect's wings move [433].

Eavesdropping on Echolocation Calls

As bat call echoes (reflections) are strongly attenuated in air, bats can hear calls emitted by other bats from much further away than they can detect echoes from their own calls. Bats approaching feeding individuals or groups, and eavesdropping on their calls, can therefore increase the active space of their prey detection range by between 10 and 50 times, depending on species, over that provided by their own echolocation ability.

An illustration of the potential benefits of eavesdropping on echolocation calls is provided in [288] which finds that little brown bats (*Myotis lucifugus*) have a self echo detection range of some 5 metres, whereas they can hear the echolocation calls of foraging conspecifics at some 50 metres. Dechmann et al. [137] report similar findings for the bat species *M. molossus*, which can detect a small insect at a range of some 2 metres but can detect calls of conspecifics at some 54 metres under the same acoustic conditions. Other documented detection ranges for echolocation calls of other bats run between 20 and 50 metres [178]. A study by Cvikei et al. [122] found that the insectivorous bat *Rhinopoma microphyllum* aggregated as a group during foraging (forming, in effect, a flying sonar array), leading to an enhanced ability to find prey. However, as the group became large, bats spent an increasing amount of their time devoting attention to avoiding conspecifics, creating a trade-off between group size and foraging efficiency.

The ability of bats to recognise and use information gleaned from echolocation calls was demonstrated by Barclay [30] in a study which showed that bats responded to the playback of recorded echolocation calls from feeding sites and roosts by approaching the source of the call. The bats also responded to the playback of calls from foraging sites of other species of bat with a similar diet. This provides an example of *associative learning*, as the bats have learnt to associate information in these calls with the availability of food resources.

In addition to the obvious information content in hearing the feeding buzz of another bat, the pause during prey consumption (termed the *post-buzz pause*) and the nature of echolocation calls immediately after the feeding buzz have additional information content, as they disclose whether the attack was successful [82, 289]. The rate of feeding buzz production by bats in a locality may also be informative to eavesdropping bats, as it implicitly provides information as to prey density.

Apart from the adaptive benefits of being able to detect prey availability at an extended distance, echolocation calls have also been shown to improve the performance of bats in finding other resources such as tree roosts. Like food, roosts are an important, patchily distributed resource. As tree-dwelling bats change roost frequently, having an efficient mechanism for locating suitable new roost sites is important. Many species of bat, particularly Vespertilionid bats (insect-eating bats of the family *Vespertilionidae*, characterised by a long tail), engage in echolocation and social calling around their roost tree. Other bats can eavesdrop on these calls in order to reduce the time and energy cost of finding suitable roost sites themselves [497]. Similar findings for communally breeding female Bechstein's bats (*Myotis bechsteinii*) have been noted, with information exchange concerning roost sites occurring [316].

6.3.1 Bat Algorithm

A recent addition to the family of foraging algorithms in the natural computing literature is the *bat algorithm*, developed by Yang [639], which draws inspiration from elements of the foraging processes of bats, specifically echolocation, in order to design an optimisation algorithm. The algorithm has produced competitive results both on benchmark optimisation problems and across a variety of applications. At a high level, the process of bat foraging has a number of elements, in that bats:

- can directly detect items within a range of space based on the echoes of their own calls;
- can eavesdrop on echolocation calls from other bats;
- may possess a memory as to the location of food sources; and
- can move stochastically.

In the bat algorithm two items are emphasised, namely, a concept of information as to the current best foraging location in a population of bats, and a local search step which exploits information in the existing population. These processes embed some parallel with real-world bat foraging, as they encompass social information transmission. The algorithm also seeks to maintain diversity in the population by means of a periodic random solution generation process.

Algorithm

Algorithm 6.9 describes a variant on the canonical bat algorithm as presented in [639]. In essence, the virtual bats in the algorithm commence by flying randomly in the search space, and each bat has a different initial call frequency, loudness and rate of pulse emission. The bats move around the search space, using a mix of social information and random search. Search is also intensified around good solutions using a local search step.

The algorithm consists of a number of discrete elements, namely initialisation, generation of new solutions, stochastic local search around existing good solutions combined with the stochastic generation of randomly created solutions and, finally, an updating of the location of the current best solution.

The objective function is denoted by $F : \mathbb{R}^d \longrightarrow \mathbb{R}$, which is to be minimised. In every iteration (time step) t, each bat i has a location vector $x_i(t)$ and a velocity vector $v_i(t)$ in the d-dimensional search space $\mathbb{R}^d$. The current best location of all bats is denoted by x^*. The general relationship between wavelength λ and frequency f for echolocation calls in air is $\lambda = v/f$, where the speed of sound v is approximately 340 metres per second. As noted above, bats can adjust the wavelength or equivalently, the frequency of their calls. In the algorithm, the frequency is varied.

Generation of New Solution

The velocity update of each bat in each iteration of the algorithm is given by

$$v_i(t+1) = v_i(t) + (x_i(t) - x_*)f_i(t), \tag{6.21}$$

and the position update is given by

$$x_i(t+1) = x_i(t) + v_i(t+1). \tag{6.22}$$

As the bat searches, the frequency of its echolocation calls at time step t is generated using

$$f_i(t) = f_{min} + (f_{max} - f_{min})\beta, \tag{6.23}$$

where f_{min} and f_{max} are the minimum and maximum frequencies, respectively, of bat calls and β is randomly drawn from a uniform distribution on $[0, 1]$. Initially, each bat i is assigned a random frequency $f_i(0)$ drawn uniformly from $[f_{min}, f_{max}]$.

In essence, the value of $f_i(t)$ controls the pace and range of the movement of bat i in each iteration, and the precise value of $f_i(t)$ is subject to a random influence due to β.

Algorithm 6.9: Bat algorithm

Define an objective function $F(x)$, where $x \in \mathbb{R}^d$;
Define maximum number of iterations, t_{max};
Set $t := 0$;
for *each bat* $i = 1, 2, \ldots, n$ **do**
  Randomly initialise the location $x_i(0)$ and velocity $v_i(0)$;
  Define pulse frequency $f_i(0) \in [f_{min}, f_{max}]$ at $x_i(0)$;
  Initialise pulse rate $r_i(0)$;
  Initialise loudness $A_i(0)$;
end
Let $x^* :=$ the x_i with best fitness;
while $t < t_{max}$ **do**
  for *each bat* $i = 1, 2, \ldots, n$ **do**
    Adjust frequency to $f_i(t+1)$ using (6.23);
    Update velocity to $v_i(t+1)$ using (6.21);
    Generate new solution $x_i^{new}(t+1)$ for bat i by updating location using (6.22);
    if rand $> r_i(t)$ rand **then**
      Generate a local solution around x^* and store it in $x_i^{new}(t+1)$;
    end
    Generate a random solution in a bounded range about $x_i(t)$ and store it in $x_i(t+1)$;
    if rand $< A_i(t)$ **and** $F(x_i^{new}(t+1)) < F(x_i)$ **then**
      Set the location of bat i, $x_i(t+1) := x_i^{new}(t+1)$;
      Increase $r_i(t)$ and reduce $A_i(t)$;
    end
  end
  Rank the bats in order of fitness and, if necessary, update the location x^* of the best solution found by the algorithm so far;
  Set $t := t+1$;
end
Output best location found by the bats;

Local Search

The local search component of the algorithm is operationalised as follows. If the condition rand $> r_i(t)$ is met for an individual bat i (note that rand is drawn from a uniform distribution whose range depends on the scaling of r as discussed below), the current best solution (or a solution from the set of the better solutions in the population) is selected, and a random walk is applied to generate a new solution. The random walk is produced using

$$x_i^{new} = x^* + A(t)\epsilon \tag{6.24}$$

where x^* is the location of the best solution found so far, ϵ is a vector with each component drawn randomly from $[-1,1]$ and $A(t)$ is the average loudness of all bats in the population at time step t.

The rate of local search during the algorithm depends on the values of $r_i(t)$ (the rate of pulse emission) across the population of bats. If this rate increases, the degree of local search activity will decrease. The average loudness $A(t)$ will tend to decrease as the algorithm iterates (discussed below), so the step sizes in the local search will reduce to become finer-grained. In order to enhance the explorative capability of the algorithm, the local search step is complemented by a random search process.

In real-world bat foraging, the loudness of calls reduces when a bat approaches a prey target, while the rate of pulse emission from the bat increases. This can be modelled using

$$A_i(t+1) = \alpha A_i(t) \quad \text{and} \quad r_i(t) = (1 - e^{-\gamma t})r_i(0), \tag{6.25}$$

where α (similar in concept to a cooling coefficient in simulated annealing) and γ are constants. For any values $0 < \alpha < 1, \gamma > 0$, the following is obtained:

$$A_i(t) \to 0, \quad r_i(t) \to r_i(0), \quad \text{as } t \to \infty. \tag{6.26}$$

The loudness and pulse emission rate of an individual bat are only updated if a solution is found by a bat which is better than its previous solution. The update process is stochastic as it only takes place if a random draw from a uniform distribution, bounded by the maximum and minimum allowable values of loudness, is less than $A_i(t)$. As the algorithm iterates, the values of loudness will tend to decay to their minimum, reducing the probability that an update will occur. A side effect of this process is that it makes the algorithm less *greedy*.

Parameters

In setting parameters for the algorithm, Yang and Gandomi [648] suggest values of $\alpha = \gamma = 0.9$, with each bat having a different initial random value for loudness $A_i(0) \in [1,2]$, with $A_{\min} = 0$. A value of $A_{\min} = 0$ indicates that a bat has just found prey and therefore stops emitting sound. Initial values for the pulse rate $r_i(0) \in [0,1]$, if using (6.25), and values of $f_{\min} = 0$ and $f_{\max} = 100$ (these values are domain specific and each bat is assigned a random value in this range at the start of the algorithm) are also suggested. The performance of the algorithm will be critically impacted by these parameters, and trial and error or a more systematic grid search will be required in order to adapt the algorithmic framework to a specific problem.

Yang and Gandomi [648] point out that there is some similarity between the bat algorithm as outlined above and the PSO algorithm as, if the loudness is set to 0 ($A_i = 0$) and the pulse rate is set to 1 ($r_i = 1$), the bat algorithm becomes similar to a variant of the PSO algorithm (without p_i^{best}), since the velocity updates are governed by the prior period velocity and a noisy attraction to g^{best}.

6.3.2 Discussion

Despite the relatively recent development of the algorithm, it has also been successfully used in a variety of applications, encompassing constrained optimisation [204, 648], multiobjective optimisation [69, 436, 640], binary-valued representations [429] and clustering [499]. A detailed review of applications of the bat algorithm is provided in [643].

There are multiple open areas of research concerning bat inspired algorithms. The canonical version of the algorithm does not explicitly include a personal detection mechanism (i.e. a real bat can detect any prey in an arc around its head) or a personal memory as to good past foraging locations. These mechanisms could be included in a variation of the canonical algorithm, as demonstrated in [121]. Another area of potential research is to embed a more complex processing of social influences (feeding buzzes) in the bat algorithm. Plausibly, a bat will be more influenced on hearing many feeding buzzes coming from a small area (indicating a heavy concentration of prey in that area) than hearing a solitary feeding buzz coming from elsewhere. It is also known that echolocating bats can consider the location of multiple prey items simultaneously and plan a flight path to consecutively capture multiple prey items [191]. Therefore, bats can process and act on sonar information for not just immediate but also subsequent prey items. This suggests the potential to design an algorithm which processes several pieces of environmental information simultaneously.

Apart from the processes of echolocation in flight, it is also noted that there is research in the foraging literature which claims that information transfer between bats can occur at roost sites [621]. This study suggested that bat species transfer information by following each other to feeding sites, with unsuccessful foragers leaving a roost and following previously successful foragers. Similar findings were reported for *Phyllostomus hastatus*, a frugivore bat species [137, 622]. Information transfer at roosts has been noted in other species such as birds and has inspired the design of optimisation algorithms including the *raven roosting optimisation algorithm* (Sect. 7.2). Information transfer at a roost bears a clear parallel to the colony-based information transfer of other central place foragers such as honeybees (Chap. 10), and the mechanisms of this process could inspire the design of optimisation algorithms.

In addition to bats, a number of other animals, including some species of rats, birds, dolphins and whales, use echolocation to assist in navigation and foraging (see Sect. 3.3.2). A number of research papers have drawn inspiration from these cases to develop alternative versions of echolocation-inspired optimisation algorithms such as the dolphin echolocation algorithm. Readers are referred to [311, 312] for further information.

6.4 Summary

In this chapter we have introduced examples of foraging algorithms inspired by three different species of mammals, specifically wolves, spider monkeys and bats.

Pack hunting behaviour is quite widespread in nature being exhibited by mammal species including primates, wolves, lions, hyenas, chimpanzees and dolphins. It is also found in other lineages, including fish, birds and insects. Two algorithms were introduced in this chapter which are loosely inspired by the foraging behaviour of wolves: wolf pack search and the GWO algorithm. The latter algorithm also embeds a concept of dominance hierarchy. This is encoded in the algorithm through wolf movements in the environment being influenced by the positions of the three best wolves (i.e. solutions) found to date, with the three best solutions representing three levels of the dominance hierarchy. In the GWO algorithm each of these three best solutions is considered equally dominant, and this simplification may present an opportunity for the development of variants of the canonical algorithm. The related WOA, inspired by the bubble-net feeding of humpback whales, takes a somewhat different approach to exploration and exploitation, and opens up the possibility of further developments in pack-hunting-inspired algorithms.

The SMO algorithm is based on foraging behaviours of spider monkeys, in particular their fission–fusion social organisation. In the design of population-based optimisation algorithms, critical issues include how best to allocate members of the population to different regions of the search space, and how best to modify these allocations using information gained from the search process. In this context, fission–fusion foraging behaviours of animals are potentially an interesting source of inspiration. Thus far, the algorithm has been applied to a limited number of test problems, so our understanding of its efficiency and effectiveness is incomplete.

The bat algorithm takes inspiration from the idea of echolocation and how bats can use this sense to perceive promising locations of prey. As implemented in the original algorithm, the search process is strongly informed by the global best solution to date (effectively exploiting the eavesdropping phenomenon), as at each iteration either bats fly towards the global best solution or, alternatively, a solution in the neighbourhood of the global best is chosen. Recently, extensions to the bat algorithm have been proposed, which explicitly add memory and an individual perception mechanism [121].

7

Bird Foraging Algorithms

Birds were long thought to have a limited intellect, with this view being supported by dissection studies which demonstrated that birds do not possess a prefrontal cortex. Accordingly, it was considered by scientists that birds were not capable of anything more complex than associative learning and that they primarily acted on instinct.

More recent research has indicated that this interpretation is simplistic, and it is now known that the forebrain of birds derives from the pallium, which was a forerunner of the mammalian neocortex [281]. It is speculated that the need to conserve mass due to the requirements of flight (flight is energetically expensive and is generally limited to relatively light animals) may have encouraged the evolution of smaller brains in birds, with structures designed to conserve space and weight [167].

On closer study, various species of birds have demonstrated notable capabilities, including tool use [272], complex food-caching behaviours [113], manipulation of other animals via simulated warning calls [184], and comprehensive navigation capabilities, with the Arctic tern, for example, making an annual migratory trip from the Arctic to either South America or Africa [167]. Magpies have also demonstrated a capability to pass the mirror mark test, a classic test which seeks to determine whether a nonhuman animal possesses the ability of self-recognition [167]. The only other animals to pass this test include the great apes, the Asiatic elephant, dolphins and orcas. Suggestive evidence has also started to emerge which indicates that some members of the corvid family of birds, including ravens, can plan for the future, a trait previously thought to be limited to humans and great apes [64, 293]. Unsurprisingly, in light of these findings, we now appreciate that bird foraging behaviours can be quite sophisticated in some instances.

Apart from food, animals also forage for other resources, including mates, shelter and nest sites. Perhaps one of the more unusual foraging strategies for nest sites is exhibited by some species of cuckoo. Cuckoos are small birds and are found in a variety of habitats across all continents. Some 60 species of cuckoo are *obligate brood parasites* in that they reproduce only by laying their eggs in the nests of other birds, relying on the host birds to raise their chicks. Most of the remaining 80 or so species of cuckoos build their own nests and raise their own young [131]. The

© Springer Nature Switzerland AG 2018
A. Brabazon, S. McGarraghy, *Foraging-Inspired Optimisation Algorithms*,
Natural Computing Series, https://doi.org/10.1007/978-3-319-59156-8_7

process of searching for a host nest which a female cuckoo can parasitise has been used to inspire the design of an optimisation algorithm, the *cuckoo search algorithm*.

A social behaviour which is exhibited by many species of birds and bats is *roosting*, where multiple conspecifics come together to rest. This naturally leads to the question of what evolutionary advantages this behaviour produces. Initial explanations centred on possible antipredatory benefits, increased opportunities for mate choice, enhanced care of young, increased opportunity for status display and thermal benefits during overnight roosting [128, 390]. An alternative explanation, the *information centre hypothesis* (ICH), was proposed in [612, 659]. This hypothesis suggested that birds join colonies and roosts in order to increase their foraging efficiency by means of the exchange of information regarding the location of food. Inspiration from the roosting and foraging behaviours of ravens have been used to inform the design of the *raven roosting optimisation algorithm*.

Both the cuckoo search algorithm and the raven roosting optimisation algorithm are introduced below.

7.1 Cuckoo Brood Parasitism

Obligate brood parasitism is a relatively rare strategy in birds, with only about 1% of bird species engaging in this behaviour. Typically, obligate brood parasites have lost the ability to construct nests and incubate eggs. Instead they lay their eggs in the nests of other bird species. Nonobligate brood parasites lay eggs in both their own nests and those of other birds.

The usual process when parasitising a nest is that the female cuckoo sits on a nearby perch observing the target nest until it is left unattended. She then removes one egg from the target nest in her beak, lays her own egg and flies away, the whole process taking some 10–15 seconds [132].

Nonparasitic cuckoos usually lay white eggs but many of the parasitic species of cuckoo lay eggs which are coloured and patterned to match those of their hosts, thereby increasing the chance that the host bird will not reject the egg or abandon the nest. The cuckoo chick may have colouration to resemble the young of the host. Commonly, the cuckoo egg hatches earlier than those of the host and it is hypothesised that obligate brood parasites have evolved unusually thick-shelled eggs (they are typically about 30% thicker than normal) in order to retain more heat for the developing embryo and contribute to the early hatching of parasite eggs [633]. The cuckoo chick grows faster and attempts to evict the eggs of the young of the host species from the nest in order to improve its share of the food provided by its foster parents.

Taking inspiration from the brood parasitism of cuckoos and combining this with the observed Lévy flight movement behaviour of some animals (Sect. 2.5.2), Yang and Deb [644, 645] developed the *cuckoo search algorithm* and initially applied this metaheuristic to function optimisation in continuous problems.

7.1.1 Cuckoo Search Algorithm

Three assumptions are made in designing the algorithm:

i. each cuckoo lays one egg (representing a solution vector);
ii. the best nests, with high-quality eggs, carry over to the next generation;
iii. the number of available host nests is fixed, and an egg laid by a cuckoo is discovered by the host bird with a probability $p_a \in [0, 1]$.

Each egg therefore encodes a vector representing a solution to the optimisation problem. In the canonical version of the algorithm, as each nest contains a single egg, the terms *nest* and *egg* can be used interchangeably.

Initially, a series of n nests (eggs) are randomly located in the search space and the fitness of each egg is determined by reference to the optimisation problem of interest. An iterative process commences, whereby a nest is randomly selected and a new nest site is generated by taking a random walk (governed by a Lévy flight) from the selected nest. If the new nest site has higher fitness than the current site, the new nest replaces the current nest.

The *discovery and rejection* step (mimicking the rejection of a cuckoo's egg by a host bird) is simulated by the replacement of a portion p_a (of the n nests) with lower fitness by new nests at randomly generated locations. The algorithm embeds elitism, in that the current best solution in the population is always maintained between iterations and therefore guides the subsequent search process. Pseudocode for the algorithm is provided in Algorithm 7.1.

Generation of New Nest Sites

A key mechanism in the algorithm is the generation of new nest sites using a Lévy flight. These flights are undertaken from an initial nest i, with the potential new location of the nest being calculated as follows:

$$x_i(t+1) = x_i(t) + \alpha \text{Lévy}(\lambda), \tag{7.1}$$

where $x_i(t)$ is the location of nest i at time t, $\alpha > 0$ is the step size, which should be scaled for the problem of interest, and $\text{Lévy}(\lambda)$ is a vector of random components. The move process corresponds to a random walk where each component of the random step is drawn from a Lévy distribution,

$$\text{Lévy}(\lambda)_k \sim t^{-\lambda}, \quad k = 1, \ldots, d, \ 1 \leq \lambda \leq 3, \tag{7.2}$$

which has an infinite variance and an infinite mean (see Sect. 16.5.4 for a description of how to simulate a random draw from a Lévy distribution based on the Mantegna algorithm [385]).

The move process has a power-law (heavy-tail) distribution of step lengths. While many of the step lengths will be small, and therefore generate local search around the current position, some will result in long jumps in the search space, helping to ensure that the algorithm is not trapped in a local optimum. A value of $\lambda = 1$ is suggested in [637].

Algorithm 7.1: Cuckoo search via Lévy flights [637]

Choose value for t_{max};
Set iteration counter $t = 1$;
Generate an initial population $\{x_i : i = 1, \ldots, n\}$ of n host nests, each located randomly in
    the search space (where each $x_i = (x_{i1}, x_{i2}, \ldots, x_{id})$ is a vector in d-dimensional space);
for *each nest i* **do**
   | Calculate its fitness $f_i = f(x_i)$;
end
repeat
   | Generate a cuckoo egg x_j by taking a Lévy flight from a randomly selected nest;
   | Evaluate its fitness using the value of the relevant objective function $f_j = f(x_j)$;
   | Choose a random nest i;
   | **if** $f_j > f_i$ **then**
   | | $x_i := x_j$;
   | | $f_i := f_j$;
   | **end**
   | A fraction p_a of the worst nests are abandoned;
   | Replacement nests are created at new locations via Lévy flights;
   | Evaluate fitness of new nests and rank all solutions;
   | Update location of best solution found so far (if necessary);
   | Set $t := t + 1$;
until $t = t_{max}$;
Output the best solution found;

Parameters

The algorithm has the benefit of simplicity, with few parameters requiring selection by the modeller (n, p_a, and α). Most applications have used relatively modest population sizes (n in the range 20–50). The settings for p_a and α determine the exploration–exploitation balance of the algorithm. High values for each parameter will focus attention on exploitation and convergence, while lower values will encourage exploration and slower convergence.

7.1.2 Extensions of Canonical Algorithm

A series of extensions of the canonical algorithm have been developed, including that in [646], which implements a version of cuckoo search for application to multi-objective optimisation.

    Walton et al. [608] modified the cuckoo search algorithm by implementing an adaptive Lévy flight step size and an information-sharing mechanism between the highest-quality solutions. The impact of these modifications is to increase the speed of convergence of the algorithm. The Lévy flight step size α, used when replacing nests to be abandoned, is reduced as the algorithm iterates in order to enhance the rate of local search over time. The value of α in each iteration of the algorithm is calculated as $\alpha = A/\sqrt{t}$, where $A = 1$ (the initial step size) and t is the iteration number.

The second modification adds information exchange between the higher-quality eggs ('nests'). A fraction of the total population of eggs, those with the best fitness, are put in a group. For each of these, a second egg is drawn at random from the same group, with a new egg being generated at an intermediate point between the two eggs. This new egg enters the population if its quality exceeds that of its initial parent.

A similar approach which adapts the parameters of the algorithm as it runs, in order to obtain a better balance of exploration earlier in the algorithm and more local search (convergence) later in the algorithm, is presented in [581]. In this approach, the parameters p_a and α are reduced as the algorithm runs based on the number of iterations. Good reviews of recent applications of cuckoo search are provided in [183, 647].

Other foraging algorithms which also employ a Lévy flight as part of their search process include the flower pollination algorithm (Sect. 16.5.4).

7.2 Raven Roosting and Foraging Behaviours

Many species of birds and bats engage in roosting where multiple conspecifics come together to rest. In this section we examine the roosting and foraging behaviours of one species of bird, the common raven, and describe how this has inspired the design of the *raven roosting optimisation* (RRO) *algorithm* [76].

Raven roosts consist of juvenile, nonbreeding common ravens. Ravens normally arrive at roosts shortly before sunset and typically leave the roost in highly synchronised groups at dawn the next day. The first comprehensive study of *information transfer* at raven roost locations was undertaken by Marzluff et al. [391], who examined roosting behaviours of the common raven (*Corvus corax*) in the forested mountains of Maine, USA. Ravens in this region are specialist feeders on the carcasses of large mammals, sometimes scavenging the kills of large carnivores such as wolves [541]. These food sources are ephemeral as they degrade or are consumed quickly, and the location of carcasses is unpredictable. Hence, the search for food locations is continuous.

The typical food discovery process commences with a small number of birds feeding at a carcass site, followed by a rapid (overnight) doubling in numbers with most of the birds arriving simultaneously at dawn. The carcass is consumed over several days and, at the final stage of carcass depletion, feeding group size declines rapidly as many birds leave the carcass in the afternoon (prior to sunset) and do not return to it the next day.

Control experiments, in which naïve birds (birds introduced into the roost with no knowledge of local food locations) were released at roosts, demonstrated that these birds found feeding locations by following their roost mates, providing evidence for the existence of information sharing. Observations by the authors of the study also indicated that the same individuals in a roost were not always knowledgeable, suggesting that information sharing rather than mere parasitism (wherein 'excellent' foragers would always be followed by less skilled conspecifics) was taking place. The study concluded that information sharing did take place at roosts and that ravens which successfully found a new food resource recruited other members of the roost to that resource.

These findings were extended in a study [630] which examined the behaviour of ravens in a large roost in North Wales in the United Kingdom. Observational evidence suggested that the birds that first discovered a food resource recruited conspecifics using preroost (evening) acrobatic flight displays and vocalisations. The discoverer birds spent the night surrounded in the roost by the group that would follow them out the next morning to the food source. Recruitment appeared to be a competitive activity which was more successful for geographically closer carcasses, consistent with the idea that the preroost displays accurately reflected the energetic state of the displaying bird and therefore the relative distance and profitability of the carcass discovered. Support for the information centre hypothesis has also been found in other bird species, with [240] finding behaviour consistent with the hypothesis in the Eurasian griffon vulture, a social obligate scavenger. When foraging, these birds are capable of travelling long distances, with food search being based on visual detection of a carcass. The estimated range of visual detection for this species is a remarkable 4 km. In this study, all the tracked birds were found to profit from following others on occasion, although individual birds differed in their tendency to be informed or uninformed.

A simulation based study by Boyd et al. [73] provided additional evidence that birds could glean useful foraging information from their roost mates, even without explicit communication mechanisms between birds. In the simulation, seabirds searching for ephemeral food resources that orientated their outbound headings in line with those of aggregations of returning birds were shown to be able to avoid areas without prey even at low population densities. Hence, even a simple observational strategy, combined with local enhancement, enhanced their foraging success, with the value of the strategy depending in turn on the predictability and persistence of resource patches relative to foraging trip duration and intervals between trips.

Other Features of Raven Foraging Behaviour

The foraging process of ravens has a number of additional features. Just like other animals, a bird may have private knowledge of food locations and may place different weights on socially acquired information depending on its private knowledge. In addition, birds can visually survey a wide terrain while flying and may decide to deviate if an alternative food source is seen whilst in flight.

In common with many animals, ravens are known to suffer from *neophobia* (a fear of new things) and have an initial reluctance to forage at new carcasses which they have not seen before, particularly if no other birds are feeding there already [132, 541]. This suggests that ravens will have some reluctance to abandon existing feeding locations for new ones, even if the new locations offer good resources.

The roosting and foraging behaviour of ravens has inspired the design of the *raven roosting optimisation* (RRO) *algorithm* by Brabazon et al. [76] and this is described in the next section.

7.2.1 Raven Roosting Optimisation Algorithm

As discussed in Sect. 5.3, there are typically four primary elements in foraging algorithm design:

  i. individual sensory capability;
 ii. individual memory as to previously successful/unsuccessful food-foraging locations;
iii. the capability of an animal to broadcast or receive social information about food locations; and
 iv. a stochastic element to animal movement when searching for new food resources.

The RRO algorithm includes all four of these components. The algorithm draws metaphorical inspiration from the process of raven foraging in that in each iteration of the algorithm, a percentage of the simulated flock of birds (denoted by c_{follow}) are recruited (mimicking social transmission of information at a roost) and follow the bird which found the best location in the last iteration of the algorithm back to that location. The remaining birds return to the best foraging location that they have found so far (simulating a memory). In both cases, if a bird perceives (sees) a better location whilst en route to the relevant location, it may decide stochastically to stop there instead of continuing with its flight. The detail of the operationalisation of the algorithm is provided in the next subsection.

Description of Algorithm

Initially, a *roosting site* is randomly located, and this roost location is then fixed for the remainder of the algorithm. Each of the population of N ravens is placed at a random initial location (a potential food location) in the search space. The fitness values of the N locations are assessed, and the bird at the location of the best solution is denoted by L, for 'leader'.

A roosting process is then simulated by mimicking an information-sharing step. In the next iteration of the algorithm, a proportion c_{follow} of the ravens are recruited to leave the roost site and follow L to the best-so-far location. Each follower bird then selects a random point within a hypersphere of radius r_{leader} around the location previously found by the leader L at which to forage.

As in real-world raven roosting, only a portion of the roost members will be recruited to a new food source, and other roost members will continue to return to a private food location. On leaving the roost, unrecruited birds travel to the best location p_{best} that they have personally found to date, and continue to forage there. The inclusion of a personal best memory for each bird embeds a concept of private information, as unrecruited birds in essence are choosing to rely on private information rather than piggyback on socially broadcast information from the leader L.

Whilst in flight to their intended destination, all birds maintain a search for new food sources en route. This process is simulated by dividing their flight into N_{steps} steps. The length of each step is chosen randomly, and the bird's position in flight is updated using

$$x_i(t) = x_i(t-1) + d_i(t), \tag{7.3}$$

where $x_i(t)$ is the current position of the ith raven, $x_i(t-1)$ is its previous position, and $d_i(t)$ is a random step size. At each step, a raven senses (perceives) the quality of its stopping point and also of its surrounding environment in the range of radius r_{pcpt}. In the latter case, it samples N_{pcpt} perceptions randomly within this hypersphere. If a better location than the bird's personal best is perceived amongst these locations, there is a probability of P_{stop} that the raven stops its flight at that point and forages at the newly found location; otherwise, it takes another flight step and continues to move to its destination. At the conclusion of the algorithm, the highest-quality location found is returned. Pseudocode is provided in Algorithm 7.2, supplemented by Figs. 7.1 and 7.2, which illustrate the following and step processes.

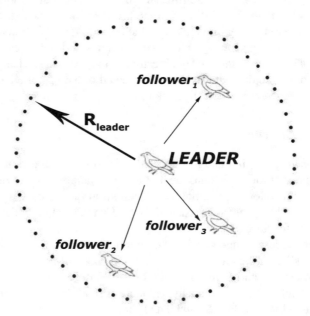

Fig. 7.1. Illustration of the following process in the RRO algorithm

---

Algorithm 7.2: Raven roosting optimisation algorithm

Set values of N, c_{follow}, r_{leader}, N_{steps}, r_{pcpt}, P_{stop};
Randomly select a roosting site;
repeat

    Assign the N foraging ravens to N random locations in the search space;
    Evaluate the fitness of each raven location;
    Update the personal best location of each raven;
    Denote the location of the best solution by L;
    Recruit proportion c_{follow} of the N foragers from the roosting site to search in the
        vicinity of L (within the range of radius r_{leader}), while the rest of the ravens will
        seek to travel to their personal best locations;
    Set $t = 0$;
    while $t < N_{steps}$ **do**
        On the way to its destination (whether the destination is the vicinity of L or its
            personal best location), each raven flies for a while and searches in the
            vicinity of its current position (within a range of radius r_{pcpt});
        if *a better solution is found than the bird's personal best* **then**
            With probability P_{stop} the raven will stop;
            Update the personal best location (p_{best});
        else
            The raven continues to fly;
        end
        $t := t + 1$;
    end
    for *each raven which finally arrives at its destination (the vicinity of L or its*
    personal best) **do**
        Update its personal best location if necessary;
    end
    Update location of the best solution found so far if necessary;
until *terminating condition*;
Output the best solution found;

Discussion

The RRO implements a specific instance of the foraging metaframework described above, embedding sensory perception, memory, social communication and a stochastic component. Perception is implemented in RRO, in that followers or birds returning to a private location can sample other locations en route and stochastically stop at these if they find a better location than their p_{best}. Memory is embedded, as each bird remembers the best resource location that it has found to date (p_{best}). Social communication is implemented through the follower mechanism, whereby the bird which has found the best location to date recruits a number of other birds (followers) and leads them back to this location. However, not all birds in the roost are recruited, and therefore the best-so-far location does not influence the search process of all birds in each iteration of the algorithm. This helps to avoid premature convergence.

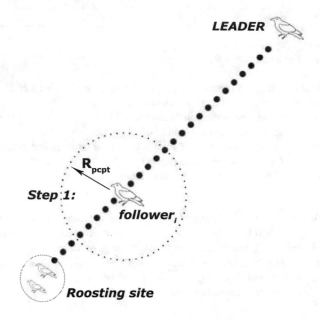

Fig. 7.2. Illustration of the step process in the RRO algorithm

A stochastic element is embedded in the algorithm, as the length of each step taken is chosen randomly as each bird flies, and the perception samplings at each step are drawn randomly from within a hypersphere at each of these locations.

7.3 Summary

From a human-centred starting point of a century ago which considered most animals as 'primitive', we have come to a deeper understanding of the true sophistication of the capabilities and behaviours of other species, including birds. In essence, we have a greater appreciation that the sensory and cognitive capabilities of an animal are critically driven by the needs of the specific environment which it inhabits and not all of these capabilities will be easily apparent to a human observer who inhabits a distinctly different ecological niche. In this chapter, two exemplars of the foraging behaviours of birds, one drawn from the brood parasitism of some species of cuckoo and the other drawn from the foraging behaviour of juvenile ravens, were used to inspire the design of two distinct optimisation algorithms. In the next chapter we introduce a number of algorithms which have been inspired by the foraging behaviours of fish.

8

Aquatic Foraging Algorithms

Marine environments cover approximately 70% of the earth's surface and encompass over 90% of its habitable biosphere [586]. In spite of the dramatic gradient in global biodiversity between marine and terrestrial environments (marine environments only contain about 15–25% of all species [586]), the former are still teeming with life, encompassing fish, plants and other organisms. The approximately 33,300 species of fish are split into teleosts or 'bony fish', comprising about 31,300 species (representing some 60% of all vertebrate species on the planet), and 1300 cartilaginous species of fish (these species have a skeleton made from cartilage instead of bone) such as sharks, rays and skates [26].

The nature of aquatic environments produces some distinctive characteristics in fish [26]. The density of water is 800 times that of air, so streamlined shapes which minimise drag during movement are favoured. Water is a good conductor of sound waves, and these are used extensively by fish for orientation and communication. Fish have good hearing (as first demonstrated in the 1930s by the Nobel Prize-winning biologist Karl von Frisch; von Frisch also decoded the dance language of honeybees; see Sect. 10.1) and have evolved many mechanisms for producing sound. Water is also a good medium for the dispersal of water-soluble chemicals and, consequently, fish are generally capable of chemical sensing. Some species of fish are thought to have magnetic sensory capabilities (particularly long-distance migratory species), and more than 300 of the teleost species are electrically sensitive. Most species of fish also have vision.

A practical issue in the study of fish is that their behaviours are more difficult to observe (being below water) than those of terrestrial animals, and hence relatively little is known about many species. However, it is well known that some species display complex group behaviours, including clustering together and executing coordinated movement in their environment. There is also evidence that groupings of fish can exhibit swarm intelligence, arising from the interactions of many individuals.

A number of optimisation algorithms have been designed taking inspiration from various aspects of the behaviours of fish, and two of these, the *fish school algorithm* and the *fish algorithm*, are overviewed in this chapter. We also discuss the *krill herd*

© Springer Nature Switzerland AG 2018
A. Brabazon, S. McGarraghy, *Foraging-Inspired Optimisation Algorithms*,
Natural Computing Series, https://doi.org/10.1007/978-3-319-59156-8_8

algorithm. Initially we provide some background on fish schooling behaviours and then discuss each of the algorithms in turn.

8.1 Fish Schooling

A particular example of an *aggregation economy* (Sect. 4.4.2) is exhibited by some social species of fish which *shoal*. Shoaling behaviour occurs when fish are observed to cluster together. If the fish also demonstrate a tendency to swim in the same direction in a coordinated manner, they are said to *school*.

These behaviours are common. Approximately a quarter of fish species, including tuna, herring and anchovy, shoal for their entire lives (this behaviour is termed *obligate shoaling*) and approximately half shoal for at least part of their lives (this behaviour is termed *facultative shoaling*). More than 4000 species of pelagic fish are known to school [549]. These aggregations can be very large, with herring forming schools of a billion or more fish on occasion [449].

8.1.1 Benefits of Shoaling and Schooling

The benefits of shoaling and schooling include enhanced defence against predators, as the shoal possesses 'many eyes' (or 'distributed sensing') and has a high level of vigilance. There is also better protection from individual capture by predators due to the *predator confusion effect*, where the presence of many moving prey overloads a predator's visual channel, making it difficult for the predator to visually isolate a single prey item.

A shoal may also exhibit enhanced foraging success, since many eyes search for food, and information on food finds can transmitted through the shoal, as the fish can visually monitor each other's behaviour. Another claimed benefit of schooling is increased hydrodynamic efficiency as the school moves through the water [549].

Schooling may also reduce or even eliminate the need for sleep, therefore reducing predation risk. During waking, the brain of most vertebrates is busy processing sensory, particularly visual, information. This conflicts with the need to refresh and consolidate memories [310]. During schooling, the need for sensory processing by fish inside the school is greatly lessened, with the burden of this processing being shifted from individuals to the entire school [310]. Schooling behaviours may therefore play a role similar to that of restful waking or sleep in nonschooling fish species.

Shoaling behaviour also carries certain costs, including oxygen and food depletion in the vicinity of the shoal.

8.1.2 How Do Fish Schools Make Decisions?

A natural question facing any modeller who is seeking to develop an optimisation algorithm using fish-school-inspired behaviours is how do fish schools actually make decisions and, critically, is there any theoretical reason to suppose that distributed sensing can generate better-quality decisions than those which could be made by an individual fish?

When we consider the dynamic environment which faces a school of fish, it is apparent that many complex decisions are faced. In which direction should it swim if approached by a predator? When should it stop and forage? When and where should it migrate?

Each of the fish in a school has similar sensing capabilities and similar behaviour patterns for acting on sensory information [549], with the collective schooling of fish creating in essence an 'informationally-connected emergent creature' [216]. A study [561] has suggested that fish schools may implement a form of consensus-based decision making employing a simple quorum rule. Under a quorum rule, an individual's probability of committing to a particular decision option increases sharply when a threshold number of other individuals have committed to it. If individuals can observe the decisions of others before committing themselves to a decision such as in what direction to swim next, a relatively naïve copying behaviour can be an effective strategy for successful decision making, without the need for individuals to undertake complicated information processing.

Distributed perception and quorum decision processes can combine to create swarm intelligence (Sect. 5.4) which can replace the need to undertake complex cognition at agent level. It also allows robust decision making to take place even when individual perceptions are noisy. The quality of the decision and the size of the group are highly correlated, so the quality of the decision improves as group size increases [561]. This suggests that fish school behaviours can form a useful platform for the development of optimisation algorithms (see Sect. 10.4 for an example of quorum-based decision making by honeybees).

8.2 Fish School Search Algorithm

One of the better-known fish school algorithms, *fish school search* (FSS), was developed by Bastos et al. [34]. In this algorithm, fish swim (search) to find food (candidate solutions) in an aquarium (search space). No explicit memory of the best locations found to date in the search process is maintained; rather, the weight of each fish acts as a memory of its individual success to date during the search process, and promising areas in the search space can be inferred from regions where larger ensembles of fish gather. The location of the centre of mass (barycentre) of the whole school of fish is considered to provide a proxy for this (Fig. 8.1). In designing the algorithm, the authors considered six design principles to be important, namely:

  i. simple computation at each agent;
 ii. creative yet simple means of storing distributed memory of past computations in the fish weights;
iii. local computations (preferably in small radii centred on each fish);
 iv. simple communication between neighbouring individuals;
  v. minimum centralised control (preferably only the barycentre information is exchanged); and

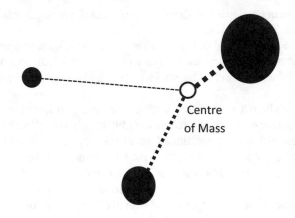

Fig. 8.1. Three objects with their associated centre of mass

vi. simple diversity-generating mechanisms among individuals.

Considering the impact of each of the above on any resulting algorithm, item (i) reduces overall computation cost, item (ii) allows adaptive learning, items (iii), (iv) and (v) keep computation costs low and allow some local knowledge sharing, promoting convergence, and item (vi) is a useful mechanism in a dynamic setting.

8.2.1 Algorithm

FSS [34, 35] has two primary operators which are inspired by fish behaviour. These are *feeding*, inspired by the natural instinct of fishes to feed (food here is a metaphor for the evaluation of candidate solutions in the search space), and *swimming*, which aims to mimic the coordinated movement of fish in a school. The operationalisation of each of these behaviours in the algorithm (Algorithm 8.1) is discussed below.

Individual Movement

A swim direction is randomly chosen. Along each dimension of the search space, a random variable r drawn from a uniform distribution $U(-1, 1)$ is multiplied by a fixed step size s_{ind}. If the food density (fitness) at the resulting location is greater than that at the fish's current location, and the new location is within the feasible search space, then the fish moves to the new location:

$$x_j(t+1) = x_j(t) + rs_{\text{ind}};\tag{8.1}$$

otherwise, no move is made.

In order to promote exploitation as the algorithm progresses, the parameter s_{ind} decreases linearly as the algorithm iterates (up to t_{max}, the maximum number of iterations):

Algorithm 8.1: Fish school search algorithm

Initialise all fish to random locations;
repeat
    for *each fish in turn* **do**
        Evaluate fitness function;
        Implement individual movement;
        Implement feeding;
        Evaluate fitness function;
    end
    for *each fish in turn* **do**
        Implement instinctive movement;
    end
    Calculate barycentre;
    for *each fish in turn* **do**
        Implement volitive movement;
    end
    Update step size;
until *termination condition is satisfied*;
Output the best solution found;

$$s_{\text{ind}}(t+1) = s_{\text{ind}}(t) - \frac{s_{\text{ind}}^{\text{initial}} - s_{\text{ind}}^{\text{final}}}{t_{\max}}. \tag{8.2}$$

Feeding

As a fish moves in the search space, it gains weight in proportion to its success in finding food (fitness). Weight is gained if the fitness at the current location of a fish is better than the fitness at its previous location; weight is lost otherwise. The value of weight is constrained to lie in the range 1 to $w_{\max}$ and all fish are initialised to a weight of $w_{\max}/2$ at the start of the algorithm. The starting weights for all fish are chosen randomly at the start of the algorithm:

$$w_i(t+1) = w_i(t) + \frac{\Delta f_i}{\max\{\Delta f\}}, \tag{8.3}$$

where $w_i(t)$ is the weight of the fish i, Δf_i is the difference in the fitness between the previous and the new location, and $\max(\Delta f)$ is the maximum fitness gain across all the fish. As discussed above, $\Delta f_i = 0$ for any fish which do not undertake individual movement in that iteration.

Collective–Instinctive Movement

After the individual movement step is completed for all fish (as above, not all fish will actually move), a weighted average of all movements in the school is computed. Fish with successful moves influence the resulting direction of movement of the entire school, and fish which did not update their position in the individual movement step are ignored. This overall movement direction is computed as

$$m(t) = \frac{\sum_{i=1}^{N} \Delta x_i \Delta f_i}{\sum_{i=1}^{N} \Delta f_i}. \tag{8.4}$$

When the overall direction is computed, all fish are repositioned, including those which did not undertake an individual movement, as follows:

$$x_i(t+1) = x_i(t) + m(t). \tag{8.5}$$

Collective–Volitive Movement

After the individual and collective–instinctive movements, a final positional adjustment is made based on the overall weight variation of the school as a whole. If the school is putting on weight collectively (and, therefore, it is engaging in successful search), then the radius of the school is contracted in order to concentrate the search. Otherwise, the radius of the school of fish is increased in order to enhance the diversity of the school. The expansion or contraction occurs by applying a small step drift to the position of every fish in the school with respect to the location of the barycentre of the entire school.

First, the barycentre b (centre of mass) needs to be calculated:

$$b(t) = \frac{\sum_{i=1}^{N} x_i w_i(t)}{\sum_{i=1}^{N} w_i(t)}. \tag{8.6}$$

If the total weight of the school has increased during the current iteration, all fish update their position (including those which did not undertake an individual movement) using

$$x_i(t+1) = x_i(t) - s_{\text{vol}} r \frac{x_i(t) - b(t)}{d(x_i(t), b(t))}, \tag{8.7}$$

or, if the total weight has decreased,

$$x_i(t+1) = x_i(t) + s_{\text{vol}} r \frac{x_i(t) - b(t)}{d(x_i(t), b(t))}, \tag{8.8}$$

where $d(\,,\,)$ is a function which returns the Euclidean distance between x_i and b, and r is a random variable drawn from the uniform distribution $U(0,1)$. The parameter s_{vol} is a predetermined step size used to control the displacement from/to the barycentre.

Looking at each of the swimming mechanisms above, individual movement is a mix of a stochastic search and personal cognition (as a fish needs to be able to assess the worth of each location). Collective–instinctive movement embeds an element of social learning as fish are assumed to move based on the success of their conspecifics in finding new high-quality feeding locations. Collective–volitive movement embeds an exploration–exploitation adaptation based on the collective success of the school. Therefore, this also embeds an element of social learning.

8.2.2 Discussion

FSS can be considered as a swarm algorithm, as the search process embeds bottom-up learning via information flow between searching agents. Other work which has extended the canonical FSS algorithm includes [35], which analyses the importance of the swimming operators and shows that all the operators have an important impact on the results obtained; [278], which examines a variety of alternative weight update strategies for FSS; [102], which develops a PSO–FSS hybrid called *volitive clan PSO*; and [362], which describes a parallel GPU implementation of FSS.

Other approaches to the design of fish school algorithms have been taken. A related strand of literature concerns the artificial fish swarm algorithm (AFSA) [243, 353]. This embeds a number of fish behaviours, including preying, swarming and following, so that the behaviour of an artificial fish depends on its current state, its local environmental state (including the quality of its current location) and the states of nearby companions. A good review of the recent literature on AFSA, including a listing of some applications, is provided in [432] and the reader is referred there for further information.

8.3 Fish Algorithm

A practical issue that arises in attempting to develop an algorithm based on the behaviour of fish schools is that we have relatively little hard data on the mechanisms which underlie schooling phenomena. At the level of the individual, agents respond to their own sensory inputs, physiological and cognitive states, and locomotory constraints [229]. It is not trivial to disentangle the relative influence of each of these. At group level, it is often difficult to experimentally observe the mechanics of the movement of animal groups or fish schools, and much previous work on developing fish school algorithms has relied on high-level observations of fish behaviour rather than on granular empirical data concerning these behaviours.

A detailed study of the schooling behaviours of *golden shiner fish* was undertaken in [52]. These fish are strongly social and form shoals of some 200–250 individuals in the freshwater lakes where they live. They display a marked preference for shaded habitats. In order to investigate the mechanism underlying the observed collective response of golden shiner fish to light gradients, fish were tracked individually to obtain information on both individual and group trajectories. The study examined the degree to which the motion of individuals is explained by individual perception (sensing of the light gradient by an individual fish) and by social influences based on distributed perception (as captured by the locations of conspecifics). The results indicated that an individual's acceleration was influenced more by the location of conspecifics than by locally perceived environmental gradients. In essence, a problem-solving behaviour (here the tracking of shade) is undertaken by the fish school (a social network), without the individual fish being aware that any computation is taking place [116].

The *fish algorithm* (FA), based on a simplified model of the above, was introduced by Brabazon et al. [75].

8.3.1 Algorithm

In the fish algorithm, the movement of each fish is governed by three factors which are described below. The movement process also embeds a stochastic element. In each iteration of the algorithm, a fish is displaced from its previous position through the application of a velocity vector:

$$p_i(t) = p_i(t-1) + v_i(t), \tag{8.9}$$

where $p_i(t)$ is the position of the ith fish at the current iteration of the algorithm (t), $p_i(t-1)$ is the position of the ith fish at the previous iteration ($t-1$) and $v_i(t)$ is its velocity.

The velocity update is a composite of three elements: the prior period velocity $v_i(t-1)$; an *individual perception* (IP) mechanism $I_i(t)$; and social influence via the *distributed perception* (DP) of conspecifics, $D_i(t)$. The update is

$$v_i(t) = v_i(t-1) + D_i(t) + I_i(t), \tag{8.10}$$

or, more generally,

$$v_i(t) = w_1 v_i(t-1) + w_2 D_i(t) + w_3 I_i(t). \tag{8.11}$$

The difference between the two update equations is that weight coefficients are given to each of the update items in (8.11). In [75], (8.10) was used for the velocity update. While the form of the velocity update bears a passing resemblance to the standard particle swarm optimisation algorithm (PSO) velocity update, in that both have three terms, it should be noted that the operationalisation of the IP and DP mechanisms is completely different from the memory-based mechanisms of p^{best} and g^{best} in PSO. The next paragraphs explain the operation of the two perception mechanisms.

Prior Period Velocity

The inclusion of a prior period velocity can be considered as a proxy for momentum or inertia. Although this feature was not described in the study of golden shiner fish [52], the inclusion of this term is motivated by empirical evidence from the movement ecology literature which indicates that organisms tend to move with a directional persistence [591].

Distributed-Perception Influence

In all social models, a key element is how the overall population influences the decisions of each agent at each time step. Typically, the actions of each agent are influenced by a subset of the population that are within an interaction range. This influence can be modelled in a variety of ways, including the fraction of an individual's neighbours taking a particular course of action or the action of their nearest neighbour. Here, the distributed-perception influence for the ith fish is implemented as follows:

$$D_i = \frac{\sum_{j=1}^{N_i^{\mathrm{DP}}} (p_j - p_i)}{N_i^{\mathrm{DP}}}, \quad j \neq i, \tag{8.12}$$

where p_i is the position of the ith fish, and the sum is calculated over all neighbours within an assumed range of interaction of the ith fish r_{DP}, that is, $0 <| p_j - p_i | \leq r_{\mathrm{DP}}$, where p_j is the position of the jth neighbouring fish and N_i^{DP} is the number of neighbours in the assumed range of interaction of the ith fish. If there are no neighbours in its assumed range of interaction, this term becomes zero. Figure 8.2 shows how the ith fish is affected by the three neighbouring fish p_1, p_2, p_3 which are within its visible range (defined by the radius r_{DP}).

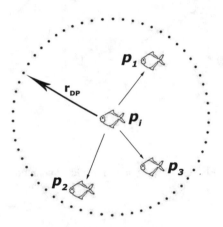

Fig. 8.2. Illustration of distributed perception

Alternative methods of modelling this social influence could be implemented, such as only considering neighbours within the angular visual range of each agent as suggested in [401]. While this would be more plausible from a biological perspective, it would impose additional computational complexity, so the canonical fish algorithm uses a simpler approach and assumes 360° vision.

Individual-Perception Influence

Individual perception is implemented as follows. In every update, each fish assesses the local light gradient surrounding it, by drawing N_i^{IP} samples within an assumed visibility region of radius r_{IP}. While a real-world fish will have a specific angle of vision depending on its own body structure, a random sampling in a hypersphere around the fish is implemented on grounds of generality. The individual-perception influence for the ith fish is determined by

$$I_i = \frac{\sum_{j=1}^{N_i^{\mathrm{IP}}} (s_j - p_i) f_j}{\sum_{j=1}^{N_i^{\mathrm{IP}}} f_j}, \quad j \neq i, \tag{8.13}$$

where p_i is the position of the ith fish, r_{IP} is the radius of the assumed range within which the ith fish can sense environmental information, N_i^{IP} is the number of samples which the ith fish generates, s_j is the position of the jth sample ($0 < |s_j - p_i| \leq r_{IP}$), and f_j is the fitness value (or quality) of the jth sample. Figure 8.3 demonstrates how the ith fish is influenced by the five random samples ($s_1 - s_5$) in the perception range with a radius r_{IP}.

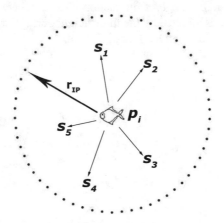

Fig. 8.3. Illustration of individual perception

8.3.2 Discussion

The key parameters which need to be set in applications of the algorithm include the radii of perception r_{DP} and r_{IP}, and the number of samples (denoted by s) used in the simulated IP component. The sensitivity of the results obtained from the algorithm to various choices of parameters was examined in [75].

As would be expected, the choice of perception radius is important. From a biological point of view, it is plausible to assume that fish have a bigger radius for DP than for IP, namely $r_{DP} > r_{IP}$. The value chosen for the two radii will be problem-specific, as it is influenced by the choice of the number of fish (N), the radius (size) of the search space (R) and the dimensionality of this space (d). In [75], the values of r_{DP} and r_{IP} were chosen after initial experimentation as $R/(1.5 \sqrt[d]{N})$ and $R/(1.8 \sqrt[d]{N})$.

Across the mechanisms of prior period velocity (momentum), individual perception and distributed perception, the results of sensitivity analysis indicate that while IP and DP provide important information for the search process, the momentum mechanism does not appear to be as significant. The best results are obtained when using a mix of information from IP and social communication. However, while social communication can usefully spread information on good locations amongst the population of agents, it needs to be supplemented by information from the IP mechanism in order to guide the search process. This finding is in concordance with Rogers' analysis of the producer–scrounger game as discussed in Sect. 4.2.1.

8.4 Krill Herd Algorithm

Krill are small marine organisms, usually about 2–5 cm in length, which are found in all oceans. About 120 species of krill exist, with some species living for several years. Krill do not have bones but possess a hard outer exoskeleton covering, and they belong to the family of crustaceans, sharing this grouping with crabs, lobsters, shrimps and barnacles. They also possess intricate compound eyes and several pairs of legs, which they use for swimming.

Krill play a critical role in the marine food chain as they feed on small organisms such as phytoplankton (microscopic, single-celled plant algae that drift near the ocean's surface) [494]. Phytoplankton play a similar role to that of plants in the terrestrial food web, being a foundational building block for the marine food web. In turn, krill are consumed by larger marine species such as whales, seals, penguins, squid and fish. Krill are filter feeders and typically follow a daily migration cycle, rising to shallower waters to feed, and then sinking after eating to deeper waters where they are safer from predation.

Most species of krill display swarming behaviours and in some cases these swarms can reach densities of 10,000 to 60,000 individuals per cubic metre. *Superswarming behaviour* has also been noted, with some swarms covering 150 km^2 and ranging up to 250 metres in depth [494]. Unsurprisingly, swarms of krill can attract attention from potential predators, especially when they form near the surface. Several rationales for the swarming behaviour of krill have been proposed, including defence (predator confusion), mate attraction, and attraction to food-rich locations.

Gandomi and Alavi [203] introduced the *krill herd algorithm* (KHA)[1] based on a simulation of the herding of krill swarms. Three aspects of krill behaviour are included in the model, namely:

i. movement induced by the presence of other individuals (V);
ii. foraging activity (F); and
iii. random diffusion (D).

The algorithm assumes that krill herding is designed to achieve two goals, (i) increasing krill density and (ii) finding areas of high food density. The developed algorithm has been applied for optimisation purposes, mimicking the process of a herd searching for resources and then concentrating around the best resource available.

8.4.1 Algorithm

In this section we outline the canonical KHA based on its exposition in [203]. The time-dependent position of an individual krill can be generalised to

$$\frac{dx_i}{dt} = V_i + F_i + D_i, \tag{8.14}$$

---

[1] As krill are invertebrates, this algorithm could be placed in Part IV of this book rather than in this chapter. However, given that krill live in a marine environment, it seems more appropriate to include the algorithm here.

where V_i is the motion induced by other krill individuals, F_i is the foraging motion and D_i is the physical diffusion of the ith krill individual. Next, we describe each of these components in some detail.

Movement Induced by Other Individuals

It is assumed that krill try to maintain a high local density and that individual i moves in a direction α_i in order to achieve this. This movement process is modelled as

$$V_i(t+1) = V_{\max}\alpha_i + \omega_m V_i(t), \tag{8.15}$$

where $V_{\max}$ is the maximum induced speed (a fixed parameter whose value is selected by the modeller, which, in conjunction with the choice of inertia weight, determines the balance placed on prior period velocity versus information from other herd members), α_i is the direction of motion induced which is estimated from the target swarm density and the local swarm density, ω_m is the inertia weight of the induced motion in the range $[0, 1]$, and $V_i(t)$ is the prior period induced motion:

$$\alpha_i = \alpha_i^{\text{local}} + \alpha_i^{\text{target}}. \tag{8.16}$$

In (8.16), α_i^{local} is the local effect due to neighbours of the ith individual and α_i^{target} is the target direction effect due to the best krill individual.

The term α_i^{target} for an individual i is calculated as follows:

$$\alpha_i^{\text{target}} = C_{\text{best}}\widehat{f_{i,\text{best}}}\widehat{x_{i,\text{best}}}, \tag{8.17}$$

where C_{best} is a coefficient which determines the relative impact of the krill individual with the best fitness on the movement of the ith individual. The coefficient C_{best} is calculated using

$$C_{\text{best}} = 2(r + I/I_{\max}), \tag{8.18}$$

where r is a random variable drawn from a uniform distribution on $[0, 1]$, I is the iteration number and $I_{\max}$ is the maximum number of iterations.

The movement induced by other individuals is critically determined by α_i^{target}, which encourages an individual krill to move in the direction of the current best individual in the herd, and also by the term α_i^{local} (8.19) which generates movement that is locally influenced by the current neighbours of the krill individual. This movement can be attractive (towards a neighbouring individual) or repulsive (away from a neighbour) depending on the relative scaled fitness of each individual. The term α_i^{local} is defined by

$$\alpha_i^{\text{local}} = \sum_{j=1}^{N_{\text{nbh}}} \widehat{f_{i,j}}\widehat{x_{i,j}} \tag{8.19}$$

where

$$\widehat{x_{i,j}} = \frac{x_j - x_i}{\|x_j - x_i\| + \epsilon} \tag{8.20}$$

and

$$\widehat{f}_{i,j} = \frac{f_i - f_j}{f_{\text{worst}} - f_{\text{best}}}. \tag{8.21}$$

In (8.19)–(8.21), x_i is the position of the ith individual, f_i is the fitness value for the ith individual, f_{best} and f_{worst} are the best and worst fitness values of the krill individuals so far, there are N_{nbh} neighbours of the ith individual and f_j is the fitness of the jth neighbour ($j = 1, 2, \ldots, N_{\text{nbh}}$).

Equations (8.19)–(8.21) contain some unit vectors and some normalised fitness values. As the normalised value can be positive or negative, an attractive or repulsive movement can be generated between individuals.

The operationalisation of this local region around a krill individual requires the selection of a specific neighbourhood definition. In [203], each krill individual is assumed to have a sensory perception range described by a hypersphere, with radius d_s defined as:

$$d_{s,i} = \frac{1}{5N} \sum_{j=1}^{N} \|x_i - x_j\|, \tag{8.22}$$

where $d_{s,i}$ is the sensory range for the ith individual and N is the number of krill individuals. The constant term (5) in the formulation is noted as being empirically determined in [203]. Therefore, all individuals whose distance from i is less than $d_{s,i}$ are considered to be neighbours of i (Fig. 8.4).

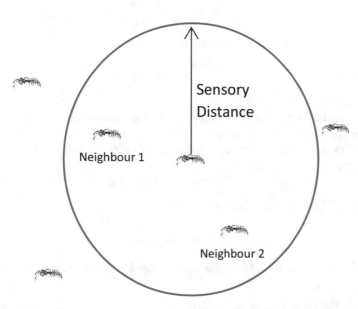

Fig. 8.4. Illustration of sensory range around an individual krill. Here, two other krill are in sensory range

Foraging Movement

The foraging behaviour is modelled using two main mechanisms. The first is the location of food, and the second is the previous experience (memory) of individual krill as to the location of food. Foraging movement is modelled as

$$F_i(t+1) = S_f\beta_i + \omega_f F_i(t), \tag{8.23}$$

where S_f is the foraging speed (a fixed parameter whose value is selected by the modeller), ω_f is the inertia weight of the foraging motion in the range $[0, 1]$ and $F_i(t)$ is the foraging motion in the prior period. The term β_i is calculated as:

$$\beta_i = \beta_i^{food} + \beta_i^{best}, \tag{8.24}$$

where β^{food} is the term which embeds attraction towards food and β_i^{best} embeds the attraction towards the best solution that the ith individual has found so far (mimicking a memory of a previous location of good resources). This formulation determines that the foraging process is a function of food location, previous experience of food location and a momentum term based on prior period movement.

The location of food in each iteration of the algorithm is assumed to be proxied by the centre of mass of the distribution of the fitness values of each krill individual, and is calculated using

$$x_{food} = \frac{\sum_{i=1}^{N} x_i/f_i}{\sum_{i=1}^{N} 1/f_i}. \tag{8.25}$$

The food attraction for the ith krill individual is

$$\beta_i^{food} = C_{food}\widehat{f_{i,food}}\widehat{x_{i,food}}, \tag{8.26}$$

where C_{food} is the food coefficient.

The effect of the foraging movement, therefore, is to encourage the herd of krill to move towards regions of supposed higher fitness. In order to avoid premature convergence in the algorithm, the impact of this term is reduced as the algorithm iterates, using

$$C_{food} = 2(1 - I/I_{max}). \tag{8.27}$$

The term β_i^{best} is determined using:

$$\beta_i^{best} = \widehat{f_{i,ibest}}\widehat{x_{i,ibest}}, \tag{8.28}$$

where $f_{i,ibest}$ is the best location previously visited by the ith individual.

Random Diffusion

Krill are subject to ocean currents and other phenomena which can result in a degree of random diffusion in their movements over time. This effect is also implemented in the model of krill movement created in [203]. In terms of design of an optimisation algorithm, this mechanism helps to maintain continual exploration by the algorithm. In contrast, the two movement mechanisms described above generally encourage the movement of krill individuals towards regions of higher fitness, therefore exploiting information uncovered by the swarm thus far. The diffusion process is modelled using:

$$D_i(t+1) = D_{max}(1 - I/I_{max})\delta, \tag{8.29}$$

where D_{max} indicates the maximum diffusion speed and δ is a random directional vector whose components are drawn from $[-1,1]$. In order to ensure convergence of the algorithm, the impact of diffusion is gradually reduced via the term $(1 - I/I_{max})$ in order to encourage more localised search.

Position Update

Bringing all the components of krill herding (movement) together, the position update of each krill during an interval from t to $t + \Delta t$ is given by

$$x_i(t + \Delta t) = x_i(t) + \Delta t \frac{dx_i}{dt}. \tag{8.30}$$

The selection of an appropriate value of Δt is important in operationalising the general model for a specific optimisation problem. A suggestion for setting the value from [203] is

$$\Delta t = C_t \sum_{j=1}^{N_{var}} (U_j - L_j), \tag{8.31}$$

where N_{var} is the number of variables, and L_j and U_j are lower and upper bounds, respectively, on the jth variable ($j = 1, 2, \ldots, N_{var}$). It is suggested that C_t be set to a value in the interval $[0, 2]$. Pseudocode for the algorithm is outlined in Algorithm 8.2.

Enhancing Algorithm Performance

In addition to the above steps, [203] suggested that the performance of the algorithm could be enhanced by adding simulated genetic operators of crossover and mutation, these steps being added after the physical diffusion step in Algorithm 8.2.

The crossover operator is controlled by a parameter C, where $C = 0.2\widehat{f_{i,best}}$, with $\widehat{f_{i,best}}$ defined as $f_i - f_{best}$. A random number r is generated uniformly in the interval $[0, 1]$, and the mth component $x_{i,m}$ of x_i is determined using

$$x_{i,m} = \begin{cases} x_{r,m}, & \text{if } r < C, \\ x_{i,m}, & \text{otherwise}, \end{cases} \tag{8.32}$$

where $r \in \{1, 2, \ldots, i-1, i+1, \ldots, N\}$. Hence, C for the global best location is zero, and the value increases as the fitness of an individual krill decreases.

Algorithm 8.2: Krill herd algorithm

Define parameters for algorithm (D_{max}, V_{max}, S_f etc.);
for *each krill $i = 1,\ldots,N$ in turn* **do**
> Generate random location x_i;
> Determine fitness value $f_i = f(x_i)$;

end
Determine values of best solution, worst solution, location of virtual food (centre of
   mass) etc.;
Save the location x^* of the best individual;
repeat
> Sort the krill population from best fitness to worst;
> **for** *each krill $i = 1,\ldots,N$ in turn* **do**
> > Implement movement induced by presence of other individuals;
> > Implement movement due to foraging;
> > Implement physical diffusion;
> > Update individual krill position;
> > Determine fitness value for krill individual;
>
> **end**
> Update best fitness location x^* if necessary;

until *termination condition is satisfied*;
Output the best solution found;

The mutation scheme suggested in [203] is controlled by a mutation parameter
M, where $M = 0.05/\widehat{f}_{i,\text{best}}$. The mutation scheme is therefore

$$
x_{i,m} = \begin{cases} x_{\text{gbest},m} + \mu(x_{p,m} - x_{q,m}), & \text{if } r < M, \\ x_{i,m}, & \text{otherwise,} \end{cases} \tag{8.33}
$$

where $p,q \in \{1,2,\ldots,i-1,i+1,\ldots,f\}$ and μ is a number $\in [0,1]$. Since $\widehat{f}_{i,\text{best}}$ is $f_i - f_{\text{best}}$, the mutation probability for the global best is zero, and it increases as the fitness
of an individual decreases.

8.4.2 Variants of the Canonical Algorithm

Following the introduction of the canonical KHA, there have been several papers
which have developed variants of the KHA and which have applied it to real-world
problems. The performance of the algorithm has been benchmarked against other
recently developed stochastic search algorithms and has generally been found to be
quite competitive. An exemplar of these studies is provided in [531], which examined
the performance of the KHA on a range of unimodal and multimodal problems,
benchmarking its performance against the firefly and cuckoo algorithms. All three
algorithms performed well on the benchmark problems, with the cuckoo algorithm
performing best, and similar levels of performance being exhibited by the KHA and
the firefly algorithm.

A number of studies have sought to increase the convergence speed of the KHA. A mutation process based on harmony search was added to the KHA in [610], replacing the random diffusion process. This was found to speed up convergence and to encourage a greater degree of local search in the resulting hybrid algorithm. The chaotic krill herd optimisation algorithm was introduced in [504]. This variant of the KHA uses chaotic maps whenever a random variable is needed in the algorithm, for example as in (8.17). The results of this study indicate that the resulting algorithm outperforms canonical KHA in terms of avoiding capture by local optima.

8.4.3 Discussion

The search process in the KHA is driven by three simulated behaviours: a tendency of krill to herd; an attraction towards food; and a random diffusion component. No claim is made by the designers of the algorithm that it fully captures all the behaviours of krill herding and movement. Nonetheless, the algorithm has shown itself to be capable of competitive results on benchmark and practical problems. Two components of the algorithm which differentiate it from others include the embedding of both attraction and repulsion, with individual krill being drawn to other krill of higher fitness but being repelled from lower-fitness regions. The algorithm also makes use of global information in the herd, via the *centre of food*, whereby the fitness information across all krill is used to estimate where regions of high fitness may lie, with individual krill being attracted to that location.

A drawback of the algorithm is that it is relatively complex, embedding multiple mechanisms which contribute to the movement process. As a result, it is not easy to ascertain which of these mechanisms are most important in driving an effective search process. A study [329] examined the impact of varying the values of V_{max} and ω and found that smaller values of these parameters were preferred when tackling unimodal optimisation problems, with larger values giving better results on more complex multimodal problems.

The KHA is a relatively new algorithm, and further research is required in order to better understand how the algorithm should be tailored for different problem settings.

8.5 Electrolocation in Fish

As discussed in Sect. 3.3.5, some animals, notably multiple species of fish, including lampreys, paddlefish, catfish, sharks, skates and rays, are capable of perceiving electric fields in their vicinity. The information from this sensory modality can be used to build up an image of their surroundings. Perhaps the best-known example of passive electrolocation is that of sharks. Although scientists had long known of the existence of the anatomical structures in sharks called *ampullae of Lorenzini* after their discoverer, the Italian anatomist Stefano Lorenzini, who first described them in 1678, their function was unknown until about 1960, when it was discovered that they are very sensitive to weak electric fields [423, 424]. The ampullae are connected to sea-water by pores on the snout and other zones of the head of sharks and rays. The use of this sensory modality for prey detection by sharks was demonstrated by Kalmijn [297]. Sharks use multiple senses in hunting. At longer distances they rely on smell and hearing but as a shark gets closer to prey, its vision and mechanosensory lateral line become more important. In the final stage of a shark attack, electroreception becomes the primary tool for the pinpointing prey location [180]. Hence, the ampullae of Lorenzini effectively act as a passive radar system, detecting naturally occurring electric fields such as those generated by living creatures.

8.5.1 Active Electrolocation

In contrast to passive electrolocation, some species of fish have evolved a capability for active echolocation by emitting an electric field and monitoring the distortion in this caused by neighbouring objects. These fish may be:

- strongly discharging, or
- weakly discharging.

Strongly electrical fish can generate discharges of up to 500 volts for predatory purposes, and examples of these include African electric catfish, electric eels and marine torpedo rays. Weakly electrical fish consist primarily of two groupings of fish, mormyrids and gymnotiforms. The family of African mormyrids is composed of more than 200 species of freshwater fish. They are usually found in sluggish, muddy water in which a visual sensory modality would be of limited benefit. The gymnotiform group of species are found in fresh water in South America and are mostly nocturnal fish [598]. Both of these groupings are capable of active electrolocation, as they can produce weak electric fields using a specialised electric organ in their tail and possess sensor organs under the skin along both sides of their bodies. By sensing distortions in the electric field reflected back onto their body from objects around them, they are able to form an impression of what surrounds them [631]. Gymnotiforms evolved their electric sense independently from the African mormyrids. While each group is capable of active electrolocation, the structures of their electric organs and electrosensory structures differ [598]. Active electrolocation is much rarer than passive electrolocation and was first discovered in the early 1950s by Hans Lissmann [363].

In active electrolocation, fish produce a series of electric signals (called *electric organ discharges*, or EODs), creating an electric field shaped like a dipole, with field lines describing a curved arc from positive pole to negative, around their body [598]. Each EOD builds up a stable three-dimensional electric field around the fish and this causes a specific spatial voltage pattern over the fish's skin surface that does not change as long as objects do not enter the electric field. EODs are species-specific and consist of either continuous sinusoidal signals or short pulses.

Neighbouring objects may conduct electricity better than or not as well as the surrounding water, or perhaps not conduct electricity at all. If an object conducts electricity better than water, the electric field discharged by the fish will be attracted to the conductor (as more current flows through the low-impedance object compared with the water it replaces) and will bend towards it, producing an area of concentrated current entering the skin of the fish opposite the object which can be detected by the receptors on the fish's body. In contrast, a less conductive object (an object with higher impedance than water) will causing a thinning of current lines in skin areas neighbouring such objects.

The resulting 'electric image' on the surface of the skin of the fish depends on the object's electrical properties, size, shape and location [598]. Electric fish detect these distortions as changes in the transepidermal electric current flow on their electroreceptive skin surface and interpret these to gain information about objects in their vicinity [600, 599]. If the object is alive, it will have complex impedances and will also change the waveform of the locally occurring EOD waveform (nonliving things will not) [600]. This information is clearly of relevance in a foraging scenario. The systems for active electrolocation and interpretation of the resulting feedback have been carefully adapted by evolution, resulting in coevolution of the mechanisms for signal production, signal perception and signal analysis by the brain [599].

In contrast to other signalling modes which use waves (for example, sound), the signal magnitude of electric discharges decreases as the inverse cube of distance $1/(r^3)$. This makes signal generation an energetically costly process and limits the range of the carrier. However, unlike sound waves, electric fields are subject to very little distortion from echo and reverberation effects. The range of active electrolocation varies according to the size of the fish, the size of the object, and the impedance of the object and surrounding water. Typically, the electric fields produced by biological organisms like fish attenuate rapidly in aquatic environments and have a field magnitude of around zero at ranges of a few metres [525, 598].

At a high level, the process of active electrolocation bears some comparison with that of echolocation. In each case the organism uses active sensing of its environment and expends energy to emit a signal, electrical in the case of active electrolocation and acoustic in the case of bats, and uses information arising from distortions in this signal to form an 'image' of its surroundings. In both cases, the signal emitted by the actively sensing animal is the carrier for the information extracted from the environment. As with echolocation, inspiration could potentially be taken from aspects of electrolocation in order to design a search algorithm.

8.6 Summary

Fish school algorithms embed a mix of local search (or sensing), group-influenced movement and stochastic movement. There are parallels with swarm intelligence in that some of these algorithms do not incorporate significant information-processing capabilities on the part of individual agents (fish), with the search process being driven by emergent effects caused by local information rippling through the population. The resulting algorithms have shown promising results when tested against benchmark and real-world problems. The fish perception algorithm embeds a greater degree of individual agent (fish) perception in the search process. The chapter has also introduced the krill herd algorithm.

A practical issue in developing these algorithms is that we still have a very limited understanding of the mechanisms which underlie schooling phenomena, the decision processes of fish schools, and the relative weight that different species of fish place on private and social information in decision making. To date, fish school algorithms have relied on very abstract models of fish behaviour. We can expect to see notable refinement of these algorithms as our knowledge of these behaviours improves.

Part IV

Invertebrates

.

Over the next five chapters we explore a series of algorithms which have been inspired by the foraging activities of a variety of invertebrate organisms. Of the estimated total of 1.4 million known species of animals, more than 1.3 million are invertebrates, including approximately 1 million species of insects [628]. Insects, arachnids (which include spiders and scorpions), crustaceans (including crabs and lobsters), and centipedes and millipedes, make up the *Arthropods*, a grouping of species that have a hard external skeleton and jointed limbs. Other examples of invertebrates include snails, clams, octopuses, starfish, sea-urchins, jellyfish, and worms.

While possessing a central nervous system, these organisms do not have a spinal column with an internal skeleton made of bone, but may have a shell or exoskeleton to protect their bodies. Central nervous systems come in varying degrees of complexity. For example, there are some 200 million neurons in the nervous system of a rat, some 3–6 billion neurons in monkeys, depending on species, and some 86 billion neurons in humans. Owing to their small size, most of the organisms considered over the next few chapters have much more limited nervous systems, with bees having some 950,000 neurons, ants and flies typically having some 250,000 neurons, and the worm *C. elegans* having only 302 neurons.

Despite the lower neural and cognitive complexity of invertebrates relative to larger animals, they are capable of quite sophisticated foraging behaviours. As discussed in Sect. 3.4, enhanced cognitive or other capabilities will not necessarily produce better net foraging outcomes. There are many ways to earn a living and not all involve the complex neural circuitry of higher animals.

The best-known and most widely applied families of algorithms inspired by the foraging activities of invertebrate organisms are ant algorithms and honeybee algorithms. These are explored in Chaps. 9 and 10, respectively. We then introduce algorithms inspired by the foraging activities of bioluminescent insects (Chap. 11), social spiders (Chap. 12) and the nematode worm *C. elegans* (Chap. 13).

9

Ant Foraging Algorithms

It is estimated that there are some 22,000 species of ants, with a total combined population numbering some 10^{15} to 10^{16} [263]. Equivalently stated, this amounts to about 1 million ants for each human being on the planet. Ants have successfully colonised almost every landmass on earth and exploit a wide range of food resources as generalist predators, scavengers or herbivores. While most ant species are *eusocial* (eusocial behaviour involves individuals living cooperatively to raise a brood, where the majority of individuals forgo reproduction in favour of a queen), colony sizes can vary considerably, with colonies in some species consisting of a few dozen members and in other cases consisting of societies of millions of individuals. Colonies consist mainly of sterile females, which are divided into castes based on work tasks. Ant societies can be quite long-lived, with queens in some species surviving for decades and (female) workers surviving for up to a year.

The activities of ant colonies have attracted close study, as they frequently exhibit collective problem-solving capabilities in colony construction, foraging and brood care. Rather than being top-down and directed, many of these behaviours are *emergent*, arising from the actions of individual ants responding to cues in their local environment. In essence, as noted in [176] the group-level cognition exhibited in many ant species involves the emergent behaviour of assemblages of workers and social networks rather than requiring a significant quantity of energetically expensive neuronal circuitry on the part of each ant in order to produce adaptive decision making.

In this chapter we introduce a number of algorithms which have been inspired by the food-foraging behaviours of ants. Initially, we provide some background on ant foraging behaviours in order to contextualise the algorithms.

9.1 Ant Foraging

Oster and Wilson [444] classify the foraging strategies of social insects into five groupings (Table 9.1). Each of these strategies can form the basis for the design of an ant-inspired optimisation algorithm. In group I, ants search individually and

© Springer Nature Switzerland AG 2018
A. Brabazon, S. McGarraghy, *Foraging-Inspired Optimisation Algorithms*,
Natural Computing Series, https://doi.org/10.1007/978-3-319-59156-8_9

Table 9.1. Oster and Wilson's classification of insect foraging [444]

Group	Description
I	Solitary insects find and retrieve prey individually
II	As in I but solitary foragers signal the prey location to other insects
III	Foragers depart the nest and follow *trunk trails* before branching off to search new terrain
IV	As in II except a group of insects assaults or retrieves prey en masse
V	Multiple insects forage as groups

do not communicate with each other. In contrast, groups II to V embed social foraging mediated by communication mechanisms. Ants possess a range of sensory mechanisms, including sight, touch, sensitivity to vibrations and sensitivity to chemicals. Communication among ants can therefore utilise any of these sensory channels. Three primary means of communication are observed in social insects:

i. indirect, or *stigmergic*, communication;
ii. direct interaction of individuals, where the actions of one individual influence those of another; and
iii. direct (nonphysical) communication between individuals.

Stigmergic communication [223] arises when individual members of a group communicate indirectly, by altering the environment faced by their peers. This alteration in the environment has the effect of influencing the subsequent behaviour of other members of the group. Examples of stigmergy include the chemical marking of a region or location as being *attractive* or *unattractive*.

Direct interaction occurs via mechanisms such as touch (for example, *antennation* in which insects rub antennae against each other in order to communicate chemical information about food, hunger levels, nestmate recognition, sexual identification etc. [44]) or phenomena such as *stridulation*, whereby individual ants use vibrational signals to communicate with other ants, for example to recruit them for a specific task [262, 511].

Although ants use a variety of communication modes for transmission of foraging-relevant information to peers, the use of pheromones for the chemical marking of trails to food sources has attracted most attention in the design of ant-inspired optimisation algorithms. In the next section we delve into this signalling mechanism in more depth.

9.2 Recruitment Pheromones and Foraging

The realisation that ants use chemical signposts to locate food resources and recruit conspecifics to foraging locations has a long history, dating at least to John Lubbock, a British banker, parliamentarian and amateur biologist in the 18th century who observed that ants followed their nestmates to food sources. Lubbock was the first to suggest that ants were actually following scent trails to these food sites [375]. Subsequent studies uncovered the workings of this mechanism.

Consider, as an exemplar, the Argentine ant *Linepithema humile*. This species of ant lays down a trail of chemical attractant (pheromone) on successfully finding food resources when returning to the colony nest. Subsequent foraging ants are inclined to follow these trails during their own foraging efforts. This simple trail-following behaviour acts to ensure that the ant colony's foraging activities are efficient and that foragers can be concentrated into a small region, in space and time, where the food is located.

If a group of ants search randomly around their nest for food, the strength of pheromone trails from the nest to nearby food sources will tend to grow at a faster rate than that of the trails to faraway food sources as ants travelling to the closest food source will return more quickly to the nest (and, therefore, further reinforce the trail). Once foraging ants have a tendency to follow stronger rather than weaker pheromone trails, *autocatalytic* behaviour will emerge, with an increasing portion of the foraging ants travelling along the strong trail, reinforcing it further (Fig. 9.1). This creates a positive feedback loop among the ants in the search for food.

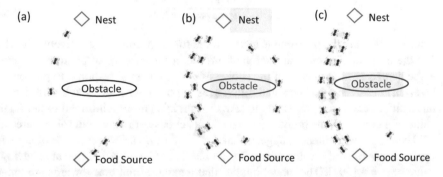

Fig. 9.1. Foraging ants reinforcing the shorter trail to a food source. In (a), three ants have discovered the food source, having gone round different sides of the obstacle. Those which travelled the shorter path (on the left-hand side) return more quickly to the nest and reinforce their trail. In (b), as the shorter trail becomes reinforced, most ants travel on this route. By time step (c), virtually all ants are using the shorter route as pheromone on the longer trail evaporates

The effect of pheromone deposit behaviour is to create an indirect communication mechanism between ants. As trails emerge over time, a collective *memory*, chemically embedded in the environment, is created as to the route to the food source. One feature of this communication mechanism is that it is scalable, as an individual ant does not need to directly communicate with every other ant in the colony in order to pass on the knowledge it uncovers during its foraging travels.

Apart from providing a guide to naïve foragers as to the location of food re-sources, pheromone trails also help ensure that ants can efficiently find their way back to the colony nest when returning from foraging, thereby increasing the net energy gain from each foraging expedition.

Recent research also suggests that the mere presence of nestmates on a trail can encourage other ants to follow it, as nestmates on a trail provide a signal as to food availability and also provides reassurance as to the safety of the trail from predation [123].

9.2.1 Danger of Lock-In

One potential drawback of a positive feedback learning mechanism is that it can lead to *lock-in*, whereby a heavily pheromone reinforced path continues to be used, even if a rich food source subsequently becomes available closer to the nest.

This phenomenon is illustrated in a *choice function model* initially developed by Deneubourg et al. [144]. In the model, an ant has a choice between two routes X and Y. The probability that the ant takes route X is proportional to

$$\frac{(x+k)^a}{(x+k)^a + (y+k)^a},\tag{9.1}$$

where x and y are the pheromone levels on the routes X and Y respectively. The term k is the intrinsic degree of attraction of a route in the absence of pheromone, and a is the degree of nonlinearity of the choice. If $a > 1$, the ants' response to pheromone levels is nonlinear and even small differences in pheromone levels will give rise to a large difference in the propensity to take the (slightly) more reinforced route. As the value of a increases, the probability that all ants choose the same route also increases.

Experimental evidence suggests that the value of a varies by species of ant with *Lasius niger* having a value of $a = 2$ (and $k = 6$) [41], and Argentine ants having a value of $a = 4$ [594]. The model implies that a small initial bias towards the longer route, depending also on the rate of evaporation of pheromone in the environment and the flow of ants from the nest, could result in the longer route becoming locked in. As the value of a increases above one, the probability of this increases.

Ant Mill

An extreme case of dysfunctional lock-in is illustrated by the well-known *ant mill* phenomenon (Fig. 9.2) which is sometimes exhibited by groupings of army ants. Army ants consist of some 200 species of ant and are best known for their aggressive foraging, in which huge groupings, sometimes consisting of up to 200,000 ants [188], can forage in the same area. Most female army ants are nearly blind and in contrast to many other social insects, which use multiple sensory inputs in foraging and trail running, these ants place strong reliance on pheromone-following behaviours. If a grouping of foraging army ants are isolated from the main foraging party, they can lose the pheromone track of the larger group and begin to follow one another, forming

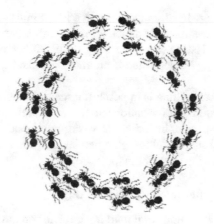

Fig. 9.2. An illustration of an ant mill, with a foraging subgroup engaged in dysfunctional pheromone trail following

a densely packed rotating circle of ants. In extreme cases, this can persist until all the ants die of exhaustion [42, 139, 510].

A number of pheromone deposit and signalling mechanisms act to reduce the risk of lock-in and these are discussed next.

9.2.2 Pheromone Signalling Behaviours

Although most optimisation applications of ant foraging algorithms rely on positive reinforcement of routes to high-quality solutions via a simulated pheromone deposit process, real-world ant-foraging pheromone signals can be much more complex. For example, some species of ants, such as *L. niger*, encode information as to food quality (sugar content and volume) in the deposited pheromone.

As well as positive feedback, pheromones can be laid down by ants which discourage other ants from following a particular trail. These can provide *negative feedback* to allow a colony to switch away from resources as they deplete [231]. An example of this is provided by Pharaoh's ants (*Monomorium pharaonis*), which can deposit repellant pheromone as a *no entry* signal to mark a foraging path which has become unrewarding [487].

The quantity of pheromone deposited when marking trails is not necessarily constant. In some species, pheromone deposition is strongly upregulated if ants are returning from a food source on an unmarked path (i.e. a new food source) [125]. Crowding at feeding sites or along a trail to a feeding site can lead to the downregulation of the deposit of pheromone by ants on that trail [124], thereby encouraging ants to forage elsewhere.

Ants are also capable of depositing more than one form of pheromone, with varying pheromones being laid down by different glands on an ant. Some trail-marking pheromones can be short-lived, evaporating in a few minutes (for example, when recruiting other ants to communally attack a prey item), whereas others can be more persistent, potentially lasting for some days [276]. Other pheromones, known as alarm pheromones, may be deposited if an ant is attacked while transiting along a trail.

While pheromone signalling is important in many ant species, ants typically use multiple sources of external information and do not usually follow trail pheromones blindly [123]. Some species, including harvester ants, place reliance on individual memory of previously successful foraging patches and accord a lower importance to pheromone-trail-following behaviours [559].

9.2.3 Limitations of Pheromone Trail Marking

The effectiveness of pheromone trail marking depends on environmental factors and colony size. The desert ants *Cataglyphis fortis* appear to use olfactory cues/odour plumes to navigate [91] rather than relying on pheromone trails. Given their environment, this makes sense as, in high-temperature desert environments, pheromone trails evaporate too quickly to be useful and there are a lack of visual landmarks. Pheromone trail marking is also ineffective for small ant colonies as there are insufficient ants to build up trails of concentrated pheromone. One such ant species is *Pachycondyla apicalis*, which lives in small colonies and relies on a combination of individual foraging and one-to-one communication mechanisms rather than mass recruitment via trail marking.

9.3 Overview of Ant Foraging Algorithms

Ant foraging algorithms constitute a family of population-based optimisation algorithms that are metaphorically based on the foraging activities of ants. They are amongst the best-known foraging algorithms, with their initial development now dating back a quarter of a century to 1992 [153].

These algorithms were first introduced to help solve network problems such as finding good paths through graphs [158]. More generally, it was noted that these approaches could be applied to any discrete optimisation problem for which a solution construction mechanism could be defined. In construction-based algorithms, the start point is an empty solution, and solution components are iteratively added until a complete solution is produced. In applications of canonical ant colony optimisation (ACO) algorithms, an artificial ant is a simple computational agent which iteratively constructs a solution for the problem at hand. In the construction step, an ant seeks to extend its partial solution by examining the set of feasible options it faces (i.e. its decision choices). It stochastically chooses among the available choices, partially based on information accumulated by the population of ants during the algorithm run as to the quality of each choice.

Subsequent to the development of ant foraging algorithms for discrete optimisation, a range of alternative ant algorithms were also developed for real-valued optimisation. In this chapter, we cover both cases. Initially, in Sects. 9.4 to 9.6, we introduce a number of ant foraging algorithms for discrete optimisation. This is followed by a discussion of a number of algorithms which can be applied to real-valued optimisation (Sects. 9.8 to 9.10).

9.4 Ant Algorithms for Discrete Optimisation

Many discrete or combinatorial optimisation problems, such as vehicle routing or arc routing problems, can be represented using a *graph* or *network* structure. Such a structure consists of two sets: a set of *nodes* (or *vertices*); and a set of *arcs* (or *edges*), each arc connecting two nodes. If two nodes i and j are connected by an arc, we denote that arc by ij. The first step in applying ACO to discrete optimisation is to map the problem onto a suitable *construction graph* so that a path through the graph (corresponding to an ant foraging route) represents a solution to the problem. The task is then to find the optimal path through the graph. For example, in the *shortest path problem*, given a start vertex and an end vertex, the objective is to find a path (a sequence of vertices and their connecting arcs, with no repeated vertex) of least possible distance or time or other cost, which goes from the start vertex to the end vertex.

Once a construction graph has been designed for the combinatorial optimisation problem of interest (for example, Fig. 9.3), a computer simulation of the activities of artificial ants can be used to uncover a good route across the graph. These artificial ants are released one at a time at the start node. Each ant builds a complete solution to the problem by traversing arcs from one node to the next until a terminating node is reached.

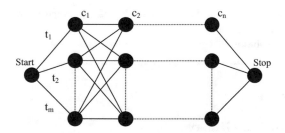

Fig. 9.3. Construction graph for a timetabling problem. Each arc corresponds to a choice of time slot (m time slots exist), and each node corresponds to a class (n in total). A walk through this graph over specific arcs produces a timetabling of each class to a specific time slot

At each node, the ant typically will have a choice of several outgoing arcs. In canonical ant colony algorithms, the ant has access to two pieces of information when making its decision as to which arc to select. The first piece of information is the quantity of pheromone deposited on each arc, which acts as a guide as to how many previous ants have traversed that arc; the second is some information on the quality of each arc (usually based on a heuristic such as 'pick the outgoing arc with the lowest cost value'). The ant stochastically selects its next arc based on a blending of these pieces of information, tending to favour arcs which are heavily reinforced with pheromone and arcs which are considered good using the heuristic guide to arc quality. The solution constructed by an ant is therefore built up based on both domain knowledge (in the form of the heuristic) and learning arising during the algorithm.

After each ant in the population has constructed a solution by walking across the construction graph, the quality of each of these solutions is assessed. The pheromone trails on the arcs are then updated, with arcs on the highest-quality solutions having additional pheromone deposited on them. Over multiple iterations of the algorithm, the arcs belonging to the better solutions become more heavily reinforced with pheromone, and consequently more ants tend to follow them. This leads to an intensification of search around those solutions.

The pheromone trails on arcs are also subject to an *evaporation* process during each iteration of the algorithm. This guides the ants by ensuring that defunct, less travelled solution fragments are forgotten over time. The high-level pseudocode for an ant foraging algorithm is outlined in Algorithm 9.1.

Algorithm 9.1: Canonical ant foraging algorithm

Initialise pheromone trails to positive value;
repeat
    for *each ant in turn* **do**
        Construct a solution;
        Measure quality of solution;
    end
    Update pheromone trails via pheromone deposit and evaporation;
until *terminating condition*;
Output the best solution found;

Pheromone Matrix as a History

The pheromone information associated with each arc in a construction graph is stored in a *pheromone matrix* ((9.2) below). The entries in the matrix correspond to the quantities of pheromone on the edges between each node (here we assume that each node is connected to each other node via a single edge), with values of 0 on the main diagonal of the matrix. The matrix is updated in each iteration of the algorithm, and the values represent a form of memory for the ant system. In (9.2) it is assumed that the pheromone levels are the same regardless of the direction that the arc is traversed. More generally, this need not be the case. The following is an example of a pheromone matrix:

$$
\begin{array}{c}
\textbf{Nodes} \quad 1 \qquad 2 \quad \cdots \quad n \\
\begin{array}{c}
1 \\
2 \\
\vdots \\
n
\end{array}
\left(
\begin{array}{cccc}
0 & 0.23 & \cdots & 0.5 \\
0.23 & 0 & \cdots & 0.33 \\
\vdots & \vdots & \ddots & \vdots \\
0.5 & 0.33 & \cdots & 0
\end{array}
\right).
\end{array}
\tag{9.2}
$$

In ant foraging algorithms, the memory of past learning (the past search experience of the algorithm) does not reside in the individual ants; instead, it resides in the pheromone matrix. These pheromone values are used to bias the construction of solutions towards higher-quality regions of the search space.

As with real-world ant foraging, most ant algorithms embed elements to control the strength of the positive reinforcement mechanism arising from pheromone trail following in order to reduce the risk of inappropriate lock-in.

9.5 Ant System

The original ant foraging algorithm, known as the *ant system* (AS), was developed by Dorigo [153]. In operationalising the general framework in Algorithm 9.1, a number of decisions must be addressed by the modeller:

  i. How are pheromone trails initialised?
 ii. How do the ants construct protosolutions?
iii. How are the pheromone trails updated (and which solutions participate in the update)?

Pheromone Initialisation

Choosing the appropriate levels of pheromone to initialise the arcs at t_0 is important, as there is a link between the level of pheromone on the arcs of the construction graph at the start of the algorithm and the rate of convergence of the algorithm to a single solution. If the initial levels of pheromone are very low, the algorithm will tend to quickly converge on the first good solution which is uncovered, before adequate exploration has occurred. This occurs because the first solution to receive reinforcing pheromone deposits will be highly favoured by subsequent foraging ants. On the other hand, if the level of pheromone used to initialise the arcs is very high, early update steps will have little effect and useful search will be delayed until sufficient evaporation has occurred to decrease the pheromone levels to the point where the pheromone deposit can begin to bias the search towards good solutions. Dorigo and Stützle [159] provide guidelines for appropriate parameter settings, including initial pheromone values, for a variety of forms of ant system.

Constructing Protosolutions

The key issue faced by the ants, when constructing a route through the graph, is which outgoing arc should be selected at each node. A simple approach would be to select the outgoing arc which has the highest pheromone level associated with it. However, this would result in too rapid convergence to a single solution, usually producing a poor result.

A more sensible approach would be to allow the ant to choose stochastically from the set of available arcs. For example, the probability of choosing arc ij from amongst the K possible feasible arc choices at a particular construction step could be determined using the following discrete probability distribution:

$$P_{ij} = \frac{\tau_{ij}}{\sum_{k=1}^{K} \tau_{ik}}, \tag{9.3}$$

where τ_{ik} is the quantity of pheromone associated with arc ik. Suppose there are three arc choices facing an ant at node i, with $P_{i1} = 0.3$, $P_{i2} = 0.4$ and $P_{i3} = 0.3$. A random draw from $U(0,1)$ producing (say) 0.2 falls inside the range $(0,0.3)$, implying that the ant follows the arc $i1$. This approach ensures that while arcs which have been part of good solutions in the past are more likely to be selected, an ant still has the potential to explore any arc with a nonzero pheromone value. However, even with the addition of this stochastic choice mechanism, ants tend to quickly lock in on a single route, resulting in poor exploration of the search space.

To combat this problem, the AS algorithm combines pheromone information with information from a heuristic when making the choice of which arc to add to the solution being constructed. Adding heuristic information to guide the solution construction process is known as adding *visibility*, or look-ahead, to the construction process.

An illustration of how heuristic information can be used is provided by the well-known travelling salesman problem (TSP). The TSP was one of the earliest applications of AS and is a standard benchmark problem for combinatorial optimisation techniques, as it is NP-complete. In the TSP, there is a network of n cities. Each road segment, or arc, between two cities has a distance (or travel time or some other cost) associated with it, and the object is to find the tour (a path including all vertices in the graph) which minimises the distance travelled or the cost for visiting all the cities exactly once and returning to the starting city. An example of a construction graph for a TSP with four cities is given in Fig. 9.4.

In the case of the TSP, a simple heuristic for assessing the possible utility of each arc choice facing an ant when it leaves city i is the distance between city i and each other city to which it is connected, with shorter distances being preferred to longer ones. The ant weighs up both the information from this heuristic and the pheromone information on each arc when selecting which city to visit next.

As the objective is to visit each city once, and once only, during a tour, the choice of which city to visit next should exclude cities already visited. The ant maintains a memory of all cities it has already visited, for this purpose. For the TSP, (9.3) is adapted to produce (9.4). Thus, the probability of ant k travelling from city i to city j at time t, where C_i^k is the set of feasible cities reachable from city i and not yet visited by ant k, is

$$P_{ij}^k(t) = \frac{\tau_{ij}(t)^\alpha \eta_{ij}(t)^\beta}{\sum_{c \in C_i^k} \tau_{ic}(t)^\alpha \eta_{ic}(t)^\beta}, \quad j \in C_i^k. \tag{9.4}$$

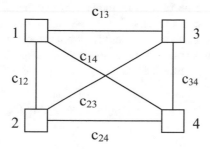

Fig. 9.4. A construction graph for a TSP with four cities. The routes correspond to the arcs/edges of the graph and the four cities are the nodes/vertices

The term η_{ij} is equal to $1/d_{ij}$, where d_{ij} is the distance between the cities i and j, and the parameters α and β represent the weights attached to pheromone and heuristic information, respectively.

Examining the form of this arc choice rule, if $\beta = 0$, then (9.4) effectively reduces to (9.3) and only pheromone information is used in determining the ants' movements. Conversely, if $\alpha = 0$, the pheromone information is ignored and only heuristic information is used to guide the search, resulting in a greedy search by the ants. While good choices for α and β are problem specific, parameters of 2 and −2, respectively, have been suggested [44].

Commonly, in TSP applications, the number of ants is equal to the number of cities, and during each iteration of the algorithm, an ant is started from each city in turn in an effort to ensure that tours starting from all cities are investigated.

Updating Pheromone Trails

After each of the ants has traversed the graph and has constructed its individual solution to the problem, the quality of each of these solutions is assessed and this information is used to update the pheromone trails. The update process typically consists of an *evaporation* step and a pheromone *deposit* step:

$$\tau_{ij}(t+1) = \tau_{ij}(t)(1-p) + \delta_{ij}. \tag{9.5}$$

In the evaporation step, the pheromone associated with every arc is degraded or weakened. The evaporation rate p crucially controls the balance between exploration and exploitation in the algorithm. If p is close to 1, then the pheromone values used in the next iteration of the algorithm are highly dependent on good solutions found in the current iteration, leading to local search around those solutions. Smaller values of p allow solutions from earlier iterations of the algorithm to influence the search process.

The amount of pheromone δ_{ij} deposited on each arc ij during the pheromone update process depends on how the deposit step is operationalised in the algorithm.

The deposit step can be performed in many ways, depending on which solutions are chosen to participate in the deposit process, what weight is applied to each of these solutions in the deposit process and how pheromone is deposited on each arc participating in the deposit process. One design of the deposit mechanism (called the *ant cycle* version of AS) for the TSP is

$$\tau_{ij}(t+1) = \tau_{ij}(t)(1-p) + \sum_{k=1}^{m} \Delta\tau_{ij}^{k}(t), \tag{9.6}$$

$$\text{with } \Delta\tau_{ij}^{k}(t) = \begin{cases} Q/L^{k}(t), & \text{if } ij \in T^{k}(t), \\ 0, & \text{otherwise,} \end{cases} \tag{9.7}$$

where m is the number of ants, ij is the arc from city i to city j, T^k is the tour done by ant k at iteration t and $L^k(t)$ is the length of this tour. The term $\Delta\tau_{ij}^{k}(t)$ represents the pheromone laid on arc ij by ant k, and Q is a positive constant.

Therefore, in (9.7) the amount of pheromone laid on an arc by an ant varies inversely with the length of the tour undertaken by that ant. Arcs on longer tours get less reinforcement than arcs on higher-quality, shorter tours. In the ant cycle version of AS, every ant lays pheromone at the end of each complete iteration of the algorithm and the total of all these deposits influences ants in the next iteration of the algorithm. Over multiple iterations of the algorithm, solution construction is the result of community learning by all the ants.

If the object is to maximise, rather than minimise, an objective function, (9.7) can be recast as

$$\Delta\tau_{ij}^{k}(t) = \begin{cases} Qf(x^{k}(t)), & \text{if } ij \in \text{path } x^{k}(t), \\ 0, & \text{otherwise,} \end{cases} \tag{9.8}$$

where $f(x^{k}(t))$ is the fitness or quality of ant k's solution.

Variations on Canonical AS

In the canonical AS algorithm (as seen above), all solutions in the current population participate in the pheromone update process. Although the canonical version of AS produced good results on test TSP problems, it did not outperform state-of-the-art algorithms on larger instances of the TSP [160].

As a result, a series of variants of the canonical AS algorithm were devised which delivered better results. Several of these variants adopted alternative pheromone deposit rules which restricted the number of ants participating in this step and, in the limit, only allowed the ant corresponding to the best solution to deposit pheromone.

The *elitist ant system* (EAS), developed by Dorigo [153], combines these approaches so that while each ant deposits pheromone, the best-so-far (elite) tour discovered is also reinforced in each iteration of the algorithm. In this case, the update step in (9.6) is amended to (9.9) through the addition of an extra term $\Delta\tau_{ij}^{*}$ (see (9.10)), which increments the pheromone on edge ij if it is on the tour traversed by the elite ant. The tour found by this ant is denoted by T^* and the length of this tour by L^*, and σ is a scaling constant which controls the degree of reinforcement of the best tour. As the value of σ increases, the ants are encouraged to search intensively around the best-so-far solution, increasing the risk of premature convergence on a locally optimal solution:

$$\tau_{ij}(t+1) = \tau_{ij}(t)(1-p) + \sum_{k=1}^{m} \Delta\tau_{ij}^k(t) + \Delta\tau_{ij}^*, \tag{9.9}$$

$$\text{with } \Delta\tau_{ij}^* = \begin{cases} \sigma Q/L^*, & \text{if } ij \in T^*, \\ 0, & \text{otherwise.} \end{cases} \tag{9.10}$$

The *rank-based ant system* (AS_{rank}) developed by Bullnheimer et al. [93] adopts a different approach. Before the pheromone deposit process, all ants are ranked according to the quality of their solution, and the amount of pheromone deposited by each ant decreases with its rank. As with elitist ant systems, the best-so-far ant also participates in the pheromone deposit step. The pheromone update (and evaporation) rule is therefore

$$\tau_{ij}(t+1) = \tau_{ij}(t)(1-p) + \sum_{k=1}^{w-1}(w-k)\Delta\tau_{ij}^k(t) + w\Delta\tau_{ij}^* \tag{9.11}$$

where only the $(w-1)$ best-ranked ants participate in the update process, along with the tour of the best-so-far ant (on tour T^*). The value of $\tau_{ij}^k(t) = 1/L^k$ and the value of $\Delta\tau_{ij}^* = 1/L^*$, where L^* is the length of the best-so-far tour. Thus, the best-so-far tour receives a weight of w in the update process, with the successive ranked tours receiving a lower weighting.

MAX-MIN Ant System

In order to reduce the risk of premature convergence of the optimisation process to a single solution, and in order to ensure that every arc has a nonzero chance of being selected during the solution construction process, the pheromone τ_{ij} associated with each arc ij may be constrained so that after the pheromone update process, $0 < \tau_{min} \leq \tau_{ij} \leq \tau_{max}$ (the values of τ_{min} and τ_{max} being set by the user). The bounds prevent the relative differences between the pheromone trails on each arc from becoming too extreme.

A variant of AS which adopts this idea is the MAX-MIN ant system (MMAS) designed by Stützle and Hoos [552]. In MMAS, all arcs are initialised with τ_{max} pheromone. During the update process only arcs on the global best *best-so-far* solution (or the best path found in the current iteration, i.e. the *iteration best* solution) are updated, and pheromone levels are constrained to $0 < \tau_{min} \leq \tau_{ij} \leq \tau_{max}$. As MMAS strongly exploits the information in the best tour found so far, it is necessary to restrict the deposit of pheromone along this tour in order to ensure that the search process does not stagnate too quickly. Another mechanism often implemented in

MMAS in order to encourage continued diversity in the search process is to periodically reinitialise the pheromone levels on all arcs in the network. This step is usually undertaken when stagnation of the optimisation process is detected.

9.6 Ant Colony System

The *ant colony system* (ACS) algorithm was developed by Dorigo and Gambardella [155, 156] in order to improve the scalability of the canonical AS algorithm. ACS differs from AS as it uses:

- a different construction rule,
- a different pheromone update rule, and
- an additional local pheromone update process.

Construction Rule in ACS

In comparison with AS, the transition rule in ACS is altered so that an ant k at city i moves to city j according to the rule

$$j = \begin{cases} \text{argmax}_{l \in J_i^k}\{\tau_{il}(t)\eta_{il}(t)^\beta\}, & \text{if } q \leq q_0, \\ J, & \text{otherwise,} \end{cases} \quad (9.12)$$

where q is a random variable drawn from a $U(0,1)$ distribution and $0 \leq q_0 \leq 1$ is a threshold parameter. With probability q_0, the ant exploits the information in the pheromone trails and the decision heuristic, selecting the best possible next arc based on this information. With probability $1 - q_0$, the ant selects its next arc $(J \in J_i^k)$ from the set of cities yet to be visited (or from a candidate list) randomly according to the probability distribution in (9.4), where $\alpha = 1$.

Therefore, q_0 is a tuning parameter which determines the degree to which the system focuses search attention on exploiting already discovered good arcs.

Pheromone Update in ACS

In ACS, only the best-so-far ant updates the pheromone trails at the end of each iteration of the algorithm (an *offline pheromone update*). In contrast, in canonical AS, all ants participate in the update process. The pheromone evaporation step is also adjusted so that arcs on the best-so-far trail are subject to an evaporation process. The level of pheromone at iteration $t + 1$ is a weighted average of the pheromone level at iteration t and the new pheromone deposited, with the parameter p governing the relative importance of both. The deposit and evaporation steps are governed by

$$\tau_{ij}(t+1) = \tau_{ij}(t)(1-p) + p\Delta\tau_{ij}^*, \quad (9.13)$$

where only the arcs traversed by the best-so-far ant (on tour T^*) participate in the pheromone deposit/evaporation process. The term $\Delta\tau_{ij}^*(t) = 1/L^*$, where L^* is the length of the best-so-far tour. As in AS, this pheromone update step is performed after all ants in the population have constructed their solutions.

Local Update in ACS

In addition to the offline pheromone update process on the best-so-far solution, a real-time local update is performed by *each* ant after it traverses an arc. Each ant applies this update only to the last arc traversed. If an ant traverses an arc ij, then the pheromone level on that arc is immediately adjusted as follows:

$$\tau_{ij}(t+1) = (1-\alpha)\tau_{ij}(t) + \alpha\tau_0, \tag{9.14}$$

where $0 < \alpha < 1$ is an evaporation parameter. The parameter τ_0 is typically set to the same value as was used to initialise the pheromone trails at the start of the algorithm. The intent of the local updating rule is to reduce the value of τ_0 each time an arc is traversed by an ant so that subsequent ants are encouraged to try other arcs. Implementing this mechanism also discourages the algorithm from stagnating.

Considering the three ACS mechanisms in their totality, they create a dynamic interplay of exploitation and exploration. Depending on the parameter settings chosen for each mechanism, ACS is capable of a wide range of search behaviours. Although the algorithm places considerable focus on the best-so-far solution, it counterbalances the convergent effect of this by means of the local updates and through the stochastic transition rule.

9.7 Ant Algorithms for Continuous Optimisation

While the canonical version of the ant algorithm was designed for discrete optimisation, many optimisation problems involve real-valued parameters. Real-world ant foraging behaviour takes place in a continuous space with pheromone trails diffusing on the landscape once they are laid down. Accordingly, it is not difficult to develop ant foraging algorithms for real-valued optimisation.

The earliest approaches applying an ant colony metaphor to continuous optimisation used discretisation, whereby the continuous range of possible options for each solution fragment is reduced to a discrete range using defined grid intervals. A refinement of this approach is to undertake an initial coarse-grained search, switching to a finer-grained search once a promising solution region is identified [56].

Following this early work, a series of ant-inspired foraging algorithms capable of working directly with a real-valued representation have been developed, and we introduce a number of these in the following sections.

9.8 API Algorithm

The API algorithm (named after the *Pachycondyla APIcalis* ant species) was first devised by Monmarché et al. [410] and is loosely inspired by the foraging behaviour of *Pachycondyla apicalis* ants. These ants are typically found in tropical forests and live in relatively small colonies of up to about 100 individuals. At any point in time, some 20–30% of these ants will be foraging [410].

A stylised description of the foraging behaviour of the ants is as follows. Initially, an ant leaves the colony nest and begins a foraging search, usually within a radius of about 10 metres around the nest. If the ant discovers a prey item, it brings it back to the nest, memorising visual landmarks on the return journey. This enables it to return to the prey capture location on a subsequent foraging mission. In contrast, if a foraging mission is not successful, an ant will search elsewhere in the future. The ants have been shown to be able to memorise a number of good foraging locations and return to these even after the passage of some time. Periodically, the area around a colony nest will be emptied of resources. In this case the colony resites the nest to a new location and recommences foraging from there.

Unlike other ant species, *Pachycondyla apicalis* ants do not interact heavily during the foraging process, although some implicit and explicit communication does take place between the ants. In cases where the population of a colony is small, mass recruitment via pheromone trail marking is usually ineffective as trails to foraging sites cannot be sufficiently strongly reinforced, owing to the relatively sparse number of foragers. Instead, foraging mechanisms in smaller colonies tend to rely on individual search or one-to-one recruitment methods such as tandem running (Sect. 4.2).

The API algorithm contrasts with previously discussed ant algorithms as it does not adopt a mass recruitment metaphor and does not embed stigmergic communication.

9.8.1 Algorithm

The stylised foraging behaviour of the *Pachycondyla apicalis* ant is adapted as follows to create the API algorithm. In this description, we draw from [23, 410].

Assume a population of n ants (denoted as $\{a_1, \ldots, a_n\}$), located in a search space S (all points $s \in S$ are considered valid solutions), where the aim is to minimise a function $f : S \longrightarrow \mathbb{R}$. As with all search algorithms, the core elements concern the process of movement around the search space in an effort to uncover good solutions to the problem of interest. In the API algorithm, two operators are defined to control the movement of the nest location and the movement of individual ants during the algorithm.

Move Operators

Let the operator O_{rand} be the generation of a random point in S using a uniform distribution. The operator O_{rand} can be defined in a variety of ways, depending on the nature of the search problem. Assume that we have no a priori information on the likely location of good solutions, that the problem is defined on a search space $S \subseteq \mathbb{R}^l$ and that the evaluation function $f(x_1, x_2, \ldots, x_l)$ to be minimised is defined on a subset of $\mathbb{R}^l$, where each dimension has lower and upper bounds b_i and B_i, respectively, for all $i \in \{1, \ldots, l\}$. Under these assumptions, a reasonable definition of O_{rand} consists of the generation of a random point in each interval using a uniform distribution.

A second operator, $O_{explo}(s, A^i)$, is also defined and is used to create a new point s' within the neighbourhood of an existing point s, according to a defined amplitude (or maximum radius of step size) A^i. This operator can be used to generate a new hunting site around the location of the colony nest (N) for an individual ant a_i (here, $A^i = A^i_{site}$ and $s = N$) or to undertake a local exploration around the location of the hunting site at which the ant (a_i) is currently located. In this latter case, $A^i = A^i_{local}$ and $s = s^i_j$m where s^i_j is a hunting site of ant a_i. If the new location s' uncovered during this local exploration is a better solution than s, then s^i_j is set equal to s'.

Assuming the same search space definition as for O_{rand}, the operator O_{explo} generates a new point $s' = (x'_1, x'_2, \ldots, x'_l)$ from an initial point $s = (x_1, x_2, \ldots, x_l)$ using

$$x'_i = x_i + rA(B_i - b_i) \quad \text{for all } i \in \{1, \ldots, l\}, \tag{9.15}$$

where r is drawn from a uniform distribution $U(-0.5, +0.5)$ on $[-0.5, +0.5]$ and A is the maximum amplitude of the step size. More generally, the operator O_{explo} need not be defined to be a random search process but can utilise any desired search heuristic.

The parameter A^i (either A^i_{site} or A^i_{local}) controls the maximum amplitude of the exploration steps during the foraging process. A^i is defined within $A^i \in [0, 1]$ such that if $A^i = 0$, then $s' = s$ (i.e. step size is zero). When $A^i = 1$, then s' can be any point in S.

In order to promote a local search process for an individual ant around an existing hunting site, the amplitude for A^i_{local} is set to one tenth of the amplitude of the search step around the colony nest location A^i_{site}. Therefore, $A^i_{local} = A^i_{site}/10$.

As the value of A^i_{site} varies for each ant (see below), the amplitude of the local search step also varies for each ant. This implies that the population of ants are using heterogeneous parameter settings.

Main Loop of Algorithm

When the algorithm commences, the nest is placed at a random location N in S as determined by O_{rand}. Every ant a_i leaves the nest in turn and, over a number of iterations of the algorithm, each ant in the population builds up a memory of hunting sites using the operator O_{explo} with a starting point of N and a search amplitude of A^i_{site} (Fig. 9.5). In [410], the value of the memory locations (p) is set to 2; therefore, each ant can memorise two locations. The values for A^i_{site} are not the same for all ants a_i but are set as follows:

$$A^1_{site} = 0.01, \ldots, \tag{9.16}$$

$$A^i_{site} = 0.01z^i, \ldots, \quad i = 2, \ldots, n-1, \tag{9.17}$$

$$A^n_{site} = 0.01z^n = 1, \tag{9.18}$$

where $z = (1/0.01)^{1/n}$. Once an ant has built up a full memory of hunting sites, either it revisits its hunting location from the previous iteration of the algorithm (if its previous exploration was successful in that it found a new, improved solution) or,

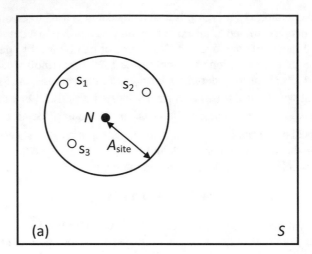

Fig. 9.5. Generation of sites s_1, s_2 and s_3 at random locations within a radius of A_{site} of nest N

alternatively, the ant randomly selects one of its p hunting sites in memory to revisit. In either case, let us designate the chosen hunting location as s. Then ant a_i applies the operator O_{explo} with an amplitude of A^i_{local} at point s and obtains a new location s'. This step, in essence, performs a local search in the neighbourhood of point s (Fig. 9.6).

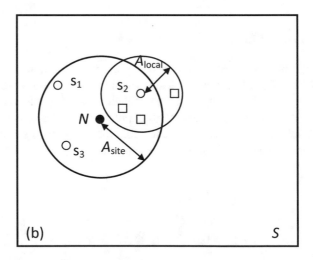

Fig. 9.6. Illustration of local-exploration locations (denoted by squares) within a radius of A_{local} around s_2

If the value of the objective function at the new location s' is better than its value at point s (i.e. $f(s') < f(s)$, in a minimisation problem) then ant a_i memorises this successful location and updates its memory to include s' and to forget s. As the search process has led to an improvement in this case, the ant will start its search from this new location in the next iteration of the algorithm.

If the local search around s did not produce a successful outcome, a counter (of failed improvement attempts) for location s is incremented by one. If the counter value exceeds a user-defined threshold parameter $P_{local}(a_i)$, then the location s is wiped from the memory of ant a_i and, in the next iteration of the algorithm, a new location is generated and inserted in the memory of ant a_i using O_{explo}, with an amplitude of A^i_{site}.

The algorithm iterates with every ant in turn exploring its environment. After each ant in the population has left the nest T_{nest} times in order to forage, the location of the nest (N) is moved to the best location found since the nest was last moved and the ants' memories of good foraging locations (p) are cleared. The ant memory vectors are again built up over several iterations of the algorithm as the ants explore the landscape around the resited colony nest.

The algorithm iterates until a terminating condition is met, and the best solution found is returned at that point. Figure 9.7 illustrates in stylised form the workings of the algorithm, and pseudocode for the algorithm is outlined in Algorithm 9.2.

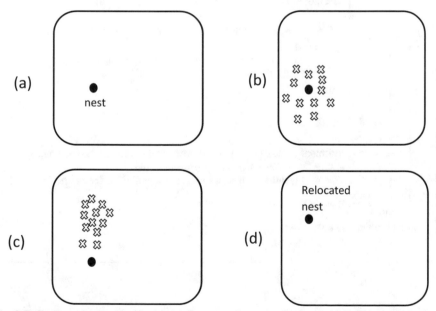

Fig. 9.7. Outline of API algorithm. Initially, a nest is randomly situated (a), with hunting sites being created around this location (b). During local exploration, the hunting sites gravitate to better areas of the search space (illustrated in (c)). The nest is then relocated to this new (better) region (d) and the search process is iterated

---

Algorithm 9.2: API algorithm [23, 410]

Select a random initial location for the nest N using $O_{rand}(S)$;
Clear memories of all ants;
repeat
    for *each ant $a_i, i \in 1, \ldots, n$* **do**
        if *a_i has less than p hunting sites in its memory* **then**
            Create a new site in the neighbourhood of N using $O_{explo}(N, A^i_{site})$;
            Explore the new site $s' := O_{explo}(s, A^i_{local})$;
            if $f(s') < f(s)$ **then**
                | Set $s := s'$;
            end
            Add s to memory of ant a_i;
        else
            if *the previous exploration by ant a_i was successful* **then**
                | Explore the same site s;
            else
                | Explore a randomly selected site s (among the p sites in memory for
                | ant a_i);
            end
            Set $s' := O_{explo}(s, A^i_{local})$;
            if $f(s') < f(s)$ **then**
                | Remove site s from memory of a_i and replace with s';
            else
                if *a_i has explored site s unsuccessfully more than P^i_{local} consecutive*
                times **then**
                    | Remove site s from memory of a_i (note: this frees up a memory
                    | slot for that ant);
                end
            end
        end
    end
    Perform recruitment (best-site copying between two randomly selected ants);
    if *more than T_{nest} iterations have been performed* **then**
        | Change the nest location to best solution found and clear the memories of all
        | ants;
    end
until *terminating condition*;
Output the best solution found;

Communication Mechanisms

Although the API algorithm does not embed a broadcast communication mechanism between individual ants via simulated pheromone trails, it does incorporate some ant-to-ant communication. The search locations of all ants are impacted by the location of their nest, forming a pseudocommunication mechanism. In the real world, the decision to move nest and the location of the new nest are collective decisions and therefore embed direct communication.

The algorithm also incorporates a simulated one-to-one communication mechanism drawn from the idea of tandem running, where one ant can recruit another ant to a favourable location by physically carrying or leading an ant there. This is implemented in the API algorithm by randomly selecting two ants a_i and a_j after all n ants have engaged in a search step. The qualities of the best sites found by each ant are compared and the location of the poorer of these is replaced by the location of the better site in that ant's memory. This process therefore enhances exploitation, as it intensifies the search process around the better site since two ants now use this information to guide their future searches.

Parameters

In order to operationalise the algorithm, the modeller needs to set a number of parameters. Monmarché et al. [410] suggest that using an ant population size of 20 produces good results, with simulation experiments on a range of benchmark problems indicating that using larger population sizes does not necessarily enhance the quality of the search process. The periodic resetting of the nest location was found to be an important factor in driving the results obtained from the algorithm, with the nest being resited after each member of the population had taken 50 search steps. A forgetting mechanism was implemented on each nest move but ant memory was otherwise maintained between nest moves. The recruitment process (simulated tandem running) was found to play a relatively minor role in determining the results of the algorithm. In testing across a small range of real-valued benchmark functions, the API algorithm performed competitively against the genetic algorithm and a multiple restart random hill-climbing heuristic.

In contrast, [23] suggested a higher value for the parameter T_{nest}, with the value being given by $2np(P_{\text{local}}^i + 1)$ in order to promote search around all sites in memory. In this implementation, the tandem recruitment mechanism was omitted.

9.8.2 Discussion

The API algorithm implements a relatively simple algorithm in which the (local) search process for every individual ant is centred around a point in each iteration of the algorithm, as is the search process of the entire population of ants. The search process of an individual ant proceeds successively to better points via random search from the current set of locations memorised by the ant. If the search from any of these points stagnates, that location is forgotten and is replaced by a new location.

At a populational level, the nest is periodically moved to the best location found since the last move, again helping to avoid search stagnation and intensifying the search process in the region where the best location so far has been found. The parameter settings for individual ants (i.e. the amplitudes of their search steps) are not fixed across the entire population. This may help promote algorithm robustness, as a priori the modeller is unlikely to know the magnitude of a good step size for a specific optimisation problem.

Above, we have described the application of the API algorithm to real-valued optimisation. The algorithm can also be applied to integer-valued or permutation problems [410] with relatively little modification. A variant of the algorithm has also been applied for clustering [328].

9.9 Continuous Ant Colony System

The *continuous ant colony system* (CACS) was introduced by Pourtakdoust and Nobahari [462]. In CACS, a continuous pheromone model is defined. Pheromone is not considered to be laid along a single discrete track; rather, it is laid down spatially. For example, a food source surrounded by multiple ants would be expected to show a high pheromone concentration, with this concentration dropping off as the distance from the food source increases.

One way of modelling this phenomenon is to use a normal probability density function (PDF), where the probability of observing a (real-valued) sample point x is defined as

$$PDF(x) = e^{-(x-x_{\min})^2/(2\sigma^2)}, \qquad (9.19)$$

where $x_{\min}$ is the location of the best point (assuming that the object is to minimise a function) found so far in the range $[a, b]$ during the search, and σ is a measure of the ants' clustering (density) around the best point.

Initially, $x_{\min}$ is randomly chosen from the range $[a, b]$ and σ is set to $3(b - a)$ in order to ensure that the real-valued strings corresponding to each ant are well distributed in the range $[a, b]$.

When the solution vector is made up of multiple elements (real values), each element will have its own normal PDF and the values of $x_{\min}$ and σ for each of these individual PDFs will alter during the optimisation run.

Algorithm

High-level pseudocode for the CACS algorithm is given as Algorithm 9.3.

Applying the CACS Algorithm

Suppose a problem of interest requires that optimal values for three real-valued parameters are to be uncovered and that a range in which the optimal values lie is known for each of the parameters. Initial values for $x_{\min}$ and σ are selected for each of the

---

Algorithm 9.3: Continuous ant colony system

---

for *each component i = 1,...,n of the real-valued solution encoding* **do**
 | Select value of x_{min} randomly from the range of allowable values for component i;
 | Set σ for component i;
 | Construct the normal PDF for component i from x_{min} and σ, using (9.19);
end
repeat
 | **for** *each ant k in turn* **do**
 | | Construct a solution $(x_1,...,x_n)$ by making random draws from $PDF_1,...,$
 | | PDF_n, respectively;
 | | Measure quality of solution;
 | **end**
 | Update the x_{min} value for each PDF if a new best-so-far solution has been found;
 | Update the σ value for each PDF;
until *terminating condition*;
Output the best solution found;

---

three parameters as above. Successive ants in turn then complete a 'tour' (construct a potential solution) by generating three random numbers and using each of these in turn to generate a sample value x using the PDFs corresponding to each locus of the solution vector. To do this, each PDF is first converted into its corresponding cumulative distribution function (CDF),

$$CDF_j(x) = \int_a^x PDF_j(y)\,dy \quad \text{for all } x \in \mathbb{R} \tag{9.20}$$

where a is the lower limit of the probability distribution. By generating a random number r from $(0,1)$, the CDF can be used to obtain a value x, where $x = CDF^{-1}(r)$.

After all ants have completed a tour, the quality of each ant's solution is found using the objective function for the problem. If any of these solutions is better than that found by the best-so-far ant, the value of x_{min} for each PDF is updated using the values from the revised best-so-far ant. The value of σ for each PDF is also updated using

$$\sigma^2 = \frac{\sum_{j=1}^k (x_j - x_{min})^2/(f_i - f_{min})}{\sum_{j=1}^k 1/(f_i - f_{min})}, \tag{9.21}$$

where k is the number of ants, and f_j and f_{min} are the objective function values for ant j and for the best-so-far ant, respectively. As the solutions encoded by each ant converge, the value of the variance σ^2 decreases.

Another way of updating the value of σ for each PDF is to calculate the standard deviation of the values of that parameter across the population of ants during the last iteration of the algorithm [588]. As good solutions are uncovered, a positive reinforcement cycle builds up. More ants are attracted to the best solution vector and, as the population of ants converges to similar sets of real values, the value of σ

automatically decreases, leading to intensification of search in the region of the best solution found so far.

The algorithm iterates until predefined termination criteria are reached, such as a maximum number of iterations or a time limit, or after a set number of iterations without any improvement in the solution.

In order to discourage premature convergence of the population of ants to a single solution vector, an explicit pheromone evaporation mechanism can be applied. Evaporation can be simulated by increasing the value of σ for each PDF, thereby increasing the spread of pheromone in the environment. One method for achieving this, suggested by Viana et al. [588], is to apply the following rule to each PDF's σ at the end of each iteration:

$$\sigma_{t+1} = \alpha\sigma_t, \tag{9.22}$$

where $\alpha > 1$ is the evaporation rate. As the value of α is increased, the level of evaporation also increases.

9.10 ACO$_\mathbb{R}$ Algorithm

Perhaps the best-known ant algorithm for real-valued optimisation, ACO$_\mathbb{R}$, was introduced by Socha [536, 537]. Like CACS, this algorithm uses continuous probability functions in searching for good real-valued solution vectors but rather than emphasising the best-so-far ant in the creation of new candidate solutions, an archive of good past solutions is maintained and potentially any solution in this archive can influence the creation of new candidate solutions.

In ACO$_\mathbb{R}$, as in the CACS algorithm, the discrete probability distributions which are used to construct the next step of a solution in canonical discrete ACO applications are replaced by continuous probability distributions, i.e. probability density functions (PDFs), when the next step in a real-valued solution is being constructed (i.e. when the value of the next element of a real-valued solution vector is being determined). The PDFs are learnt as the algorithm iterates, with a history or memory of good solutions being maintained in a *solution archive*. This archive represents the algorithm's *pheromone model*.

The algorithm uses this archive to construct a probability distribution of promising solutions in the search space, this distribution being generated using a sum of weighted Gaussian functions.

Algorithm

The ACO$_\mathbb{R}$ algorithm commences by randomly generating k complete solutions for the solution archive (Fig. 9.8). Each of these solutions consists of a real-valued n-dimensional vector with the ith component $x_i \in [x_{\min}, x_{\max}]$, where $i = \{1, \ldots, n\}$ (note that $k > n$ in order to ensure that the algorithm can rotate the coordinate system as needed during the solution process [537]). Unlike the pheromone model (pheromone

S_1	s_1^1	s_1^2	$\cdots$	s_1^i	$\cdots$	s_1^n	$f(s_1)$	ω_1
S_2	s_2^1	s_2^2	$\cdots$	s_2^i	$\cdots$	s_2^n	$f(s_2)$	ω_2
$\vdots$	$\vdots$	$\vdots$	$\vdots$	$\vdots$	$\vdots$	$\vdots$	$\vdots$	$\vdots$
S_j	s_j^1	s_j^2	$\cdots$	s_j^i	$\cdots$	s_j^n	$f(s_j)$	ω_j
$\vdots$	$\vdots$	$\vdots$	$\vdots$	$\vdots$	$\vdots$	$\vdots$	$\vdots$	$\vdots$
S_k	s_k^1	s_k^2	$\cdots$	s_k^i	$\cdots$	s_k^n	$f(s_k)$	ω_k

$$G_1 \quad G_2 \quad \cdots \quad G_i \quad \cdots \quad G_n$$

Fig. 9.8. Representation of solution archive. Each solution S_j consists of a vector of real numbers $(s_j^1, \ldots, s_j^n)$ and an associated objective function value $f(s_j)$, and has an associated weight ω_j. The solutions are stored in rank order from best to worst. The PDF G^i is constructed using the ith element of all k solutions. The k solutions in the archive represent a *pheromone model* and alter as the algorithm iterates

matrix) in combinatorial optimisation applications of ACO which maintains an implicit memory of the search process, the solution archive maintains an *explicit* memory of previous solutions.

After initialisation, the k solutions in the archive are ranked according to their quality (objective function value) from best to worst. Each solution S_j is then accorded a weight ω_j based on its rank, calculated as follows:

$$\omega_j = \frac{1}{qk\sqrt{2\pi}} e^{-(\text{rank}(j)-1)^2/(2q^2k^2)}, \tag{9.23}$$

where rank(j) is the rank of solution S_j in the sorted archive, and q is a user-defined parameter. The best solution in the archive receives the highest weight, with the weight value declining exponentially as the rank increases. The value of q determines the degree of selection pressure in the process. As $q \to 0$, the likelihood that the Gaussian function associated with the best solution is selected increases. As the value of q increases, a greater number of solutions in the archive may be selected. Therefore, larger values of q slow the convergence of the algorithm.

In each iteration, m new solutions are generated in a two-step process. First, a guide solution is selected from the solution archive, and then a new solution is generated from this. The selection process is stochastic, based on the weights associated with each solution in the archive. The probability of selecting a specific solution S_j is given by

$$P(S_j) = \frac{\omega_j}{\sum_{a=1}^k \omega_a}. \tag{9.24}$$

The higher the rank of a solution, the greater the chance it is selected.

After a guide solution (S_j) is selected, a new candidate solution is generated around it, variable by variable, by means of sampling draws from a Gaussian PDF with a mean equal to that of the guide solution for each variable i in turn $(\mu^i_{\text{guide}} = s^i_{\text{guide}})$, and a standard deviation for each variable i $(\sigma^i_{\text{guide}})$ calculated as:

$$\sigma^i_{\text{guide}} = \xi \sum_{r=1}^{k} \frac{|s^i_r - s^i_{\text{guide}}|}{k-1}, \tag{9.25}$$

corresponding to the average distance between the value of the ith component of S_{guide} and the values of the ith component of the other solutions in the archive, multiplied by a user-defined parameter $\xi > 0$ which has the same value for all dimensions. The parameter ξ acts in a similar manner to the evaporation mechanism in ACO, with higher values of ξ producing slower convergence in the search process.

A total of m new candidate solutions are generated (corresponding to m simulated ants). The best k of the $k + m$ solutions survive into the next iteration of the algorithm, with the other solutions being discarded. This broadly corresponds to the pheromone matrix update process in canonical, discrete-valued ACO. The insertion of new solutions into the archive bears parallel with the deposit of pheromone on good solution fragments, and the removal of poorer solutions from the archive bears parallel with an evaporation process.

Thus, the search process is refined in each iteration of the algorithm towards the best solutions found thus far. The pseudocode for the $\text{ACO}_{\mathbb{R}}$ algorithm is outlined in Algorithm 9.4.

Algorithm 9.4: $\text{ACO}_{\mathbb{R}}$ algorithm

Select values of k, m, n, q and ξ;
Generate k random solutions and evaluate quality of each;
$T = \text{Sort}(S_1, \ldots, S_k)$ (Sort k solutions in archive);

repeat
    for $l = 1$ *to* m **do**
        for $i = 1$ *to* D **do**
            Select Gaussian g^i_j according to weights;
            Sample Gaussian g^i_j with parameters μ^i_j, σ^i_j;
        end
        Store and evalute newly generated solution;
    end
    $T = \text{Sort}(S_1, \ldots, S_{k+m})$;
    Maintain best k solutions in archive, discard the rest;
until *terminating condition*;
Output the best solution found;

Parameters

While the parameter values will be application specific, the values applied in [537] on a range of test problems included $m = 2$, $\xi = 0.85$, $q = 10^{-4}$ and $k = 50$.

A number of variants of the canonical ACO_R algorithm have been developed, including the unified ant colony optimisation algorithm (UACOR) [357] which develops a metaheuristic based on three distinct versions of ACO_R. A variant of ACO_R which can be applied to mixed-variable optimisation problems [356] has also been developed.

9.11 Ant Nest Site Selection

In many species of social organisms, decision making is communal. Crucial decisions such as where to forage, how to allocate foraging resources between food sites and where to nest are made at community level. In some cases, the population may split into discrete groups, each of which reaches its own decision. In other cases, a single option must be chosen by the entire group.

In the latter case, an interesting question is how the group can converge on a single option from amongst several alternatives. During *consensus-based decision making*, individuals in groups balance private information with social information, allowing the group to reach a single collective choice [401]. In most cases, a form of voting occurs and a group decision is reached once the number of individuals voting for a particular option reaches a threshold or *quorum* level. The voting behaviour can be subtle, and examples of voting behaviour in animal groups include body posture, movements and vocalisations [364].

An advantage of group or quorum decision making is that better decisions can result from the pooling of even imperfect individual estimates. If each organism reaches an independent decision, a relatively small group can make correct choices via majority voting even when individuals vote on the basis of noisy signals [401].

Nest site selection (i.e. foraging for a new nest site) in colonies of social ants provides an example of group decision making. While some ant species form long-lived nests, others such as army ants move very frequently [465]. An intermediate example is provided by ants of the genus *Temnothorax* (formerly *Leptothorax*), which typically live in small colonies of no more than a few hundred workers and inhabit fragile nest sites such as rock crevices, resulting in frequent nest migrations due to the disturbance or destruction of their nesting sites. Their nest site foraging process was initially described in [408] with later contributions in [464, 465, 466, 467].

Nest Site Selection by *Temnothorax* Ants

When searching for a new nest site, only about 30% of the colony scout for new potential locations [466]. The nest selection process has four stages (Fig. 9.9). Initially, ants search the neighbouring environment individually. Once a potential nest site is found, the ant assesses the site using criteria such as cavity size, degree of darkness

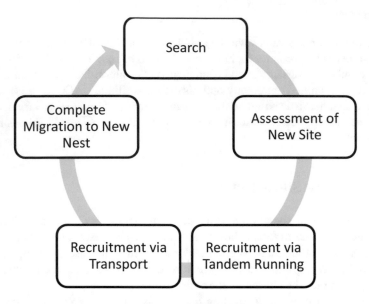

Fig. 9.9. Nest selection process of ants of the genus *Temnothorax*

and size of entrance (smaller entrances being preferred as they are easier to defend). The assessment phase may entail visits (and revisits) to one or a number of potential nest sites, with the length of the assessment phase being inversely proportional to the quality of the site [384]. If the site is accepted by the scout, she recruits another scout at the old nest to the location and leads the single follower to the location by tandem running. In turn, the recruited ant makes its own assessment of the site. The final phase in the process occurs when an ant which has accepted the site encounters a critical number of nest mates at the new location. At this point she enters a committed phase and actively transports (carries) the passive majority of the colony (both adults and brood members) from the old nest to the new site.

The site selection process has two methods of recruitment with the first stage being the 'recruitment of the recruiters' [408]. By delaying recruitment to poorer-quality sites, the colony creates time for the discovery of better locations and avoids too quick lock-in to any site [384]. Conversely, better-quality sites are accepted more quickly, with tandem runs to them beginning sooner than for poorer sites.

The transport phase only commences when a quorum threshold of nest mates at the site has been reached, which helps ensure that an individual ant's assessment of the quality of the site is supported by similar assessments of a sufficient number of other nest mates [466]. In turn, this generates a positive feedback loop as once a small number of ants commence transport recruitment, the likelihood of other ants doing so is increased. Transporting runs are some three times faster than tandem runs [465] and therefore quickly build population size at the new site.

The nest site choice does not stem from global information, as few of the ants will have visited all potential nest sites. Rather, it emerges from the local information gathered by each ant and their interactions. This allows the colony to choose between multiple sites of varying quality even in the absence of top-down control.

Although the search process could be multistage, potentially with transport activity to more than one new location (followed by subsequent transports to the final winning location), the quorum threshold required before transport activity commences lessens the chance of this happening. Although the exact mechanism by which individual scouts decide that their quorum threshold has been reached is not known, it is thought that the rate of encounter by an ant with other scouts at the new location is the trigger (the mechanism may also be pheromone based, with the concentration of pheromone increasing as the density of ants at the new site rises). Experimentally, the probability of transitioning to transport behaviour rises substantially when the new nest site has 5–10 ants present [464].

The nest site selection process of these ants bears some parallel with that of honeybees (described in Sect. 10.4) in that both involve a search and recruitment phase, with a final decision being dependent on a quorum threshold being attained. An optimisation algorithm arising from the honeybee site selection process is outlined in Sect. 10.4.1, and a variant on this algorithm could be designed drawing on inspiration from the nest site selection behaviour of *Temnothorax* ants.

Group Decision Making When Individual Ants Make Differing Quality Assessments

A recent study by O'Shea-Wheller et al. [443] of the nest site selection behaviour of *Temnothorax* ants provides further insights into their group decision-making process. It was found that while individual ants were consistent in spending more time when investigating higher-quality nest sites, the absolute amount of time spent investigating a new site was quite variable between individual ants.

As site quality improves, ants will generally spend longer periods of time investigating the site and are more likely to subsequently accept the site. However, from empirical investigation, some ants are more easily satisfied with potential nest sites and will accept a site even after a relatively short visit, whereas other ants need to spend a considerable period of time investigating a location before deciding whether to accept it. Each of these groupings of ants plays a role in the colony's site selection. Those that accept quickly, recruit other ants from the colony to the new location. Those that spend longer in the new site before deciding whether to accept it facilitate the selection of the site by contributing to collective quorum attainment at the site. Some of the searching ants are harder to please, rarely being satisfied even with a high-quality nest site and continuing to search for alternatives. This suggests that individual ants can have quite different acceptance thresholds and leads to the question as to how variation in individual assessments could impact on group decision making.

This heterogeneity of ant assessments can in fact contribute to enhanced group decision making by increasing the colony's flexibility in adapting to differing environments. With some ants being 'hard to please' a continual search for new sites is maintained, reducing the chance of too early lock-in to a poor-quality site when better alternatives may exist nearby. The 'easier to please' grouping of ants facilitates site selection in the case where only poor-quality sites are available in the environment. The different levels of time spent investigating a new site by ants also generate a subtle balance between recruitment, which is facilitated when ants quickly accept a site, and quorum attainment, which is facilitated when ants spend longer investigating new sites as their presence at the site increases the propensity of visiting ants to transition to transporting behaviour.

O'Shea-Wheller et al. [443] point out that group (colony) decision processes can be influenced by differences in individual behaviour akin to a personality, and indeed that heterogeneity of assessments can enhance colony decision-making flexibility. As noted in the section above, a variety of optimisation algorithms could be designed based on features of the nest site selection behaviour of these ants.

9.12 Summary

The many species of ants inhabit a wide range of foraging environments and have consequently evolved varying physical capabilities, grouping behaviours and foraging behaviours in order to survive. At one extreme, individual small food items may be located randomly in time and space, as is the case for desert ants. In this scenario, ants typically engage in solitary foraging and have developed notable skills for search and for navigation back to their colony. This environment would not tend to produce selective pressure for food recruitment communication and collective foraging [176].

In contrast, other food environments may be much more predictable, with large quantities of food being available at specific locations for longer periods of time. This setting would produce much more selective pressure for mass-recruitment cooperative food retrieval but would not necessarily require individual ants to have strong personal navigation capabilities [176]. In this scenario, an ant colony can be considered as akin to a distributed sensory system, with information on desired resources being transmitted to other colony members.

The diversity of ant foraging behaviours provides a plentiful source of inspiration for the design of optimisation algorithms but we are reminded that each foraging behaviour is closely tied to the nature of the underlying environment and the physical capabilities of that species of ant. Hence, algorithms modelled on a set of specific behaviours will likewise produce their best results in problem environments for which those behaviours are well suited. Notably, only about 50 species of ants have been studied in detail [216] and consequently, much remains to be uncovered concerning the complete repertoire of foraging behaviours of this family of organisms.

In this chapter we have introduced a number of ant-foraging-inspired algorithms which can be applied for both discrete and real-valued optimisation. Many of these algorithms emphasise the sharing of information via mechanisms such as trail marking by simulated pheromone deposit, and the subsequent autocatalytic reinforcement of high-quality solutions. Real-world ant foraging behaviours can use information from other senses as well although few of these mechanisms have yet been incorporated into ant-foraging-inspired algorithms. For example, some species of ants use vibrational signalling for nest mate recruitment, with foraging leaf-cutting ants of the species *Atta cephalotes* using substrate-borne vibratory signals to attract other ants to areas of leaf harvesting (i.e. the process bears parallel with a signal being broadcast from the food source) [488, 489]. Ant species also have varying capabilities concerning visual learning and memory, and place differing focus on private and public information when foraging. Some species are also known to be able to modulate their use of each type of information according to state information such as degree of starvation. The potential to design more sophisticated optimisation algorithms which embed some of these ideas is noted in [604].

10

Honeybee Foraging Algorithms

The last decade has seen the development of a family of powerful optimisation algorithms inspired by the foraging behaviours of the European honeybee, *Apis mellifera*. Like ant algorithms, a key component of most honeybee algorithms is the concept of recruitment. In the case of honeybees, successful foragers directly broadcast information to other colony members by means of a *dance language*. This language is a symbolic system of communication and can be used to transmit information on (amongst other things) the location and quality of resources in the vicinity of a colony's hive.

In this chapter we provide an introduction to aspects of honeybee foraging behaviours, particularly the honeybee recruitment dance. We also describe a number of optimisation algorithms which have been developed based on the food and nest site foraging behaviours of honeybees.

10.1 Honeybee Recruitment Dance

The foraging activities of bees involve searching for resources such as pollen (a source of protein), water, waxy materials for hive building and nectar from flowers. Nectar is converted by bees into honey, which serves as a durable energy store in the hive. When a scout or explorer bee discovers a food source of sufficient quality, she may subsequently undertake a dance on her return to the hive once she has unloaded her harvest. The objective of the dance is to inform and potentially recruit other foragers to the resource. In turn, the newly recruited bees may also undertake a dance when they return to the hive if the food resource that they have been recruited to is of sufficient quality, creating a positive feedback cycle.

The dance language consists of repetitive patterned movements that are intended to communicate, in the context of foraging, information about the location and desirability of a food source. In essence, the dance can be considered as a reenactment of the flight from the hive to the location of interest. The dance is undertaken in a specific location of the hive near the entrance called the *dance floor*. The dance floor

© Springer Nature Switzerland AG 2018
A. Brabazon, S. McGarraghy, *Foraging-Inspired Optimisation Algorithms*,
Natural Computing Series, https://doi.org/10.1007/978-3-319-59156-8_10

consists of a vertical comb in the hive, and typically this area contains multiple potential foraging recruits. The dance is social in that it is never undertaken without an audience [117].

Nature of the Dance

The nature of the dance movements depends on the location of the food source relative to the hive. If the food source is close by (up to about 100 metres from the hive), the bee undertakes a *round dance*, with circular movements predominating. If the food source is further away, a *waggle dance*, in which the movement pattern resembles a figure of eight, is undertaken. The direction to the resource on leaving the hive (relative to the sun) is indicated by the angling of the bee's body during the dance. The desirability of the location is communicated by the dance's *liveliness* or *enthusiasm*, with more desirable locations producing livelier dances [516]. The duration of the waggle portion of the dance is a proxy for the distance to the location of further-away food sources.

The dance floor of a hive is a *marketplace* for information about the state of the colony and the external environment. At any point in time there may be several bees on the dance floor, with multiple food locations being advertised simultaneously. Recruited foragers tend to travel to richer food sources in greater numbers as dances for high-quality food sources are more conspicuous and repeated, thereby creating a positive feedback loop resulting in the amplification of the exploitation of those food sources. However, unlike the mass recruitment mechanisms of some ant species as discussed in Sect. 9.2, the positive feedback mechanism is not sufficiently strong to lock a honeybee colony into harvesting from a single location. This helps ensure that the colony forages at a variety of locations, and this permits quick adaptation by the colony in the event that a particular food resource becomes exhausted and therefore needs to be abandoned.

Regulation of the Dance

The dance recruitment process is regulated by a number of mechanisms which allow for the integration of internal- and external-state information. Internally, the resource requirements of the colony are indirectly communicated to foragers via their unloading time. If a successful forager is unloaded quickly by receiver bees on her return to the hive (as the resource is needed urgently by the colony), she is more likely to dance for the relevant foraging location. Conversely, if the resource is not in short supply, the unloading time is likely to be longer, as fewer receiver bees will prioritise its unloading. Therefore, the unloading time acts as a regulatory feedback mechanism for recruitment as it influences dance propensity [559].

External-state information can be incorporated into the recruitment process by means of communication of a negative or inhibitory signal for foraging at particular locations, i.e. a *stop signal*. A stop signal is a brief vibrating signal made by a bee that lasts for about a tenth of a second. It is frequently delivered by a sender bee butting her head into a recipient, a dancing bee on the dance floor. This signal generally causes the recipient to freeze and therefore cease dancing for a location. Stop signals may result from crowding at a resource location. A honeybee that has suffered a predatory attack at a resource location being danced for is also more likely to produce a stop signal, particularly if the resource location is distant. Bee colonies are therefore capable of tuning the numbers of follower bees recruited in response to crowding and predation risk [274, 434].

Other Elements of the Recruitment Process

The above description of the dance does not incorporate the role of odours, sounds and taste, which are also relevant to the recruitment process [515, 565, 602]. For example, *trophallaxis* (the transfer of a nectar or another resource sample from one honeybee to another) plays an important role in foraging, as a bee will abandon its own source of nectar more quickly if it is exposed to higher-quality nectar by other honeybees.

It is speculated that electric fields may also play a role in the recruitment process. Bees have been found to possess electroreception capabilities [225] and, indeed, have also been found to emit both constant and modulated electric fields when flying, landing and walking and during the waggle dance. The fields have both low- and high-frequency components and can induce passive antennal movements in neighbouring stationary bees. It is suggested that these electric fields may be biologically relevant stimuli and could play a role in social communication between conspecifics. Social communication amongst bees may be more nuanced and multimodal than has been previously realised [225].

10.2 Honeybee Foraging Algorithms

Broadly speaking, most honeybee foraging algorithm variants can be decomposed into three iterated activities:

i. explore the environment for good food locations;
ii. recruit foragers to the better of the discovered locations; and
iii. harvest these locations.

Typically, a random element is also added to the algorithm in order to prevent premature convergence of the search process.

10.2.1 Bees Algorithm

The *bees algorithm* was developed by Pham et al. [457, 458]. The algorithm simulates a search process by virtual bee scouts who in turn recruit other bees to good locations or 'flower patches' that they have found. A differential recruitment process is simulated whereby the best sites found by the scouts recruit the greatest number of followers.

The parameters for the algorithm are a population of n scout bees, a number (m) of patches selected out of the n visited points (i.e. the best m locations found by n searching scouts) and e elite patches (i.e. the best e of the m patches), n_{ep} being the number of bees recruited to search in the e best locations and n_{sp} being the number of bees recruited to the other ($m-e$) patches.

Initially, all n scout bees are placed at random locations in the search space. The fitness of each bee (location) is determined. The best e locations are considered as elite patches and additional bees (n_{ep} bees per patch) are recruited to search in the neighbourhood of each these locations. For each patch, the best location found is memorised and the bee corresponding to that location survives into the next iteration of the algorithm.

A similar local search process is undertaken in the next $m-e$ best patches, with n_{sp} search trials being allocated to each of these patches. Again, the patch fitness is determined as the fitness of the best location found, and the bee corresponding to this location survives into the next generation. As the values of n_{ep} and n_{sp} are selected so that $n_{ep} > n_{sp}$, a greater number of bees are recruited to search in the patches around the elite sites; therefore, the search process is more intensive around the better locations.

The local search process can be operationalised in a variety of ways but in the canonical version of the bees algorithm, it is implemented as a randomly selected location in a neighbourhood around the original location. In the simplest case, the neighbourhood can be determined using a parameter r which defines a hypersphere of radius r around the original location.

A global search process is then undertaken using $n-m$ bees, with each of these bees being randomly located in the search space. The fitness of each of these locations is then determined.

The search process iterates until a terminating condition is hit. The algorithm consists of an exploitative phase (the local search phase) and an explorative phase (the global search process). Pseudocode is outlined in Algorithm 10.1.

10.2.2 Improved Bees Algorithm

The canonical bees algorithm was modified by the addition of two mechanisms in [456] in order to enhance its efficiency.

Neighbourhood Shrinkage

The neighbourhood size for each of the n patches $a^1, a^2, \ldots, a^n$ is initially set to a fairly large value. For each dimension a_i of a patch location a ($i = 1, 2, \ldots, d$),

$$a_i(t) = r(t)(b_i^{\max} - b_i^{\min}), \tag{10.1}$$
$$r(0) = 1.0, \tag{10.2}$$

where t is the current iteration of the bees algorithm, and $b_i^{\max}$ and $b_i^{\min}$ are the maximum and minimum values on dimension i of the search space.

Algorithm 10.1: Bees algorithm [458]

Randomly assign each of n bees to initial food sources (locations) in the search space;
Evaluate the quality of each of these locations;
$t := 1$;
while *stopping criteria not met* **do**
> Select sites for neighbourhood search;
> Recruit bees for selected sites (more bees for best e sites) and evaluate fitness for these searches;
> Select fittest bee from each patch;
> Assign remaining bees to search randomly and evaluate fitness of these searches;
> Sort locations by their fitness values;
> Set $t := t + 1$;

end
Output the best solution found;

The size of the neighbourhood remains unchanged as long as the local search process produces locations of higher fitness. The local neighbourhood is initially defined over the entire range of the search space. If no better location is found, the size of r is decreased (10.3). If no forager finds a solution of higher fitness, the size of the flower patch is shrunk. As a result, the scope of the search becomes more intensively focused in the neighbourhood surrounding the local fitness best.

$$r(t+1) = \begin{cases} 0.8r(t), & \text{if no improvement,} \\ r(t), & \text{otherwise.} \end{cases} \quad (10.3)$$

The above process bears parallel with the cooling schedule in simulated annealing in that the search process alters from being explorative to local (exploitative) over time.

Site Abandonment

If there is no fitness improvement in a patch during the neighbourhood-shrinking process over a preset number ($t_s^{\lim}$) of search attempts within the patch, the local maximum of fitness is considered found. The patch is then abandoned and a new scout is located at a randomly generated location in the search space.

Parameters

While the parameters for the successful application of the bees algorithm will be application-specific, the parameter settings in [658] included $n = 50$, $m = 15$, $e = 3$, $n_{ep} = 12$, $n_{sp} = 8$, $r = 1$ and the maximum number of iterations was set as $t_{max} = 5000$. This implementation also included a neighbourhood-shrinking and site abandonment process.

10.2.3 Artificial Bee Colony Algorithm

Around the same time as the development of the bees algorithm, another bee-foraging-inspired algorithm, the *artificial bee colony* (ABC) *algorithm*, was independently proposed by Karaboga [302]. There are similarities between the two algorithms due to their common biological inspiration, although the implementations of the recruitment process and of the local search process differ between them.

In the artificial bee colony algorithm the population of bees is divided into *employed bees* (those which are currently exploiting an already discovered food resource) and *unemployed bees* (those which are looking for a food resource). The latter group can be further decomposed into scout bees and onlookers, where scout bees search around the nest for new food sources, and onlooker bees wait at the nest for recruitment to food resources. As before, a specific location in the search space represents a potential solution to the optimisation problem of interest. The pseudocode for the artificial bee colony algorithm is outlined in Algorithm 10.2.

The search process is undertaken by a population of $2S_N$ artificial bees, where S_N of these are designated as employed bees and the remainder are onlooker bees. Initially, each of the S_N employed bees is located at a randomly selected location in the search space (simulating the process whereby the corresponding initial food resources have been found by scout bees from the hive), corresponding to the starting food locations from which the search process develops. Each of these locations (or potential solutions) is a d-dimensional vector, with d being the dimension of the search space. The quality of each of these locations is then assessed using (10.5) below, where fit$_i$ is the fitness of the ith location ($i \in \{1,\dots,S_N\}$). The effect of (10.5) is to scale the fitness into the range $[0,1]$ and it is assumed that all raw fitness values are positive ($f_i \in [0,\infty)$). Next, the search process commences.

Local Search Step

In this step, each of the S_N employed bees seeks to locate a better food source in its vicinity, simulating the process of a bee using visual cues in order to uncover better sources of nectar. Assuming that the bee is initially located at $x_{i,j}$, it takes a randomly generated step from this location to a new location $v_{i,j}$.

The taking of this step could be operationalised in a variety of ways. For example, in [306] the process is governed by (10.4) where for $i \in \{1,\dots,S_N\}$, j is a randomly generated number in the range $[1,\dots,d]$ where $\phi_{i,j}$ is a randomly generated number $\in [-1,1]$ and $k \in [1,\dots,S_N]$ is the index of a randomly chosen solution ($k \neq i$). In essence, the new solution is obtained by mutating the current solution using a stochastic difference vector ($\phi_{i,j}(x_{i,j} - x_{k,j})$) [445]. The difference vector is obtained by comparing the current solution with another solution in the population:

$$v_{i,j} = x_{i,j} + \phi_{i,j}(x_{i,j} - x_{k,j}). \tag{10.4}$$

The quality of the resulting location v_i is compared with that of x_i and the bee exploits the better location. If the new location ($v_{i,j}$) is better than the old one ($x_{i,j}$), then the

---

Algorithm 10.2: Artificial bee colony algorithm [302, 304, 306]

Randomly assign each of S_N employed bees to initial food sources (locations) in the
search space;
Evaluate the quality of each of these locations;

Set $t := 1$;
while $t < t_{max}$ **do**
 for *all employed bees* $\in \{1, \ldots, S_N\}$ *in turn* **do**
  Search for better solutions in the vicinity of their current location (food patch)
   using (10.4);
  Evaluate the quality of the new solution;
  if *the new solution is better* **then**
   Relocate the bee to the new position;
  else
   Retain the bee at its current location;
  end
 end
 Calculate the probability of an onlooker bee choosing each of the S_N food patches
  to harvest using (10.6);
 for *all onlooker bees* $i \in \{1, \ldots, S_N\}$ *in turn* **do**
  Select a food patch to harvest using (10.6) (this chooses a value for j);
  Randomly select $k \in S_N$;
  Search for better solutions in the vicinity of that food patch using (10.4);
  Evaluate the quality of the new solution using (10.5);
  if *the new solution is better* **then**
   Relocate the bee to the new position;
  else
   Retain the bee at its current location;
  end
 end
 if *the location of any of the S_N food patches has remained unchanged for more*
  than t_{lim} iterations of the algorithm **then**
  Abandon that location and replace it with a new randomly generated location;
 end
 Record location of best solution found so far;
 Set $t := t + 1$;
end
Output the best solution found;

new location replaces the old one in the memory of S_N food sources. Otherwise, the location remains unchanged in the memory. A critical aspect of this search process is that as the positions of the best food patches begin to converge during the algorithm (and, therefore, $x_{i,j}$ and $x_{k,j}$ get closer together), the step size in the search process self-adapts, becoming smaller in order to promote more intensive exploitation around the already-discovered better solutions.

Recruitment

After all the employed bees have undertaken a local search step, the onlooker bees are recruited to the S_N food patches and, in turn, they search around the food patch to which they have been recruited using (10.4). In choosing which food patch to investigate, each onlooker bee is driven by (10.6), simulating a dance process, where p_i is the probability that any given onlooker bee is recruited to food patch i, $i \in \{1, \ldots, S_N\}$. This roulette wheel selection mechanism implies that the best food patches are more likely to recruit onlooker bees. If a food source is not improved after a predetermined number of iterations (parameterised as t_{lim}), it is abandoned and replaced by a new randomly created food source, thereby maintaining an exploration capability in the algorithm and helping to prevent premature convergence. This simulates the explorative search behaviour for new food sources by scout bees:

$$\text{fit}_i = \frac{1}{1 + f_i}, \tag{10.5}$$

$$p_i = \frac{\text{fit}_i}{\sum_{n=1}^{S_N} \text{fit}_n}. \tag{10.6}$$

The critical parameters of the algorithm include the value of S_N, the number of iterations before an unchanged food source location is abandoned (t_{lim}), and the maximum number of iterations (t_{max}) of the algorithm. The canonical ABC algorithm as described above can exhibit slow convergence to the global optimum and a number of variants have been proposed which speed up convergence. The reader is referred to [5, 318] for details of two of these.

10.2.4 Discussion

At a high level, there are similarities between the bees algorithm and the ABC algorithm as both embed a populational search process which has a stochastic component, a locally driven search component and a recruitment/social communication mechanism between individual members of the population.

However, the operationalisation of these mechanisms is different in each algorithm. Recruitment is based on a probabilistic process in ABC whereas a fixed number of bees are recruited to each elite patch in the bees algorithm. The implementation of the local search mechanism also differs between the algorithms. In ABC, the local search point is determined as a weighted average between the location of the current bee and another randomly selected member of the population. In contrast, in the bees algorithm, the local search point is determined using a random selection in a neighbourhood around the current bee.

Both algorithms have shown good performance on a range of benchmark and application studies, with ABC being the more commonly applied variant. A comprehensive review of applications of the ABC algorithm is provided in [305].

10.3 Developments in the Honeybee Literature

In most foraging-inspired honeybee algorithms, the core concept is that of recruitment, whereby bees which have found good food sources recruit colony members which travel to, and harvest, the food resource (or 'good' location on the fitness landscape). In order to avoid premature convergence and maintain diversity in the search process, forgetting mechanisms are typically included in algorithmic implementations. These can be as simple as the maintenance of continual random search by some foragers.

Whilst the resulting algorithms have proven to be highly effective for optimisation, they incorporate a limited number of features of the behavioural repertoire of real honeybees. These behaviours have been extensively studied in recent decades and we now possess a much more comprehensive understanding of the foraging process of honeybees. Drawing on this literature, three items in particular are illustrated in Fig. 10.1 and subsequently discussed.

10.3.1 Individual Perception

Foraging by honeybees does not just rely on the transmission of social information. Individual honeybees are capable of processing multiple sensory inputs during their foraging trips in order to return to desirable foraging locations and to find new ones. An interesting aspect of the memory process involved in learning the location of the hive and of good foraging locations is *look-back behaviour* whereby the departing bee will turn around and 'look back' at the location and its surrounds a number of times during initial flights from the hive or initial return flights from the foraging location. It is assumed that during this behaviour the bee is memorising an image of features associated with the location, which can subsequently be used for return navigation.

Another feature of vision in bees (and many other insects) is their experience of an *optic flow* whilst in flight [118]. In this, features directly ahead of the insect appear to be motionless, whereas environmental features to the side of the bee's flight path move past as the bee moves. The rate of optic flow experienced by the bee during a foraging trip can act as a navigational guide for subsequent flights.

In contrast to bumblebees which process information in parallel from a wide visual field, honeybees process visual information using a serial search behaviour, with search terminating the moment the first target is discovered. Honeybees have a

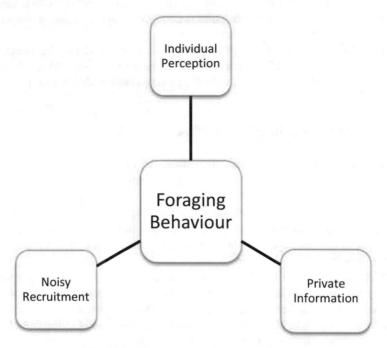

Fig. 10.1. Three aspects of honeybee foraging

smaller visual field than bumblebees and their visual search is more akin to 'moving a small spotlight step by step over the search area' [412, p. 2522]. In addition to visual sensory capabilities, chemoreception, sound and touch also provide important environmental information to honeybees. Given these abilities, honeybees are capable of identifying promising food sources at a distance and altering their flight trajectory to forage at these resources.

Other sensory modalities of honeybees include electroreception capabilities [225]. It is speculated that this may allow them to assess whether a flower has been recently visited by another bee and therefore should be avoided as its resources have already been harvested. Flying insects tend to collect a positive charge when flying through the air, whereas flowers often have a negative potential. This combination offers direct benefits as it promotes pollen transfer and adhesion [112]. Perhaps less obviously, electric fields can provide information cues for bees during foraging. When a bee visits a flower, it alters the electrical field associated with the flower. Potentially, this alteration could serve as an information cue to inform subsequently foraging bees as to whether a flower has recently been visited by another bee and therefore is likely to have a reduced concentration of pollen. As electrical fields can change in seconds (much faster than other floral reward cues), this sensory modality can facilitate rapid and dynamic communication between flowers and their pollinators [112]. Honeybees are also capable of sensing magnetic fields and using these for orientation during navigation [605]. It is thought that this capability arises from a magnetite-based magnetoreceptor located in their abdomen [345].

In summary, bees make use of multiple sensory modes when foraging and are well capable of perceiving their environment.

10.3.2 Noisy Recruitment

A second issue is that the recruitment dances of honeybees are much noisier in the real world than is typically suggested in honeybee algorithms. A dancing bee will repeat a dance multiple times (sometimes up to 100 times [576]), with higher-quality sources tending to induce more dancing behaviour. Observational evidence indicates that repeated dances for a specific resource by the same bee (or by different bees recruiting for the same resource) often vary in both directional and distance information [232, 576]. Therefore, dances only recruit to an *approximation* of the location of the food resource. Dance followers observe several iterations of the dance to compute an average travel vector, with some 5–10 performances being observed at a minimum. Despite this, most recruits have to undertake several trips before finding the advertised food source.

Counterintuitively, observational evidence indicates that the noise concerning the directional component of the vector is larger for nearby food sources (up to 1 km from the nest) than for more distant food locations. The distance component of the vector is also noisy, with the variance of the distance information being scale invariant [140], again indicating greater noise for close-by food sources.

Although, prima facie, a noisy communication mechanism would appear suboptimal, it has been suggested that the imprecision in the honeybee dance could in fact be adaptive as it would allow the discovery and exploitation of food sources which are near to the resource originally recruited for [222]. In essence, it injects a stochastic element into the foraging process, and may be well suited to an environment in which food resources are found in patches.

10.3.3 Private Information

A third issue is that in spite of the importance accorded to recruitment in most honeybee algorithms, real-world bees place substantial reliance on personal (private) information. Over the foraging lifetime of a honeybee (approximately 99.5 ± 27.3 foraging trips), only some 25% of foraging flights on average are preceded by dance observation by the forager and in less than half of these cases is a bee recruited to a new food source [55].

This study also considered whether the recruitment propensity was dependent on the level of foraging experience of the bee and found that even in the case of novice foragers, only about half made use of information acquired from observing a dance rather than searching independently. On most foraging trips, bees rely on personal, previously acquired knowledge, with even inexperienced foragers relying on trial and error learning [232, 629].

The propensity of a bee to use socially acquired information varies depending on context. If the current food location that a bee is harvesting becomes unprofitable, the forager will retire from it and subsequently look for a new resource either by trial and error search, or by following a dance and being recruited to a new resource location.

There is emerging, tantalising evidence that insects may have emotional-like states which could at least partially explain the above observed behaviours (i.e. persistence in harvesting already-located resources). Research concerning bumblebee behaviours indicates that their decisions are influenced by prior events which generated the neurochemical dopamine, a chemical which plays a key role in the reward system of humans [455].

Experienced bees can employ a flexible strategy—called *copy if dissatisfied* (see Sect. 4.2.2)—which combines both personal and social learning, rather than blindly following recruitment dances regardless of the payoffs of their current behaviour. This strategy is relatively simple to implement, as it does not require complex cognition such as a precise comparison of the relative costs and benefits of several alternatives.

Some forager bees also maintain a *memory* of old food sources which they have previously harvested but from which they retired when the source became unprofitable. These *inspector bees* continue to make occasional trips to the old location to check on its quality and will resume foraging at that location if it again becomes profitable [55]. Memory information about the route to previously harvested resources can be quite persistent and allow a foraging bee to return to food locations even after a gap of some weeks [230]. These inspector bees act as short-term memory for the bee colony and facilitate the quick reactivation of previously abandoned food sources.

Internal-state information also plays a role in determining the foraging activities of honeybees [309]. Bees in a higher nutritional state forage less often than those in a lower nutritional state, and higher individual nutritional states are also associated with a greater degree of explorative (rather than exploitative) foraging activity. A similar effect is found if the colony's nutritional state is high, with a greater degree of exploration taking place during foraging.

10.3.4 Discussion

From the above, it is evident that the use of social and private information is nuanced in real honeybee foraging behaviours and that current honeybee algorithms are based on very simplified models of honeybee foraging.

The development in our understanding of the subtleties of honeybee foraging opens up avenues for the design of new algorithms. In particular, while social information is important in honeybee foraging, the majority of foraging bees are actually using private information at any point in time. Social information does not have the dominant role which it is assigned in most honeybee algorithms.

As described by Seeley [515, p. 252] the foraging activities of a colony of honeybees bears more similarity to 'an ensemble of largely independent individuals that rather infrequently exchange information with one another'. This nuanced perspective contrasts with the suggestion in some of the honeybee optimisation literature that honeybee foraging provides an example of swarm intelligence.

10.4 Nest Site Selection

Another example of a honeybee foraging behaviour is provided by nest site selection (see Sect. 9.11 for a related discussion of the nest site selection behaviour of ants of the genus *Temnothorax*). Typically, in late spring or early summer as a colony outgrows its current hive, the colony will *fission* (i.e. divide), whereby the queen bee and about half of the population of worker bees will leave the hive and seek to establish a colony at a new nest site, leaving behind a young queen bee and the remainder of the worker bees in the existing hive. Having left the current hive, the swarm usually does not fly far and within about 20 minutes it forms a football sized cluster of bees, often on the branch of a tree [43]. From this location, *scout* bees begin a search for a new nest site and the search process can last for some days. An ideal home for a bee colony is located several metres off the ground, has a cavity volume of more than 20 litres and has a south facing entrance hole (in the northern hemisphere) smaller than 30 square centimetres which is located at the floor of the cavity [518, p. 222].

Site Discovery

During the site selection process, scout bees leave the cluster and search for a new nest site. These scout bees are the most experienced forager bees in the swarm. Usually the swarm will have several hundred scouts. As potential nest sites of satisfactory quality are uncovered, the returning scout bees communicate their location to other scout bees by doing a waggle dance on the surface of the swarm. The length of the dance depends on the quality of the site found, with longer dances being undertaken for better-quality sites. If a bee finds a good site, it visits it a number of times, eventually becoming *committed* to it. However, the length of its recruitment dance for the site decreases after each visit. This phenomenon is known as *dance attrition* [518].

A scout bee will only dance for a location if its quality exceeds a threshold value. A scout may undertake several trips before uncovering a site of the desired quality. Alternatively, a scout bee may be recruited by a returned dancing scout bee and may therefore visit a promising location found by another scout bee. In turn, if the recruited bee also considers the location to be of satisfactory quality, she will dance for that location on her return to the swarm. The net effect of the recruitment and the dance attrition phenomenon is that higher-quality sites attract more attention from the searching scouts, creating a positive reinforcement cycle. Dance attrition facilitates the forgetting of nest site locations that are not continually reinforced. While multiple nest sites (if several of sufficient quality exist) will be considered in the early stages of the search process, these will be quickly whittled down to a limited number of choices, from which one is finally chosen. Unlike the foraging process, whereby several food locations may be harvested simultaneously, the nest site selection problem produces a single winner.

Selection of Final Site

The final decision as to nest site location results from a *quorum sensing* process [518] (see Sects. 8.1.2 and 9.11). Once scout bees at a potential nest site sense that the number of other scout bees there has reached a threshold of approximately 20 bees they undertake *buzzing runs* at the new nest site, which triggers the return of all the scouts to the swarm [517]. The scouts then excite the swarm into getting ready to move by doing buzzing runs across the swarm's surface and by moving through the swarm, producing a high-pitched acoustic signal from their wing vibrations (known as *piping*).

When the swarm lifts off, an evident practical problem is that less than 5% of the swarm has previously visited the new nest site [517], so most of the swarm do not know the new site's location. The swarm is guided to the correct location by the scout bees who have visited the new site, and they signal the correct trajectory by flying rapidly through the swarm in the direction of the new nest site [43]. In addition, some scouts shoot ahead to the nest site, where they release marker pheromones.

The nest site selection process of honeybees presents an interesting example of a high-stakes exploration–exploitation trade-off [279]. If too fast a decision is made (thereby exploiting information acquired early in the search process), the swarm runs the risk of selecting a poor location. On the other hand, if the decision making process is too slow, the swarm is left exposed to the risk of bad weather and/or predation. The selection process also presents an example of decentralised decision making, as the final site selection is determined by the actions of multiple, independent agents (bees). Even where a scout bee is recruited for a potential nest site location, she will travel to the site and inspect it for herself in order to decide whether she will dance (or *vote*) for it. As noted by Passino and Seeley [452], it is plausible that evolution has tuned the process of nest site selection in order to balance its speed–accuracy trade-off, balancing the chance that the swarm selects a poor site against the energy and time cost associated with searching for a new site.

Recent years have seen a number of studies, such as [452], which have used simulation in order to examine the relative importance of various elements of the nest site selection process in determining the success of the search process. As would be expected, the values for parameters such as the number of scout bees, the quorum threshold required to make a site decision, the decay rate of the dance length following revisits to a site by scouts and the propensity of bees to search by themselves as opposed to being recruited by other scouts have all been found to be important.

The potential for drawing inspiration from the bee nest site selection process in order to design general-purpose optimisation algorithms was indicated in [149]. This framework was further developed and applied in [150] with the creation of the *bee nest site selection optimisation* (BNSO) *algorithm*. This algorithm is outlined below.

10.4.1 Bee Nest Site Selection Optimisation Algorithm

The essence of the algorithm is that the swarm of bees searching for a new nest seeks to iteratively improve the quality of its location. The swarm is composed of two types of bees, namely scouts and followers. The scouts seek new potential nest sites in the vicinity of the swarm's current location and, if a scout succeeds, it recruits a number of followers based on the fitness of the location that it has found. The followers then go to the location found by the scout and search locally around that location. If a better site is found than the swarm's current location, the swarm moves to the new location and the search process begins anew. Pseudocode for the BNSO algorithm is outlined in Algorithm 10.3.

Algorithm 10.3: Bee nest site selection algorithm (from [150])

Place swarm on random location p (set $p_{\text{swarm}} := p$);
Let $t := 0$;
Set values for d_{scout}, d_{follower}, and number of scouts and followers;
repeat
    Set value of f_{range} using (10.7);
    for *each scout in turn* **do**
        Choose new location p_s with a maximum distance of $d_{\text{scout}} \cdot f_{\text{range}}$ to the nest;
        $\text{fit}_s = \max\{0, (F(p_{\text{swarm}}) \cdot f_q) - F(p_s)\}$;
    end
    for *each follower in turn* **do**
        Choose a scout s according to (10.8);
        Choose a new location p_{follower} with a maximum distance of d_{follower} to
          chosen scout's position p_s;
        Sample search space between p_s and p_{follower} in m equally spaced flight steps;
    end
    if *a new location p was found which is better than p_{swarm}* **then**
        Relocate swarm to p ($p_{\text{swarm}} = p$);
    else
        if $t \geq t_{\text{max}}$ **then**
            Place swarm on new random location p ($p_{\text{swarm}} = p$);
            Set $t := 0$;
        else
            Set $t := t + 1$;
        end
    end
until *termination condition is satisfied*;
Output the best solution found;

In the algorithm, there is a swarm of n bees comprising n_{scout} and n_{follower} bees ($n = n_{\text{scout}} + n_{\text{follower}}$). These are searching in a real-valued space and the location p_{swarm} of the swarm in this space corresponds to a solution to the maximisation or minimisation problem of interest. During the algorithm, the swarm is initially placed at a random location, and each scout s chooses a random location p_s which is within $d_{scout} \cdot f_{range}$ of the swarm's current location (hence, $|p_{swarm} - p_s| \leq d_{scout} \cdot f_{range}$ for all scouts). This simulates the real-world phenomenon that scouts are more likely to search within a few minutes' flight time of the swarm's current location.

In order to encourage convergence of the search process, $f_{range} \in [0, 1]$ is decreased as the algorithm iterates. A simple mechanism for implementing this is presented in (10.7), whereby t_{max} is the maximum number of iterations for which the algorithm will run and t is the current iteration number:

$$f_{range} = 1 - \frac{t}{t_{max}}. \tag{10.7}$$

If the new location found by a scout is of sufficient quality (i.e. assuming a minimisation problem, $F(p_s) \leq F(p_{swarm}) \cdot f_q$, where $f_q \in [0, 1]$ is a 'quality improvement' factor), then the location is a candidate for recruitment of follower bees. The relative fitness of each scout is calculated using $fit_s = \max\{0, (F(p_{swarm}) \cdot f_q) - F(p_s)\}$. Each follower then selects one scout to follow using (10.8); the fitter locations uncovered by the scouts will tend to recruit a greater number of followers.

Each follower then proceeds to choose a random location p_f in the neighbourhood of the scout's location (p_s) subject to the constraint that the selected location is not greater than $d_{follower}$ away from p_s so that $|p_s - p_f| \leq d_{follower}$. As with d_{scout}, the value of the parameter $d_{follower}$ is set by the user of the algorithm. The follower bee then samples the search space between p_s and p_f in m equal-sized steps (calculating the fitness at each of the m steps) on a straight-line flight between the two locations.

During the scout and follower search phase, a record of the location of the fittest point found (p_{best}) is maintained and, if this is better than the fitness of the current location of the swarm (p_{swarm}), then the swarm moves to the new location and restarts the process. Otherwise the nest search process is restarted from the swarm's current location. If the swarm cannot improve its location after t_{max} attempts, it is moved to a new random location in the search space, t is reset to 0 and the search process recommences from that location:

$$P_s = \frac{fit_s}{\sum_{k=1}^{n_{scout}} fit_k} \tag{10.8}$$

The bee nest site selection optimisation algorithm as described above was applied in [150] to the molecular docking problem. The study compared the results from the algorithm against those from PSO and random search on the same problem, finding that the bee nest site selection optimisation algorithm was reasonably competitive. Further development of the algorithm and further testing will be required before its utility can be fully assessed. However, the algorithm does represent an interesting new avenue for honeybee-foraging algorithmic research.

10.5 Summary

Honeybee behaviours are a rich set of sources of inspiration for the design of computational algorithms. Thus far, most algorithms incorporate a limited number of features of the full behavioural repertoire of honeybees. Scope exists to refine these algorithms further. In addition to real-valued optimisation, there have been applications of honeybee-inspired algorithms for binary-valued optimisation problems [445], for combinatorial optimisation and for clustering purposes. Readers are directed to [37, 303] for relevant references.

In the last two chapters we have introduced a number of algorithms which have drawn inspiration from the foraging activities of some species of ants and honeybees. A feature of the foraging activities in both cases is that the organisms are central place foragers, in that they return to a colony or hive in order to store food, and therefore they can interact with colony members and potentially pass on information about food finds. Information can also be passed on indirectly by means of chemical trails in the environment. A notable aspect in both cases is that successful foragers seek to recruit other conspecifics to food resources that they have found.

In the next chapter we introduce another form of communication mechanism, whereby organisms use light signals which are self-generated in order to advertise information concerning themselves or their location.

11

Bioluminescence Foraging Algorithms

Bioluminescence refers to the generation of light by living organisms, usually by means of a chemical process [235]. Bioluminescent organisms have a variety of light generation behaviours, with some emitting light continuously and others emitting light in flashes, typically of 0.1 to 1 second in duration. Flashing is controlled by the organism by rapidly turning a light-generating chemical reaction on and off [625]. Bioluminescence is used for a variety of purposes. It can be used for signalling in scenarios as diverse as startle defence, mate attraction or territorial warning. Another use of bioluminescence is to engage in aggressive mimicry for the purposes of attracting prey.

The phenomenon of bioluminescence is particularly widespread in marine environments with multiple species of fish and other marine organisms (including jellyfish, crustaceans and cephalopod molluscs) displaying this capability [618]. For many marine animals, their primary visual stimulus comes from biologically generated light rather than from sunlight [235], as ocean depths below about 1000 metres (termed the *aphotic zone* and comprising some 90% of total ocean volume) receive no sunlight at all. A small number of terrestrial creatures such as some species of insects are also capable of bioluminescence, as are some microbes and fungi [40].

Inspiration has been drawn from the phenomenon of bioluminescence to design several optimisation algorithms. We describe three of these in this chapter, the *firefly algorithm*, the *glow-worm algorithm*, and the *bioluminescent swarm optimisation algorithm*. All of these draw metaphorically on the bioluminescent capability of fireflies.

11.1 Firefly Flash Signalling

Most bioluminescent insects are beetles, with the best-known examples being fireflies, winged beetles of the family *Lampyridae* [625]. This family consists of over 2000 species and are found in a wide range of temperate and tropical regions. The majority of fireflies are nocturnal and have evolved a bioluminescent visual sig-

© Springer Nature Switzerland AG 2018
A. Brabazon, S. McGarraghy, *Foraging-Inspired Optimisation Algorithms*,
Natural Computing Series, https://doi.org/10.1007/978-3-319-59156-8_11

nalling mechanism, being able to produce light from special light-emitting organs called *lanterns* in their lower abdomen as a result of a chemical process.

Curiously, the emitted light is circularly polarised, this being the only known example of the emission of such light by a living creature [118]. In fireflies, the enzyme luciferase acts on a light-emitting molecule (generically called *luciferin*) in the presence of ATP and oxygen to produce light. The chemical process produces *cold light*, with little or no heat being generated by the reaction. The light produced is entirely in the (human-)visible spectrum, containing no infrared or ultraviolet components.

Uses of Light Signals by Fireflies

Fireflies use light signals for two main purposes, as a warning to predators and for mate attraction in the darkness. Firefly larvae and many firefly adults contain chemicals that are toxic or unpleasant to predators, and being able to advertise their nature by emitting distinctive visual light patterns lessens their predation risk. In mate attraction, a complex interplay of flash signals between male and female fireflies has evolved as part of the mating process. Typically, the first signallers are flying males which try to attract the attention of flightless females on the ground by emitting a species-specific flash pattern while in flight. Once a female responds by emitting a response flash or continuous signal, a courtship dialogue begins in which light signals are exchanged [350]. In some species, females can flash spontaneously as well as in response to male signals.

In addition to predator deterrence and mate attraction, flash signalling is sometimes used for attraction of prey, with the females of the *Photuris* group of fireflies engaging in *predation by aggressive mimicry* [367]. These predators can mimic the response flashes of females from the *Photinus* group of fireflies. When they detect the courtship signal from a male member of the *Photinus* group, they respond by producing signal flashes containing the same response delays and flash patterns as *Photinus* females. When the male lands, it is attacked and eaten by the female. Therefore, although the underlying purpose of the signalling mechanism is for mate detection and attraction, the female *Photuris* fireflies engage in signal deception and use the mechanism for foraging purposes.

A related deceptive behaviour is exhibited by a number of species of deep-sea fish including the anglerfish and dragonfish. These fish have an appendage that contains bioluminescent bacteria whose glow the fish can control. Anglerfish are thought to be ambush predators that seek to entice prey to approach them by moving this appendage and then attacking the prey when it gets too close [118]. This potentially provides an example of a biological mechanism which encourages the movement of food *to* the predator, rather than the more usual case where the predator moves around an environment *looking for* food.

Readers wishing to learn more about the signalling processes of fireflies are referred to [350], which provides an excellent review of the literature concerning firefly biology and signalling.

11.2 Firefly Algorithm

The *firefly algorithm* (FA) was initially proposed by Yang [636] and was subsequently developed in a series of papers by the same author [637, 638, 642]. The algorithm is broadly inspired by the flash signal behaviour of fireflies.

In designing the algorithm a number of simplifying assumptions are made:

- all fireflies are unisex; consequently a firefly is attracted to all other fireflies;
- the attractiveness of a firefly is proportional to its brightness and this brightness decreases as the distance from a firefly increases;
- for any pair of fireflies, the less bright one will move towards the brighter one;
- if there are no fireflies brighter than a given firefly in visual range, it will move randomly; and
- the brightness of a firefly is determined by the landscape of the objective function.

In the canonical version of the algorithm, the initial positions for the population of n fireflies are randomly selected and an appropriate choice of objective function for the problem is determined. This function plays an important role in determining the brightness, and therefore the attractiveness, of each firefly. Once the initial light intensity for each firefly has been determined, an iterative process commences. In every iteration of the core algorithm, each pair of fireflies i, j is compared, and if firefly i is less bright than firefly j, then i is moved towards j. Otherwise, i takes a random step. After this movement step, the light intensity of the firefly at its new location is determined. The algorithm iterates until a terminating condition is satisfied.

Pseudocode for the canonical version of the FA is provided in Algorithm 11.1. In operationalising the algorithm, two key decisions are required, namely: the measurement of attractiveness between fireflies, and the definition of how movement takes place in the algorithm. The approach to each of these as outlined in [636, 637] is described below.

Attractiveness

In the simplest case, a firefly's attractiveness can be modelled as being proportional to its light intensity as perceived by adjacent fireflies. In turn, this is related to the objective function evaluation for that location. Hence, $I(x_i) \propto f(x_i)$ (assuming a maximisation problem), where $I(x_i)$ is the brightness of firefly i at location x_i. In designing a mapping from a measure of brightness to a metric for attractiveness, two factors considered in [636, 637] are:

i. the distance to the light source; and
ii. absorption characteristics of medium through which the light travels.

The first factor considers the fact that light intensity will diminish as the distance from the light source increases. Denoting the distance between two fireflies i and j as r_{ij}, the perceived intensity of a light source (I_{source}) at a distance r is modelled using the inverse square law $I(r) = I_{\text{source}}/r^2$.

Algorithm 11.1: Firefly algorithm [636]

Generate an initial population of n fireflies $i = \{1, \ldots, n\}$ with each firefly i located
  randomly at x_i in the search space;
Define the absorption coefficient γ;
Choose value for maximum number of generations t_{max} (see below);
Set generation counter $t = 1$;
repeat
    for $i = 1$ **to** n *(for each firefly)* **do**
        Determine light intensity I_i at x_i by the value of the relevant objective function
          $f(x_i)$;
        for $j = 1$ **to** n *(for each firefly other than i)* **do**
            if $I_j > I_i$ **then**
             | Move firefly i towards j using (11.4);
            else
             | Move firefly i randomly;
            end
            Vary attractiveness with distance r according to $e^{-\gamma r}$;
            Evaluate new solutions and update light intensity;
        end
    end
    Rank fireflies and find the current best;
until $t = t_{max}$;
Output the best solution found;

The second factor considers the fact that the medium through which the light travels may partly absorb or scatter the light signal. If the absorption coefficient is assumed fixed at γ, the effect of absorption on light intensity can be modelled using $I(r) = I_{source}e^{-\gamma r}$.

Combining the effect of the inverse square law and absorption, the impact of the two factors can be modelled using

$$I(r) = I_{source}e^{-\gamma r^2}. \tag{11.1}$$

Of course, variants of this function could be employed, depending on the assumptions made concerning the relationship between intensity, distance from source and rate of absorption of the medium. If we define the attractiveness of one firefly towards another using a function β (where $\beta \propto I$), then a general form of (11.1) is

$$\beta(r) = \beta_0 e^{-\gamma r^m}, \tag{11.2}$$

where r is the distance from the light source, β_0 is the attractiveness of that source at $r = 0$ and $m \geq 1$.

Movement Step

Assuming that the search space is Euclidean, a natural distance metric between any two fireflies is the Cartesian distance,

$$r_{ij} = \|x_i - x_j\| = \sqrt{\sum_{k=1}^{d}(x_{ik} - x_{jk})^2}, \tag{11.3}$$

and the update step for a pair of fireflies x_i and x_j, when firefly i is moving towards a more attractive (brighter) firefly j, can be formulated as [636, 637]

$$x_{ik}(t+1) = x_{ik}(t) + \beta_0 e^{-\gamma r_{ij}^2}(x_{jk}(t) - x_{ik}(t)) + \alpha(t)\epsilon(t). \tag{11.4}$$

The movement of each firefly i is determined by its current position, the attractiveness of firefly j and a stochastic component. The parameter $\alpha(t)$ controls the step size of the stochastic component, and this can be varied as a function of t in order to encourage enhanced exploitation behaviour by the algorithm later in the run.

Yang [642] suggests an operationalisation of the parameter $\alpha(t)$ using:

$$\alpha(t) = \alpha_0 \delta^t, \quad 0 < \delta < 1 \tag{11.5}$$

where α_0 is the initial step size and δ is essentially a cooling factor ($\delta \in [0.95, 0.97]$). The value of α_0 should be related to the range of the search space on each dimension in order to ensure that the step size is appropriate. The term $\epsilon(t)$ is a vector of random numbers drawn from a Gaussian, Lévy or other distribution [636].

11.2.1 Parameter Setting

The movement step plays a crucial role in determining the behaviour of the algorithm, and changes in the values of parameters for this step can produce a variety of search processes.

If the value of β_0 is set to zero in (11.4), the movement step generates a search process similar to that of a parallel version of simulated annealing, with the cooling schedule being determined by the value of α [182].

The value of γ plays an important role in the determination of attractiveness and, in turn, impacts on the speed of convergence of the algorithm. In the limiting case, if $\gamma \to 0$ in (11.4), the degree of attractiveness is not affected by distance and a flashing light can be seen by all fireflies. If the inner loop (for j) in Algorithm 11.1 is removed and I_j is replaced by the current g^{best} (i.e. global best, the best location found so far during the search process), the FA is equivalent to the canonical PSO algorithm [638]. If $\gamma = 0$ and $\beta_0 = 1$ in (11.4), the resulting search process bears some parallel with a simplified version of the differential evolution (DE) algorithm [550, 551] which omits mutation [182].

In the case where $\gamma \to \infty$, the attractiveness falls to zero as the distance to a light source becomes positive and the search process effectively becomes random, as no information as to good regions of the search space is transmitted between the fireflies via the attractiveness mechanism.

While good choices for the parameter values will be application-specific, values of $\beta_0 = 1$, $\alpha \in [0,1]$ and $\gamma = 1$ are suggested in [638], with a range of studies [637, 638, 642] suggesting that values for n in the range from 15 to 50 can produce good results.

11.2.2 Discussion

The FA is a relatively compact algorithm. Depending on how it is parameterised, it can present some similarities to aspects of well-known optimisation heuristics, including PSO, simulated annealing and DE. Since its introduction in 2008, many variants of the FA have been developed, including versions for application to discrete-valued problems [386, 506] and for clustering [520]. A significant literature concerning variants of the FA and its applications has emerged, and comprehensive reviews are provided in [182, 611].

A particular characteristic of the FA is that it may be suited for multimodal landscapes [642]. As the movement of fireflies in the FA is based on an attractiveness metric, which decreases over distance, the population of fireflies will tend to divide into subgroups, clustering around local optima. This could be of particular advantage in problems where several good solutions exist. The value of γ will determine the number of clusters that can arise. In the case where $\gamma = 0$, the population will not cluster into subgroups.

11.3 Glow-Worm Swarm Algorithm

The term *glow-worm* is a general name for a number of groups of insect larvae and adults that glow through bioluminescence. Indeed, there is some confusion as to the precise meaning of the terms *firefly* and *glow-worm*. In the US the term 'firefly' commonly refers to all members of the *Lampyridae* family of beetles and the term 'glow-worm' refers to bioluminescent beetles of the family *Phengodidae* (railroad worms). In contrast, in European usage, the term 'glow-worm' commonly includes larvae of *Lampyridae* and lampyrid species with flightless females (in these species, the male is capable of flight but does not emit light), and the term 'firefly' refers to lampyrid species where both males and females can fly and emit light.

Leaving definitional matters aside, a second family of algorithms based on inspiration from the process of bioluminescence, the *glow-worm swarm algorithm* (GWSA), was introduced by Krishnanand and Ghose (2005) [333, 334, 335, 336] as an optimiser for multimodal environments.

Like the firefly algorithm, the brightness of a glow-worm is determined by the value of the objective function at its location. In contrast to the firefly algorithm, the GWSA incorporates a dynamic decision range within which information as to good locations is gleaned from neighbours.

In the algorithm, each individual (glow-worm) has an associated luminescence which indicates the quality of its current location. It also has a sensor range within which it can detect the location of other light-emitting glow-worms. The essence of the algorithm is that glow-worms in the best locations shine most brightly and thereby attract other glow-worms towards them. This leads to more intensive searching of regions around the brighter glow-worms. As the interactions between glow-worms are local (i.e. within each glow-worm's sensor range), the population of glow-worms can divide itself into disjoint subgroups, with each converging to a local optimum of the function. The algorithm can be of particular use in multimodal environments when the modeller wishes to capture information on multiple optima. Algorithm 11.2 describes the GWSA as presented in [335].

Algorithm 11.2: Glow-worm swarm algorithm

Select number of glow-worms and radius of sensor range;
for *each glow-worm* **do**
$\quad\mid\quad$ Place glow-worm randomly in search space;
$\quad\mid\quad$ Initialise glow-worm's luminescence level and radius of its local decision domain;
end

repeat
$\quad\mid\quad$ **for** *each glow-worm i* **do**
$\quad\quad\mid\quad\mid\quad$ Update glow-worm i's luminescence value using (11.6);
$\quad\quad\mid\quad\mid\quad$ Find the set of glow-worm i's brighter neighbours;
$\quad\quad\mid\quad\mid\quad$ Calculate probability of glow-worm i moving towards each of these
$\quad\quad\mid\quad\mid\quad$ $\quad$ neighbours using (11.7);
$\quad\quad\mid\quad\mid\quad$ Stochastically select a peer glow-worm (j) for i to move towards;
$\quad\quad\mid\quad\mid\quad$ Update glow-worm i's position using (11.8) and (11.9);
$\quad\quad\mid\quad\mid\quad$ Update glow-worm i's local decision range using (11.10);
$\quad\mid\quad$ **end**
until *terminating condition*;
Output the best solution found;

Sensor Range

Each glow-worm i has two ranges defined around its current location, its *local decision range* r_i^{dec} and its *sensor range* r_i^{sens}, where $0 < r_i^{\text{dec}} \leq r_i^{\text{sens}}$. The size of the sensor range remains fixed during the algorithm but the size of each glow-worm's local decision range varies as the algorithm executes. Each of these ranges is illustrated in Fig. 11.1.

The local decision range defines a neighbourhood around each glow-worm which is used to control how the glow-worm moves through the search space. In each iteration of the algorithm, every glow-worm stochastically selects one of its neighbours

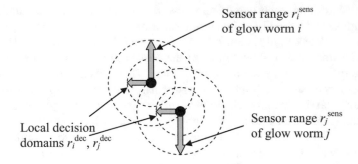

Fig. 11.1. Glow-worms i and j have sensor ranges with radii of r_i^{sens} and r_j^{sens}, respectively, and each has its own local decision domain, with radii of r_i^{dec} and r_j^{dec}, respectively. The size of the local decision domain for each glow-worm adapts during the course of the algorithm

which is brighter than it and moves towards that neighbour. One particular feature of using a local decision range is that the size of the neighbourhoods can be tuned in order to facilitate the use of the GWSA for the detection of multiple local optima.

At the start of the algorithm, the individual glow-worms are initially located randomly in the search space, each is given an equal quantity of luminescence $\tau_i(0) > 0$ and each is given the same initial local decision range $r_i^{dec}(0)$. Subsequently, in each iteration of the algorithm, the location and the luminescence value of each glow-worm are updated. This location update rule implies that the brightest glow-worm in the population in a given iteration of the algorithm will not move; therefore, the algorithm embeds elitism as the best-so-far solution cannot be lost between iterations of the algorithm.

Luminescence Update

The luminescence update depends on the value of the objective function at the current location of the glow-worm. The previous luminescence value of the glow-worm is updated by adding a component which is proportional to the objective function value at the glow-worm's current position at time t (i.e. $f_j(t)$). Just as in the case of ant foraging algorithms, an evaporation mechanism is also implemented such that a portion of the luminescence value at the end of the previous iteration is eliminated:

$$\tau_j(t+1) = \tau_j(t)(1-p) + \gamma f_j(t+1), \tag{11.6}$$

where $\tau_j(t+1)$ is the updated value for the luminescence, p is the evaporation or decay constant ($0 < p < 1$) and γ is the fixed portion of the value of the objective function $f_j(t)$ at glow-worm j's current position which is added to its luminescence value. If the glow-worm moves towards ever-improving regions of the search space, its luminescence value will tend to increase (assuming that the value of p is not excessive). On the other hand, if it moves away from a good region, its objective function value will decrease, as will its luminescence.

The term $\tau_j(t)(1 - p)$ can be considered as playing a smoothing role, as once $p < 1$ the luminescence value at $t + 1$ is influenced by its previous value, embedding an implicit *memory* of the quality of past locations visited by the glow-worm. If $p = 1$, this memory is turned off. The term γ determines the contribution the quality of the glow-worm's current location makes to its luminescence update. Once again, setting $\gamma = 0$ turns off this component of the update. The values of p and γ control the balance between current and past information about the glow-worm's trajectory in influencing the search process.

Although there is an implicit memory in the luminescence value for each glow-worm, no explicit memory is maintained by individuals as to the locations, and associated quality of these, that they have visited in the past.

Location Update

The location update is driven by the relative luminescence of nearby peers, i.e. local rather than global information. A glow-worm can only move in the direction of peers which are within its local decision radius r_i^{dec} and which are brighter than it. The choice as to which of these neighbours a glow worm tries to move towards is made stochastically. The probability that glow-worm i moves towards a brighter neighbour j is given by

$$P_j(t) = \frac{\tau_j(t) - \tau_i(t)}{\sum_{k \in N_i(t)}(\tau_k(t) - \tau_i(t))},\qquad(11.7)$$

where $\tau_i(t)$ is the luminescence of glow-worm i at iteration t, $N_i(t) = \{k : \|x_k(t) - x_i(t)\| < r_i^{dec}(t)$ and $\tau_i(t) < \tau_k(t)\}$ denotes the neighbourhood set of all glow-worms which are brighter than i and which are within a threshold distance r_i^{dec} of i, and $j \in N_i(t)$. Here, the term $\|x_k(t) - x_i(t)\|$ is the Euclidean distance between glow-worms i and k at iteration t.

By generating a random number and associating this with the calculated probability ranges, a decision is made as to which of its brighter peers to move towards. For example, suppose glow-worm i has three neighbours a, b and c, all of which are brighter than it (Fig. 11.2), and that $\tau_a(t) = 30$, $\tau_b(t) = 70$, $\tau_c(t) = 30$ and $\tau_i(t) = 10$. Plugging these values into (11.7) produces $P_a(t) = 0.2$, $P_b(t) = 0.6$ and $P_c(t) = 0.2$. If (for instance) 0.31 is drawn from the uniform distribution $U(0, 1)$, this falls into the interval $[0.20, 0.79)$; therefore, glow-worm i moves towards neighbour b.

The glow-worm takes a step of size s from its current location $x_i(t)$ towards the location of this peer $x_j(t)$, leading to a *line-of-sight* move towards a neighbouring glow-worm

$$x_i(t + 1) = x_i(t) + sd_{ij}(t),\qquad(11.8)$$

where $d_{ij}(t)$ is a unit vector in the direction from glow-worm i to glow-worm j:

$$d_{ij}(t) = \frac{x_j(t) - x_i(t)}{\|x_j(t) - x_i(t)\|}.\qquad(11.9)$$

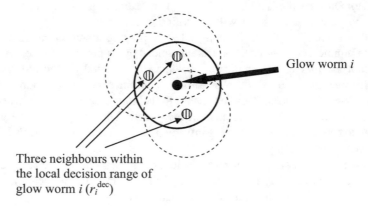

Three neighbours within
the local decision range of
glow worm i (r_i^{dec})

Fig. 11.2. Glow-worm i has three neighbours which are brighter than it within its local decision range

If a glow-worm uncovers a very good location, it may remain at that location for several iterations of the algorithm. As other neighbouring glow-worms converge on that location, the fixed step size ensures that there will be some local search around the location of the best-neighbourhood glow-worm.

Local Decision Range Update

The local decision range, in effect the visual range, for each individual glow-worm i does not stay constant during the algorithm; rather, it adapts:

$$r_i^{\mathrm{dec}}(t+1) = \min\{r^{\mathrm{sens}}, \max\{0, r_i^{\mathrm{dec}}(t) + \beta(n_t - |N_i(t)|)\}\}. \tag{11.10}$$

In (11.10), r^{sens} is the fixed range of a luminescence sensor and this acts as a hard limit on the possible size of the local decision domain. The parameter β is a constant which parameterises the relationship between the size of $r_i^{\mathrm{dec}}(t+1)$ and the number of neighbours of a glow-worm, and n_t (a user-chosen parameter) is used to control the number of neighbours. The higher the value of n_t, the larger the local decision range.

Considering specific values of the terms in (11.10), setting $\beta = 0$ implies that the radius of the local decision domain cannot change over time. If $\beta > 0$ and there are many neighbouring glow-worms ($n_t < N_i(t)$), then the value of $r_i^{\mathrm{dec}}(t+1)$ is reduced, tending to lessen the number of close-by neighbours in subsequent iterations of the algorithm. On the other hand, if $\beta > 0$ and there are few neighbouring glow worms ($n_t > N_i(t)$), then the value of $r_i^{\mathrm{dec}}(t+1)$ is increased, typically leading to an increase in the number of neighbours.

A consequence of the local-decision-range update and of the movement of each glow-worm during the algorithm is that the neighbourhood topology for individual glow-worms is not fixed during the algorithm but will adapt dynamically.

Potential Shortcoming

One potential shortcoming of the GWSA is that the formulation of the location up-date rule implies that the brightest glow-worm in the population in a given iteration of the algorithm will not move. This can result in an inefficient search around the best-so-far location as the examination of this region depends on the movement of other glow-worms.

The next algorithm, the bioluminescent swarm optimisation algorithm, addresses this issue by embedding a self adaptive search step size and by implementing local search around the best-so-far location.

11.4 Bioluminescent Swarm Optimisation Algorithm

A variant on the GWSA called the *bioluminescent swarm optimisation* (BSO) *algorithm* was developed by de Oliveira et al. [145]. Unlike the GWSA, which is designed to find multiple peaks in multimodal optimisation environments, BSO is designed to undertake a global search which can converge on a single optimal location. The algorithm is a hybrid of the GWSA and the particle swarm optimisation (PSO) algorithm [314, 315]. It also includes a number of additional mechanisms and, therefore, is only partly motivated by the foraging and attraction behaviours of glow-worms.

As in the GWSA, individual members of the population in the BSO algorithm are drawn to other individuals with a high luminescence value. The steps concerning the calculation of the luminescence update, the determination of the set of neighbours of each individual and the selection of which neighbour to move towards are very similar in both algorithms.

The position update step is modified in BSO as the update step size is stochastic, unlike the GWSA where the step size is fixed. The position update is also impacted by the location of the global best solution ever found by the population, similarly to the concept of the g^{best} location in PSO.

Two additional mechanisms are added in BSO, a local search in the region around the global best solution and a periodic *mass extinction* event. The local search step allows the algorithm to intensify its search around best-so-far locations, enabling exploitation of this information. In contrast, the mass extinction mechanism, in which every individual is reinitialised to a new random location with search then recommencing, aims to reduce the chance that the algorithm converges prematurely (i.e. it promotes exploration of the search space).

One other distinction between the GWSA and the BSO algorithm is that the latter does not use a local-neighbourhood concept, in that the radius of the neighbourhood around each individual is assumed to be infinite. Therefore, each individual i in the population is aware of the luminescence values of all other individuals in the population. The set of neighbours used in deciding which other individuals i will move towards during its position update is the set of individuals in the population that have a *higher* luminescence value than i.

Stochastic Step Sizing and Position Update

As in the GWSA, the location of each individual i in the d-dimensional search space is denoted by $x_i = (x_{i1}, x_{i2}, \ldots, x_{id})$. The step size for each move in the search process is calculated using:

$$s = s_0 \cdot \frac{1}{1 + c_s \tau_i(t)}, \tag{11.11}$$

where $\tau_i(t)$ is the luminescence value of individual i at time t, s_0 is the maximum step size and c_s is a *slowing constant*. Therefore, the step size is related to the luminescence value of an individual, with individuals having a high value (and therefore being in good search regions) moving slowly, and individuals with low values moving faster.

The step size during the search process is adaptive, depending on the current and recent performance of an individual. As with the GWSA, the value of $\tau_i(t)$ must be nonnegative. The value of c_s is the weight attached to $\tau_i(t)$ in the determination of the step size, with high values of c_s tending to produce smaller step sizes.

The position update for each individual during the BSO algorithm is:

$$x_i(t+1) = x_i(t) + u \cdot s \cdot \frac{x_j(t) - x_i(t)}{\|x_j(t) - x_i(t)\|} + c_g \cdot u \cdot s \cdot \frac{g(t) - x_i(t)}{\|g(t) - x_i(t)\|}. \tag{11.12}$$

The term $x_i(t)$ is the current position of particle i, u is a random number drawn from the uniform distribution $U(0, 1)$ on the interval $[0, 1]$, s is the current step size for particle i and c_g is a weight which determines the influence of the global best location, denoted by $g(t)$, in the update process.

Therefore, in each update, an individual moves from its current position based on its location relative to the selected individual in its neighbourhood with the highest luminescence value (individual $x_j(t)$), and the location of the global best-so-far solution ($g(t)$).

Local Search

If we examine (11.12), it is apparent that the individual at the global best location will not move during an iteration of the BSO algorithm. This individual is the brightest in its neighbourhood (by definition) and the right-hand side of (11.12) will therefore produce no change in the location of the global best individual.

In order to improve the local search capability of the BSO algorithm around the global best solution, two local search mechanisms are included. These mechanisms seek to improve the global best solution by sampling the region around it. One of the mechanisms is applied to the global best solution in every iteration of the algorithm (*local unimodal sampling*) and the other (*single-dimension perturbation search*) is applied every t_{IR} (a user-defined parameter) iterations. While both methods produce a local search, more computational effort is devoted to the periodic single-dimension perturbation search process.

The mechanism used to undertake the local unimodal sampling process was initially proposed in [453] and is described as in Algorithm 11.3 in [145]. Starting from

---

Algorithm 11.3: Local unimodal sampling algorithm

Set initial location of global best individual x_0;
Set parameters n_w (iteration limit), r_{0w} (initial sampling radius) and q (radius reduction
    parameter);
Set $x = x_0$;
Set $r = r_{0w}$;
for $i = 1$ to n_w **do**
    Randomly generate the movement vector a within r of x;
    if $f(x+a) > f(x)$ **then**
        Set $x = x + a$;
    else
        Set $r = qr$;
    end
end
Output x;

---

the initial position of the global best solution, locations within a radius of r of the
current global best position are randomly sampled. If a better solution is found, the
global best location is updated (here, we assume that the object is to maximise fitness). If the sampled position is not better than the current one, there is no change
in the location of the global best solution and the sampling radius is decreased. A
total of n_w samples are taken in each application of the local unimodal sampling
algorithm.

The single-dimension perturbation search process randomly selects an index i
corresponding to one of the dimensions in the solution ($i \le d$). A new solution is generated by randomly selecting a new value for one of the dimensions of the global best
solution, and if the new solution is better, it becomes the global best solution. The
process is repeated a total of n_s times. The details are provided in Algorithm 11.4.

Mass Extinction

In order to reduce the likelihood of premature convergence, a reset mechanism is
included in the BSO algorithm which reinitialises all the individuals to new random
locations if there has been no improvement in the global best position for t_{eT} iterations (this is a user-defined parameter) of the algorithm. The location of the global
best individual is not reset and is carried forward in the algorithm to the new population.

Algorithm

Bringing the above mechanisms together with those of the canonical GWSA discussed in Sect. 11.3, the pseudocode for the BSO algorithm is provided in Algorithm 11.5.

Algorithm 11.4: Single dimension perturbation search

Set initial location of global best individual x_0;
Set parameters n_s (iteration limit) and r_{0s} (initial sampling radius);
Set $x = x_0$;
for $i = 1$ **to** n_s **do**
  Set $C = x$ (candidate solution);
  Set $r = r_{0s} \cdot (n_s - i)/n_s$;
  Randomly select a dimension j to perturb;
  Compute step $S = ur$, where u is drawn from uniform $U(-1, 1)$ on $[-1, 1]$;
  Set $C_j = C_j + S$ (alter dimension j);
  if $f(C) > f(x)$ **then**
    Set $x = C$;
  end
end
Output x;

Parameter Settings

The parameter settings used in [145] for the BSO algorithm included population size $n = 500$, decay constant $p = 0.4$, $\gamma = 0.6$, $t_{eT} = 100$, $t_{IR} = 5$, $c_g = 0.03$ and $c_s = 5$. For the local search processes, the suggested parameters in [145] include $n_w = 10$, $n_s = 100$ (hence, there are ten times more iterations allocated to the single-dimension perturbation search process than to the local unimodal sampling process), $r_{0w} = 0.1$, $r_{0s} = 1$ and $q = 0.6$.

11.5 Summary

Bioluminescence in nature serves multiple purposes. The light emitted serves as an advertisement or lure, communicating or mimicking the presence of an organism. It is a small leap to recast this mechanism as an optimisation algorithm, where the brightness of the light emitted provides an indication of the quality of the location occupied by an individual. While most bioluminescent organisms are marine, the best-known algorithms developed from the inspiration of bioluminescence are based on an insect narrative. Both the firefly algorithm and the glow-worm/bioluminescent swarm algorithms are relatively new, and further work remains to be undertaken to develop these families of algorithms theoretically and to assess their performance thoroughly across a variety of problem domains.

It is notable that there are overlaps between some elements of these algorithms and elements of other optimisation algorithms. Similarly to ant foraging algorithms (Chap. 9), there is an attractant, chemical in the case of ant algorithms and light in the case of the algorithms in this chapter. Two of the bioluminescence algorithms (GWSA and the BSO algorithm) also make use of the idea of evaporation. However,

---

Algorithm 11.5: Bioluminescent swarm optimisation algorithm

Set values for parameters n, p, γ, s_0, t_{max}, t_{eT}, t_{IR}, c_g and c_s;
Randomly generate the population of bioluminescent individuals, $x_i : i = 1,\ldots,n$;
Set luminescence value for each individual to zero;
Determine location of global best $g(t)$;
Set $t = 1$;
repeat
   for *each individual i* **do**
      Update individual's luminescence value using (11.6);
      Find set of each individual's brighter neighbours;
      Calculate probability of individual moving towards each of these neighbours
        using (11.7);
      Stochastically select which brighter neighbour (j) to move towards;
      Update step size for individual i using (11.11);
      Update individual i's position using (11.12);
      Update location of global best ($g(t)$) if necessary;
      if t mod $t_{IR} = 0$ **then**
        Perform single-dimension perturbation search on $g(t)$ using
          Algorithm 11.4;
      else
        Perform local unimodal sampling on $g(t)$ using Algorithm 11.3;
      end
      if *number of iterations since $g(t)$ has been improved* $= t_{eT}$ **then**
        Implement mass extinction;
        Reinitialise all individuals to new locations but retain $g(t)$;
      end
   end
   Set $t = t + 1$;
until $t = t_{max}$;
Output the best solution found;

---

there are also clear distinctions; as in canonical ant algorithms, memory of good locations/solutions is maintained in the environment rather than being a property of individual ants. The communication mechanisms also differ, as in ant algorithms information from the environment at each decision point when a solution is being constructed is visible to all ants traversing that point. In contrast, some bioluminescence algorithms implement a sensory range, with individuals only being able to detect environmental information in their immediate vicinity.

Spider and Antlion Foraging Algorithms

While we commonly associate foraging with active search of the environment for prey, an alternative approach is *sit-and-wait*, where a predator remains in one location and waits for a prey item to approach. As sit-and-wait predators are expected to have lower chances of encountering prey than active foragers [379, 459], a sit-and-wait strategy is often complemented by behaviours such as trap building, careful patch choice and the use of camouflage or lures in order to increase the probability of successful prey acquisition [206].

Use of traps is found in multiple species, including spiders and antlions. In the case of spiders, many species build webs in order to capture prey. These webs can be constructed by individual spiders or, less commonly, by a colony of cooperating spiders. In contrast, antlion larvae dig a cone-shaped pit in the ground in order to capture insect prey.

In this chapter we introduce two families of optimisation algorithms which have been designed using a trap foraging metaphor, namely the *social spider optimisation algorithm* and the *antlion optimisation algorithm*.

12.1 Spider Foraging

Spiders, along with scorpions, mites and ticks, are members of the *arachnid* group. This group comprises arthropods (invertebrates with an exoskeleton) which have four pairs of legs and a body with two parts, the cephalothorax and the abdomen. The arachnid group contains over 100,000 species, around 47,000 of which are species of spider. Spiders are different from insects, as the latter have six legs and three body compartments.

One curious aspect of spider morphology is that they devote a higher proportion of their body mass to central nervous system tissue than many animals of their size, with this tissue occupying some 80% of the body cavity, even extending into the legs of a number of species of very small spiders [470]. In contrast, human brains occupy some 2–3% of our mass. It is speculated that spiders need fairly complex brains in order to control eight limbs and to undertake tasks such as weaving webs.

A. Brabazon, S. McGarraghy, *Foraging-Inspired Optimisation Algorithms*,
Natural Computing Series, https://doi.org/10.1007/978-3-319-59156-8_12

Central nervous system tissue is metabolically expensive, so much of the food intake of spiders goes to support this tissue.

Most species of spiders are solitary predators. Their usual prey is insects, but they are frequently cannibalistic and will attack other spiders if the opportunity arises. Some species of spiders occasionally supplement their insect diet with plant products such as nectar, sap or seeds [431, 440]. Spiders cannot eat solid food, and engage in preconsumptive food handling by releasing digestive fluid into or onto their prey, which dissolves tissue. The liquefied material is then ingested.

Spiders can be grouped on the basis of their foraging strategy. Perhaps the best-known strategy is that of web-building spiders, which create webs in which small insects can become enmeshed and trapped. Several types of web design exist, including spiral orb webs, funnel webs, tubular webs and sheet webs. Not all spiders forage using webs. *Cursorial spiders*, such as the wolf spider, hunt by actively searching for and chasing after prey on the ground. As would be expected, the physical and sensory characteristics of each group of spiders vary, with web-building spiders typically having poorer eyesight than cursorial spiders.

One interesting strategy employed by a subgroup of flower dwelling spiders is the use of a light-emitting lure. These spiders reflect strongly in the ultraviolet spectrum (i.e. they present an ultraviolet visual lure) attracting honeybees to the flower which they are inhabiting, whereupon the bee is preyed on by the spider [206]. It is thought that the bees mistake the spiders for *floral guides*, natural bright spots on flowers that guide pollinators to land on them.

Other Organisms Using Web-like Structures for Foraging

While foraging using webs is most commonly exhibited by spiders, a number of other organisms adopt related strategies. Perhaps the most unusual example of web-like foraging is exhibited by a small number of fungal species that supplement their diet by preying on nematodes or other microscopic animals in the soil [613]. The fungus *Arthrobotrys oligospora* has evolved sticky networks of hyphae for trapping nematodes [553, 613]. The range of morphological adaptations for prey capture in nematode-trapping fungi is quite diverse, encompassing constricting rings (as the name suggests, when a nematode moves into the ring, it triggers a response so that the cells composing the ring rapidly swell inward and close around it), sessile adhesive knobs, stalked adhesive knobs, adhesive networks, adhesive columns and non-constricting rings. Once a worm is immobilised in these hyphae, the fungus invades and consumes its body. This bears some parallel to the use of webs by spiders, except that in this case it is the 'web' that consumes the prey item.

12.1.1 The Sensory World of Spiders

In addition to a visual sense, spiders are capable of chemoreception and of mechanoreception of vibrations in their environment. Many of the receptors for these two senses are in the legs of spiders, so—unusually to us—spiders primarily smell via chemoreceptors in their legs [431].

Vibrations are sensed via mechanoreceptors called slit senillae, and by means of vibration-sensitive hairs, both located on the legs of spiders. Spiders also employ vibratory communication for a diverse range of functions, including courtship and to mediate interactions with prey and predators [254].

In the case of web-building spiders, valuable sensory information is obtained from vibrations transmitted across their web. Apart from passive detection of vibrations, spiders can actively seek information by plucking strands of their web and waiting for echo information, which will be modified if the web is damaged or if prey or other organisms (potential mates or predators) are on the web. In essence, a web provides a sensor array which serves as an external extension to a spider's senses.

A small number of species of spider use aggressive mimicry by creating vibrations on webs which simulate the vibrations caused by captured prey. When the resident spider approaches the supposed prey, it is attacked by the predatory spider.

12.1.2 Social Spiders

Although most spiders are solitary, a small number of species engage in group living. This occurs in two forms, namely cooperative breeding and colonial behaviour [57]. While these groups can contain thousands of individuals, their social structure is very different from that of social insects as there are no role specialisations and each spider acts and breeds individually.

Potential advantages of group living by spiders include economies in web building and web maintenance, and enhanced predator defence. Grouping can also facilitate the capture of larger prey, either by producing better-quality communal webs, leading to enhanced prey capture capabilities, or by facilitating the immobilisation of prey by communal attack of prey by multiple spiders once it is caught in the web [57]. Empirical evidence indicates that group living spiders capture larger-sized prey items than do solitary spiders [392, 653].

Cooperative Breeding

These species of spider typically construct a nest or communal web, capture prey together, feed communally, cooperate in raising young and show cooperative defensive behaviour. Less than 30 of the 47,000 known species of spider display these behaviours [25], with nearly all of these species being found in the tropics, typically constructing dense communal three-dimensional webs [25].

Colonial Behaviour

Colonial group living is (very) slightly more common and this form of social organisation may range from short-lived to long-lived aggregations. Colonial species do not generally cooperate in prey capture and feeding, nor in brood care. However, webs may be interconnected and, in active hunting species, the spiders may share a communal nest. Approximately 50 species of colonial spiders are known to exist [57].

Examples of Social Species of Spider

Species of spider which engage in communal prey capture include *Agelena consociata* and *Mallos gregalis*. *Agelena consociata* is found in the tropical forests of West Africa, living in colonies of up to several thousand individuals, building a huge communal web. When prey is caught in the web, a number of spiders will attack it (up to 25 in the case of large prey items). When the prey item has been subdued, it is transported to a retreat, where up to 40 spiders have been observed feeding on a single prey item.

Spiders of the species *Mallos gregalis* are found in Mexico. This species lives in groups of up to 100,000 individuals [570]. The group cooperates to build a sheet-like spider web, on which they communally capture and subsequently feed on prey.

The behaviour of this species has been well studied, and one interesting aspect is the way that the spiders construct and tune the silk strands in their webs in order to enhance the transmission of vibrations at frequencies which correspond to those produced by a captured prey item struggling to free itself from the web, while dampening frequencies corresponding to the vibrations caused by other spiders walking on the web [94]. This allows spiders to more easily distinguish vibrations caused by the capture of prey, as distinct from those of other spiders traversing the web. It is speculated that this attribute of communal webs also contributes to the emergence of social behaviour. As vibrations from conspecifics are dampened, the stimulus to predation behaviour—which could otherwise arise owing to the vibrations from other spiders walking on the web—is lessened [94].

The foraging behaviours of social spiders have been used to inspire the development of an optimisation algorithm, and this is introduced in the next section.

12.2 Social Spider Optimisation Algorithm

The *social spider optimisation algorithm* (SSA) was developed by Yu and Li [654, 655, 656, 657] and is inspired by foraging behaviours of social spiders. The SSA uses a vibration-based *information loss communication system* between spiders which acts to coordinate cooperative movement towards good food sources, corresponding to good potential solutions in an optimisation problem. In this section we describe the variant of the SSA outlined in [656], which has been applied for real-valued optimisation. An alternative version of the algorithm has also been applied to binary-valued problems, and readers are referred to [655] for details of this.

The search space for the optimisation problem is conceptualised as a multidimensional spider web on which a population of simulated spiders (search agents) are moving. Each position on the web (search space) corresponds to a solution to the optimisation problem of interest.

Every spider in the population has a current position on the web and this position has an associated fitness value. Each spider stores (remembers) a number of additional items of information, including the vibration it followed in the previous

iteration of the algorithm (both the source location and the intensity of this vibration are stored) and the dimension mask for that spider (which was used to guide the
movement of the spider in the last iteration of the algorithm).

When a spider changes its position on the web (i.e. its location), a *vibration* is
generated as a result. This vibration propagates across the web and can be sensed
by all the other spiders. It forms the basis of the communication between members
of the population, contributing to a collective social knowledge. The intensity of
the vibration is determined by the fitness value of the associated position on the
web, with fitter positions generating stronger vibrations. If a spider moves to a new
position x, it generates a vibration with an intensity I, determined as follows in the
case of a minimisation problem:

$$I = \log\left(\frac{1}{f(x)-C}+1\right),\tag{12.1}$$

where $f(x)$ is the fitness value at x and C is a constant smaller than the minimum
value which the fitness function can attain. In the case of a maximisation problem,
the denominator of the fraction in (12.1) becomes $C - f(x)$ and C is assigned a value
larger than the maximum fitness possible.

The formulation of (12.1) implies that all vibration intensities must be positive
and helps ensure that as the global optimum is approached, the rate of increase of the
intensity level is controlled so that it is consistent with the design of the attenuation
mechanism, which is described next.

As a vibration propagates across the multidimensional spider web, it is attenuated. The attenuation process is simulated using:

$$I_d = I \cdot \exp\left(-\frac{d}{\bar{\sigma} \cdot r_{att}}\right),\tag{12.2}$$

where I_d is the attenuated intensity after a vibration with a source intensity of I has
travelled a distance of d between its source and reception point, and $\bar{\sigma}$ is the average
of the standard deviation of all spider locations along each dimension. The parameter
$r_{att} \in (0, \infty)$ is user defined and controls the attenuation rate.

Rather than adopting a Euclidean metric, the SSA uses a Manhattan metric (also
called the city block or 1-norm distance), where the distance between two locations
is the sum of the absolute differences of their Cartesian coordinates. This reduces the
computational overhead in the algorithm, as the calculation of Manhattan distance is
not computationally intensive. Hence, the distance between spiders a and b, located
at x_a and x_b, respectively, is calculated as

$$d(x_a, x_b) = \|x_a - x_b\|_1 = \sum_{i=1}^{N} |x_{ai} - x_{bi}|,\tag{12.3}$$

where N is the dimension of the optimisation problem.

12.2.1 Algorithm

The population of spiders is initially located randomly in the search space. The initial vibration that a given spider is following (see below) is set to its own position and is assigned an intensity of zero. The algorithm then iterates over the following elements.

Fitness Calculation

In each iteration, the fitness values of all spiders' current positions are initially determined and the location of the global optimum is updated if necessary.

Social Influence

Next, each spider generates a vibration at its current position, based on the fitness of that location. The vibrations of all spiders propagate across the web and are attenuated to differing degrees depending on how far away a receiving spider is from the source of the vibration. Each spider selects the strongest (attenuated) vibration that it has received (denoted as v_{best}). This is compared with the strongest vibration it received in the last iteration of the algorithm, denoted by v_{target} (also termed the *following* vibration), which influenced its movement in that iteration.

If the new vibration is not greater than v_{target}, the spider increments an inactivity counter c_s by one. If the received vibration is stronger, v_{best} replaces the value in v_{target} and the inactivity counter c_s is reset to zero.

The position of v_{target} influences the position update of each spider and acts as a form of social information in the algorithm.

Position Update

When all spiders have determined their following vibration for this iteration, their positions on the web are updated.

Each spider has a *dimension mask*, which is used to induce some randomness in the position update step. The dimension mask for each spider is a binary string with N elements, the optimisation problem having N dimensions. In each iteration the probability that the dimension mask is changed is $1 - p_c^{c_s}$, where $p_c \in (0, 1)$ is a user-defined parameter. The term c_s denotes the inactivity degree of the spider, corresponding to the number of iterations since the last update of its v_{target}. As the value of c_s increases, the probability that the dimension mask is altered also increases.

If the mask is to be changed from its setting in the last iteration, each bit is assigned a value of 1 with a probability of p_m (a user-defined parameter); otherwise, the bit is assigned a value of 0. In the case where all bits are set to zero, one bit is selected randomly and set to 1 in order to avoid premature convergence of the search process.

After the dimension mask is determined for the current iteration, a new following position $x_{following,s}$ for spider s is determined for each dimension i using

$$x_{\text{following},s,i} = \begin{cases} x_{\text{target},s,i}, & m_{s,i} = 0, \\ x_{r,s,i}, & m_{s,i} = 1, \end{cases} \tag{12.4}$$

where r is a random integer value $\in \{1,\ldots,n\}$, with n being the population size, and $m_{s,i}$ is the ith dimension of the dimension mask m for spider s. The random integer r is generated independently for different dimensions, where $m_{s,i} = 1$.

The final element of the algorithm is the updating of the position of a spider, wherein each spider follows a random walk towards the source location of its v_{target}. Using the location of v_{target}, the calculated following position $x_{\text{following},s}$ and the spider's current position on the web, a new location is determined as follows:

$$x_s(t+1) = x_s(t) + (x_s(t) - x_s(t-1)) \cdot u + (x_{\text{following},s}(t) - x_s(t)) \odot R, \tag{12.5}$$

where $x_s(t)$ is the position of spider s at time t, $\odot$ denotes componentwise multiplication[1] and R is an N-dimensional vector of random numbers generated from the uniform distribution $U(0,1)$.

Looking at the position update in (12.5), it can be observed that spider s initially moves a random amount in the same direction as it did in the previous iteration and then approaches $x_{\text{following},s}$ along each dimension, with the degree of movement being determined randomly by $u \in U(0,1)$. The required values of u are generated independently for each dimension i.

The pseudocode for the complete algorithm is outlined in Algorithm 12.1.

12.2.2 Parameter Setting

The performance of the algorithm is influenced by four parameters: population size, attenuation rate, mask change probability and mask bit mutation rate. A study [657] examined the performance of the algorithm for different settings of these parameters on a series of benchmark functions. Although the best parameter settings varied across problems, general settings where population size $n \in \{20,\ldots,40\}$ and values for $r_{\text{att}} \in (0.5, 2)$, $p_c \in (0.5, 0.7)$ and $p_m \in (0.1, 0.3)$ were suggested.

An interesting aspect of the algorithm is that the social transmission of information is *lossy* in that vibration signals are attenuated [656]. As a result, a spider will not necessarily use information from the best location found by the entire population, as, if the corresponding vibration is travelling from a distant location, its attenuated strength could be low by the time it is received by a particular spider.

As the attenuation rate r_{att} gets larger, the degree of attenuation in (12.2) becomes *smaller*, and therefore the best positions found in the entire population are more likely to influence the movement of other spiders. Conversely, small values for r_{att} make the transmission of social information via the web more lossy, and consequently spiders will be more influenced by the activities of nearby, local spiders.

Setting a large value for p_c will produce more frequent mask changing and will increase the exploratory characteristics of the algorithm. Smaller values of p_c will

---

[1] As in Chap. 6, if $x = (x_1,\ldots,x_n)$ and $y = (y_1,\ldots,y_n)$ are vectors, then the componentwise product is $x \odot y := (x_1 y_1,\ldots,x_n y_n)$.

Algorithm 12.1: Social spider optimisation algorithm [656]

Set parameters for algorithm including population size n, r_{att}, p_c, p_m;
Generate an initial population of n spiders each located randomly in the search space;
Initialise v_{target} for each spider;
repeat
    for $s = 1$ **to** n *(for each spider)* **do**
        Evaluate fitness of current location of s;
        Generate a vibration at current location of s using (12.1);
    end
    for $s = 1$ **to** n *(for each spider)* **do**
        Calculate intensity of the vibrations generated by all spiders using (12.2);
        Select the strongest vibration v_{best} from all of these;
        if *the intensity of* $v_{best} > v_{target}$ **then**
            Replace value in v_{target} with v_{best} and this becomes the new following
            vibration for spider s;
        end
        Update the inactive counter c_s;
        Generate a random number $r \in [0, 1)$;
        if $r > p_c^{c_s}$ **then**
            Update the dimension mask m_s;
        end
        Generate $x_{following,s}$;
        Perform a random walk using (12.5);
    end
until *terminating condition*;
Output the best solution found;

---

enhance exploitation. A large value for p_m will result in more 1s in the dimension mask, leading to a noisier following position and a greater degree of exploration.

12.2.3 Discussion

The SSA draws inspiration from the foraging activities of social spiders to develop an optimisation algorithm. The algorithm has shown promising results in a number of benchmark studies to date.

There are some similarities between the SSA and the firefly algorithm (Sect. 11.2). In both, the searching agent (a firefly and a spider, respectively) broadcasts information concerning the quality of its current location, and the intensity of the related light or vibrational signal is *attenuated* as the distance between the source and the receiver increases. In both algorithms, this potentially allows the search process to be influenced by multiple members of the population, thereby reducing the risk of premature convergence in the early stages of the algorithm. However, the position update steps in each algorithm are quite different. Further work is required to comprehensively compare and contrast the search characteristics of the two algorithms.

The SSA is not the only optimisation algorithm which has been inspired by the cooperative behaviour of social spiders. The social spider optimisation algorithm developed in [120], while not drawing on a food-foraging metaphor, simulates a cooperative search process by a population of male and female social spiders with distinct search behaviours which interact with each other to jointly solve an optimisation task.

12.3 Antlion Foraging

Spiders are not the only organisms that use a trap strategy. Most species of antlions, an insect group in the order Neuroptera, further classified into the family Myrmeleontidae, adopt a similar strategy. There are some 2000 species of antlions found worldwide, usually in sandy, sheltered habitats.

The life cycle of members of this family typically includes a larval stage of some two to three years and, technically, the term *antlion* refers to the larval stages of the life cycle. Antlion larvae are approximately 1–2 cm in length and are generalist predators feeding on ants, other small insects and spiders. They are also cannibalistic and will prey on other antlions if the opportunity arises [264]. During the three-week adult stage of the life cycle, reproduction takes place, with the insects resembling dragonflies as they gain wings and feed on nectar and pollen.

Foraging Process

While a small number of antlion species engage in sit-and-pursue foraging, burying themselves in sand and emerging to chase prey if it gets close [508], most species of antlion construct cone-shaped pits in the ground, in which they seek to trap their prey (Fig. 12.1). The pits are 2–5 cm in diameter and about the same depth. The antlion

Fig. 12.1. Antlion pit with antlion sitting at bottom

digs out the pit so that the walls of the cone are steep, effectively being at the critical angle of repose. This makes it difficult for prey to escape if they fall into the pit, as the side of the pit destabilises beneath them owing to their weight and movement.

The antlion buries itself at the bottom of the pit with only its jaws protruding. On seizing prey, the antlion bites it, injecting a paralysing toxin. Next, digestive enzymes are injected into the prey to break down its internal tissues and then the antlion sucks out the liquefied contents from the prey's body. After feeding, the carcass is ejected from the pit, which is then repaired if necessary. If a prey item attempts to escape the pit, the antlion throws showers of loose sand at the side of the pit in order to further destabilise it, in an attempt to cause the prey to slip back down [179]. It is considered that this behaviour represents an example of tool use, as the antlion employs the sand particles to alter the physical properties of another object (here, the stability of the sand beneath the prey item) [378]. Long thought to be unique to humans, tool use is now known to be widespread, with over 50 instances of tool use in insects being catalogued in [50].

Commonly, antlions cluster in suitable habitats, forming a minefield of predatory traps for the local ant population [218]. Densities in excess of 100 antlion larva traps per square metre have been observed. There is evidence that ants can detect the presence of antlion traps and that they have a preference to forage outside rather than inside these areas [217, 218].

In spite of the apparent simplicity of their foraging strategy, antlions have been found to be capable of associative learning [265], displaying the ability to learn cues predicting prey arrival [264]. Antlions have highly sensitive mechanoreceptors located all over their bodies. It is thought that they can detect prey once it approaches within 6–10 cm of their pit and therefore get ready to attack the potential prey item, enhancing their chances of a successful capture.

Although antlions use pits for trapping purposes, this behaviour is relatively uncommon in the natural world, which presents a puzzle as pits are fairly simple constructions and allow the capture of prey items that may be faster than the predator [179]. A very similar pit-building foraging strategy is also adopted by wormlions, which are insect larvae of the family Vermileonidae [152]. One of the few other species that uses this approach is the trapdoor spider of the family *Ctenizidae*. These are medium sized nocturnal spiders that construct burrows with a trapdoor made of soil, vegetation and silk. The spider waits under the door and if prey disturb trip wires around the door, or if nearby vibrations in the ground are detected, the spider suddenly emerges from under the trapdoor and captures the prey item.

12.4 Antlion Optimisation Algorithm

The *antlion optimisation* (ALO) *algorithm* was developed by Mirjalili [403] and mimics the interactions of ants moving around an environment in which there are predatory antlion larvae, in order to create an optimisation algorithm. In the following sections we outline the algorithm as described in [403]. The design of the optimisation algorithm incorporates several elements:

- ants move stochastically around their environment;
- their movements are affected by the location of the traps of antlions;

- antlions are considered to build pits proportional in size to their fitness (i.e. a higher fitness of current location implies a bigger pit);
- antlions with larger pits have more chance of trapping an ant prey item;
- the boundary range of an ant's movement is decreased adaptively during the algorithm, simulating ants falling towards the bottom of an antlion trap;
- if an ant is fitter than an antlion, it is considered to be caught and pulled under the sand by that antlion, and in this case
- the antlion repositions itself to the position of the latest prey item caught.

As with all metaphorically inspired algorithms, the design is not a completely faithful representation of the real-life foraging process of antlions, although it does draw several ideas from this.

The ants are assumed to move stochastically in the environment. A simple random walk process could be simulated as follows:

$$X(t) = [0, s_{\text{cum}}(2r(t_1) - 1), s_{\text{cum}}(2r(t_2) - 1), \ldots, s_{\text{cum}}(2r(t_{\max}) - 1)], \quad (12.6)$$

where s_{cum} is the cumulative sum to the current iteration of the algorithm (denoted by t), $t_{\max}$ is the maximum possible number of iterations and $r(t)$ is a stochastic function as follows:

$$r(t) = \begin{cases} 1, & \text{if } u > 0.5, \\ 0, & \text{if } u \leq 0.5, \end{cases} \quad (12.7)$$

with u being drawn from a $U(0,1)$ uniform distribution. This basic random walk process is modified in the algorithm (as discussed below) in order to:

i. allow for cases where the search space is bounded in some dimensions; and
ii. allow the random walks to be influenced by the location of antlion traps (simulating the trapping of ants by antlions).

As the algorithm runs, the current location (corresponding to a solution vector for the problem of interest) of each of n ants is stored in a matrix M_{Ant}, where

$$M_{\text{Ant}} = \begin{pmatrix} A_{11} & A_{12} & \ldots & A_{1d} \\ A_{21} & A_{22} & \ldots & A_{2d} \\ \vdots & \vdots & \ddots & \vdots \\ A_{n1} & A_{n2} & \ldots & A_{nd} \end{pmatrix}. \quad (12.8)$$

The term A_{ij} refers to the value of the jth dimension for the ith ant. It is assumed that the problem has d dimensions. The objective function (fitness) values corresponding to each location are stored in M_{OA}, where

$$M_{\text{OA}} = \begin{pmatrix} f(A_{11}, A_{12}, \ldots, A_{1d}) \\ f(A_{21}, A_{22}, \ldots, A_{2d}) \\ \vdots \\ f(A_{n1}, A_{n2}, \ldots, A_{nd}) \end{pmatrix}. \quad (12.9)$$

In addition to ants, the environment also contains n antlions and their locations are stored in a matrix M_{Antlion}, where

$$M_{\text{Antlion}} = \begin{pmatrix} L_{11} & L_{12} & \ldots & L_{1d} \\ L_{21} & L_{22} & \ldots & L_{2d} \\ \vdots & \vdots & \ddots & \vdots \\ L_{n1} & L_{n2} & \ldots & L_{nd} \end{pmatrix} \tag{12.10}$$

and L_{ij} refers to the value of the jth dimension for the ith antlion. As with ants, the objective function value corresponding to the current location for each antlion is stored in an n-dimensional vector, denoted by M_{OAL}.

12.4.1 Algorithm

At the start of the algorithm, all rows in the antlion and ant location matrices are initialised to randomly selected positions. In each iteration of the algorithm, ants are assigned to antlions using a roulette selection process, with the probability of being assigned to a particular antlion being determined by its fitness relative to the fitnesses of all the antlions. The range within which the position update step takes place for each ant is governed by an adaptive process. This is driven by the current iteration number, which encourages increased exploitation of the area around an antlion as the algorithm progresses.

The position update process for each ant proceeds using this information, information as to the location of its selected antlion, and the position of the elite (i.e. best so far) antlion. The location of antlions is updated during the algorithm as better locations are uncovered by ants during their search. The location of the elite antlion is also updated as better locations are uncovered.

Algorithm 12.2 provides pseudocode for the ALO algorithm. Each of the key processes in the algorithm is described below.

Operationalising the Algorithm

The core component of any search algorithm is how the search agents update their positions as the algorithm runs. In the ALO algorithm, the simulated ants move stochastically, with their position updates being influenced by the location of antlions and the elite antlion.

To ensure that the movements of ants remain in the valid search space for the problem of interest (in the case where the search space is bounded), the random walk on each dimension (corresponding to a variable in the problem) as described in (12.6) needs to be normalised into the valid search space. One way of doing this is to use a max-min normalisation (the formulation of this normalisation is taken from the MATLAB source code of Mirjalili, which can be found in [404]):

$$X_i(t) = \frac{(X_i(t) - a_i) \cdot (d_i(t) - c_i(t))}{(b_i - a_i)} + c_i(t) \tag{12.11}$$

Algorithm 12.2: Antlion optimisation algorithm [403]

Position the initial population of ants and antlions (two populations of size n) randomly in the environment;
Calculate the fitness of the ants and antlions;
Find the fittest antlion and designate this as the *elite* antlion E;

repeat
    for *each ant in turn* **do**
        Select an antlion using roulette wheel selection;
        Update c and d using (12.15) and (12.16);
        Create a random walk and normalise it using (12.6) and (12.11);
        Update the position of the ant using (12.18);
    end
    Calculate fitness of all ants;
    Replace an antlion with its corresponding ant if it becomes fitter (12.19);
    Update elite if any antlion becomes fitter than the elite;
until *terminating condition*;
Output the best solution found (the position of the elite antlion E);

---

where the parameters a_i and b_i are, respectively, the minimum and maximum of the random walk of the ith variable. The terms $c_i(t)$ and $d_i(t)$ are the minimum and maximum of the ith variable at iteration t.

The selection of ants for assignment to antlions is undertaken using roulette wheel selection. Given a list of each of the n antlions and their associated fitnesses f_i, roulette wheel selection can be implemented by generating a random number $r \in \left[0, \sum_{j=1}^{n} f_j\right)$ and then selecting the individual antlion i such that

$$\sum_{j=1}^{i-1} f_j \leq r < \sum_{j=1}^{i} f_j. \tag{12.12}$$

This implies that fitter antlions are more likely to be assigned ants. Each ant is assumed to be trapped by a single antlion and moves in the vicinity of that antlion during its position update step. As ants move around the environment, their walks are affected by the position of the antlion to which they have been assigned. This effect is modelled using

$$c_i(t) = L_j(t) + c(t), \tag{12.13}$$

$$d_i(t) = L_j(t) + d(t). \tag{12.14}$$

The term $c(t)$ is the vector of the minimum of all variables at the tth iteration, $d(t)$ is the vector of the maximum of all variables at the tth iteration, $c_i(t)$ is the minimum of all variables for the ith ant, $d_i(t)$ is the maximum of all variables for the ith ant and $L_j(t)$ indicates the position of the jth antlion at iteration t.

Under (12.13) and (12.14), ants move randomly in a hypersphere defined by the vectors c and d around a selected antlion location, simulating an ant being trapped by a specific antlion.

To encourage convergence of the search process over time, the radius of the ants' random walk hypersphere around the location of an antlion is reduced over time, using

$$c(t) = \frac{c(t)}{I}, \quad (12.15)$$

$$d(t) = \frac{d(t)}{I}. \quad (12.16)$$

The term I is calculated as

$$I = 10^w \frac{t}{t_{max}}, \quad (12.17)$$

where t is the current iteration, t_{max} is the maximum number of iterations that the algorithm will run for and w is a constant which is adapted as the algorithm runs. A schedule of $w = 2$ when $0.1t_{max} < t \leq 0.5t_{max}$, $w = 3$ when $0.5t_{max} < t \leq 0.75t_{max}$, $w = 4$ when $0.75t_{max} < t \leq 0.9t_{max}$, $w = 5$ when $0.9t_{max} < t \leq 0.95t_{max}$ and $w = 6$ when $t > 0.95t_{max}$ is suggested in [403]. This adaptive process is loosely inspired by the process of ants falling towards the bottom of an antlion trap.

The movement process of each ant during the algorithm is also impacted by the location of the fittest antlion discovered so far (the elite antlion). Therefore, the movement of each ant in each iteration is determined by both the location of the antlion to which it is assigned using the roulette selection process and the location of the elite, as follows:

$$A_i(t) = \frac{R_A(t) + R_E(t)}{2}. \quad (12.18)$$

Here, $R_A(t)$ is a random walk around the antlion selected using roulette selection and $R_E(t)$ is a random walk around the elite antlion E. $A_i(t)$ is the resulting location of the ith ant at iteration t.

While the roulette selection process in the ALO algorithm is easy to implement, it can produce a high *selection pressure* in the early stages of the algorithm when there is likely to be a high variance in the quality of the antlion locations. In this scenario, some antlion locations could be selected multiple times for ant assignment, potentially leading to the premature convergence of the search process.

Bringing all the components of the position update step together, $R_A(t)$ and then $R_E(t)$ are calculated (and normalised into the search space using (12.11)) using (12.13) and (12.14) for each ant (in the case of $R_E(t)$, these equations are altered to $c_E(t) = E_j(t) + c(t)$ and $d_E(t) = E_j(t) + d(t)$, respectively), and then the two positions are averaged.

The final stage of antlion foraging is when the ant is caught by the antlion and consumed. This process is simulated in the algorithm when an ant reaches a fitter location than that of its antlion. When this occurs, the antlion updates its location to the fitter location discovered by the ant. Therefore,

$$L_j(t) = A_i(t), \text{ if } f(A_i(t)) > f(L_j(t)) \quad (12.19)$$

where $L_j(t)$ is the position of the jth antlion at iteration t and $A_i(t)$ is the location of ant i at iteration t.

The above step (updating the position of antlions) could be implemented in a number of ways. A simple implementation which ensures that the locations of all antlions can be improved is to concatenate the matrices M_{Ant} and $M_{Antlion}$, sort these locations by fitness and then assign the n antlions in the next iteration to the fittest n of these locations.

12.4.2 Parameter Setting

The ALO algorithm has few parameters, the main ones being the choice of population size and the maximum number of iterations. In testing the ALO algorithm on a series of benchmark problems, [403] employed a population of 30 antlions, increasing this to a population size in the hundreds when tackling larger 200-dimensional problems.

In the algorithm, exploration is encouraged by using a random initialisation of the locations for ants and antlions, and through the implementation of a stochastic position update process for the ants. Exploitation is encouraged by the adaptive reduction of the walks of ants around the location of antlion pits. As antlions relocate to better regions of the search space during the algorithm, the search process will tend to converge, therefore encouraging exploitation of the information concerning good regions already captured in the population. The cooling schedule which governs the intensification of the search process is controlled by adapting the value of I, with larger values corresponding to greater locality of search.

12.5 Summary

In this chapter we have introduced two novel families of optimisation algorithms the designs of which have been inspired by trap-foraging behaviours. These behaviours form a contrast with the more commonly seen active foraging behaviours which inspire many other foraging algorithms. Both social spider optimisation algorithms and antlion optimisation algorithms are very recent additions to the literature of biologically inspired algorithms, and further research is required to ascertain their efficiency and effectiveness.

13

Worm Foraging Algorithm

In this chapter we provide an introduction to *Caenorhabditis elegans* (*C. elegans*), a nonparasitic, soil-living nematode worm. The phylum Nematoda comprises a wide array of different species of smooth-skinned roundworms which are found in most terrestrial and marine environments.

C. elegans is one of the simplest creatures which still shares many of the essential characteristics of more complex organisms. It is conceived as a single cell, undergoes a process of development, has a nervous system and simple brain (the circumpharyngeal nerve ring), is capable of immune responses and displays a sleep-like state. It is also capable of rudimentary learning and memory [651]. Because of these characteristics, *C. elegans* is frequently used as a model organism for research in developmental biology and neurology, with over 15,000 research articles having been devoted to various biological aspects of the organism [177]. *C. elegans* was the first organism to have its connectome mapped, providing a wiring diagram for its 302 neurons and approximately 7000 synaptic connections [585, 616]. It was also the first multicellular organism to have its genome completely sequenced. The importance of *C. elegans* as a model organism was recognised by the award of the 2002 Nobel Prize in Physiology or Medicine to Sydney Brenner, Robert Horvitz and John Sulston in respect of their work on *C. elegans*.

Despite the relative simplicity of *C. elegans*, it is capable of some interesting foraging-related behaviours and has inspired the development of the *worm optimisation algorithm*, which is outlined in this chapter.

13.1 Description of *C. elegans*

C. elegans is transparent, about 1 mm in length, and feeds on bacteria such as *E. coli*. It is found in two genders, one consisting of female hermaphrodites which can self-fertilise, the other being a male gender. Its lifespan is about 2–3 weeks under ideal conditions [177]. *C. elegans* has a distinct head, a tubular digestive system and a locomotory system. It can sense and react to several stimuli, including chemicals, touch, heat, magnetic fields and light (although it possesses no visual system).

© Springer Nature Switzerland AG 2018

A. Brabazon, S. McGarraghy, *Foraging-Inspired Optimisation Algorithms*,
Natural Computing Series, https://doi.org/10.1007/978-3-319-59156-8_13

Information from its sensory modes guides various behaviours, including feeding, mating and egg laying [177]. It is capable of movement via 81 muscle cells by generating dorsal–ventral waves along its body. Therefore, it can move in response to environmental stimuli. *C. elegans* shares its environment with a diverse range of organisms, some of which feed on the same microbes and some of which are predators.

C. elegans develops through a number of larval stages before progressing to adulthood. If conditions during larval development are stressed, such as insufficient food being available or the local population density of *C. elegans*—determined by the local density of secreted conspecific pheromones called ascarosides [224]—exceeding a threshold level (which would in turn lead to an eventual shortage of nutrients in that area), the developing larva can enter a dormant state called the *Dauer state*. In this condition, its mouth is sealed and it has no food intake. The larva can remain dormant for up to a few months and recover if environmental conditions improve.

13.2 Foraging Behaviour of *C. elegans*

C. elegans makes a remarkable investment in neural tissue. Although it has only 302 neurons, these plus 56 support and glial cells comprise nearly 40% of its entire number of cells [548]. *C. elegans* is capable of several behaviours while traversing its environment, including foraging for food and avoiding toxins. Some 32 neurons are associated with chemoreception, and these enable *C. elegans* to alter its behaviour by reacting to numerous repulsive and attractive chemicals, altering its locomotion [65, 632] by switching between dwelling and roaming states.

Dwelling and Roaming

C. elegans nematodes can differentiate between high- and lower-quality food and tend to *dwell* for longer in regions with good food resources, this state being characterised by more frequent stops and path reversals during their locomotion in such regions. Conversely, if food resources are poor, *C. elegans* is more inclined to engage in *roaming* behaviour, characterised by rapid movement in straight lines. This behaviour is influenced by specific *AIY* neurons (a type of interneuron), with lower *AIY* activity levels producing local search (dwelling) behaviour [526]. It is known that *AIY* neurons exhibit a memory function, which may enable them to remember the level of nervous system activity related to previous chemical (food) concentration levels [632, 651].

Learning

There is also evidence that *C. elegans* is capable of nonassociative learning (such as habituation), associative learning, and imprinting [15]. Several genes have been identified in *C. elegans* with no known role other than mediating behavioural plasticity [15]. One example of learning in the context of foraging occurs when worms presented with medium-quality food, having been initially exposed to high-quality food resources, are found to be much more likely to move elsewhere than are worms which were initially exposed to low-quality resources [526], thereby modifying their behaviour on the basis of previous experience. In addition to being attracted to the chemical signatures of food resources such as bacteria [15], *C. elegans* is also capable of associative learning and subsequent repulsion in response to chemical signatures associated with predators such as the flatworm *Dugesia gonocephala* and toxic soil bacteria.

Foraging Styles

The various strains of *C. elegans* exhibit two distinct foraging styles, social foraging and solitary foraging [369]. Social strains of *C. elegans* cluster in tight swarms when feeding, whereas solitary strains move more slowly on food and feed alone. The difference in foraging style is governed by variations in the gene *nrp-1*, which encodes the NRP-1 receptor (found on the surface of neurons) [135]. Solitary strains express high-activity versions of *nrp-1*, with social feeders expressing a low-activity version. The level of *nrp-1* activity in turn affects the action of a specific neuron known as *RMG*.

13.3 Worm Optimisation Algorithm

The *worm optimisation* (WO) *algorithm*, inspired by elements of the foraging behaviours of *C. elegans*, was devised by Arnaout [17]. The behaviours are used to design a search process which consists of a simulated population of worms searching an environment for the best solution to the problem of interest. Specific features embedded in the algorithm include the capability of *C. elegans* to:

- distinguish food quality and to prefer higher-quality resources to poorer resources;
- engage in dwelling (local search) or roaming (global search) behaviour depending on the quality of food;
- have a social or solitary foraging style;
- have a preference for avoiding toxins (corresponding here to previously discovered low-quality solutions); and
- engage in dormancy if environmental conditions are poor (i.e. there is a low quality of food or a high local concentration of conspecifics) or, alternatively, reproduce if conditions are good.

13.3.1 Travelling Salesman Problem

As described in Sect. 9.5, the object of the travelling salesman problem (TSP) is to find the shortest tour of N cities where each link—or arc—between two cities has a distance or cost associated with it. Each city must be visited exactly once, and the tour must finish in the same city in which it started. The problem can be represented by a connected graph, the object being to find the sequence of arcs which gives the shortest distance required to visit all the nodes of the graph.

The TSP is an example of an NP-complete problem. The difficulty of solving it using brute force (by exhaustive enumeration of all possible tours) increases factorially (more than polynomially) as the number of cities in the problem grows. While small TSP problems can be solved using complete enumeration, larger examples require the application of heuristics to obtain an approximate solution. These heuristics are generally either constructive or modificative in that they either build a proposed tour from a starting point or, alternatively, take an existing (perhaps randomly created) tour and seek to improve it incrementally. One well-known example of a tour modification heuristic is the *k-opt heuristic*, where *k* is a user-defined parameter. In the 2-opt case, two edges not incident with the same node are selected and deleted, and the nodes at the ends of the two resulting segments of the tour are reconnected in the unique way possible to obtain a new tour. Among all pairs of edges whose 2-opt exchange decreases the length, we choose the pair that gives the shortest tour. This procedure is then iterated until no such pair of edges remains. The resulting tour is called 2-*optimal*. The process is illustrated in Fig. 13.1. More generally, in *k*-opt, *k*

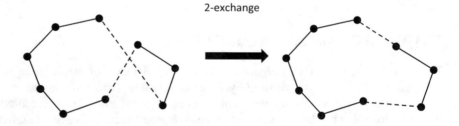

Fig. 13.1. Illustration of 2-exchange move. As the proposed move reduces the total tour length (measured by Euclidean distance), it is accepted

noncontiguous arcs are deleted and the resulting *k* segments are rejoined using arcs different from those deleted (there are, in general, *k*! ways to do this). In addition to traditional heuristics for the TSP, a large literature has emerged applying biologically inspired algorithms such as ant colony algorithms to this class of problem. Below we outline an application of the WO algorithm to the TSP.

13.3.2 Algorithm

The following description of the WO algorithm is drawn from [17]. In the algorithm, each artificial worm seeks to traverse a sequence of arcs between nodes (corresponding to cities). The complete tour associated with an individual worm will have a total cost or length. The choice of arc from each node by a worm is assumed to be determined by a number of factors, including the amount of simulated pheromone on an arc (denoted by τ_{ij} for the arc ij between nodes i and j), the social or solitary foraging style of the worm (modelled by the visibility η_{ij} of an arc ij), and a preference, ADF, for avoiding bad arcs (a proxy for the avoidance of toxic chemicals in the environment).

As the objective is to visit each city once, and once only, during a tour, the choice of which city to visit next should exclude cities already visited. The worm, when constructing a tour, (implicitly) maintains a memory of all cities it has already visited for this purpose. Thus, the probability of worm k travelling from city i to city j at time t, where V_i^k is the set of feasible cities reachable from city i and not yet visited by worm k, is

$$P_{ij}^k(t) = \frac{\tau_{ij}(t)^\alpha \cdot \eta_{ij}(t)^\beta \cdot \text{ADF}_{ij}}{\sum_{c \in V_i^k} \tau_{ic}(t)^\alpha \cdot \eta_{ic}(t)^\beta \cdot \text{ADF}_{ic}}, \text{ for } j \in V_i^k. \tag{13.1}$$

From (13.1) it can be observed that three factors determine the probability of choosing each arc:

i. the amount of pheromone (τ) deposited on the arc;
ii. the visibility (η) of the arc, which is impacted by whether a worm is a social or a solitary forager; and
iii. a parameter called a *bad solution factor* (ADF).

Two critical factors in (13.1) are the parameters α and β, and the values for these weight the relative importance of pheromone and visibility information in determining a worm's movements.

If a worm finds a better tour than the best solution found so far, then the pheromone levels on arcs of the newly found tour are updated using

$$\tau_{ij}(t+1) := \tau_{ij}(t) + \rho, \tag{13.2}$$

where ρ is the amount of pheromone laid down when the arcs on a good tour are reinforced. As with ant colony optimisation (ACO) (Sect. 9.5), the level of pheromone on arcs indicates their quality, as only arcs on the best tour found so far get updated.

At first glance, the formulation of (13.1) bears some similarity to the arc selection process in ACO. As will be seen below, the implementation of the arc selection process in WO is quite different, as the use that individual worms make of pheromone information and the way that $\eta_{ij}(t)$ is defined both depend on the foraging style of a worm. Another distinction is that (13.1) embeds a bad-arc avoidance behaviour (determined by ADF) not found in canonical ACO. It is also notable that the pheromone update step does not include the evaporation mechanism found in ACO. The WO algorithm also bears some similarity to the bacterial foraging optimisation algorithm (BFOA) (Sect. 14.3).

Foraging Style

In modelling the foraging style of each worm, a parameter RMG is used, with the value of this parameter determining whether a worm exhibits social or solitary foraging behaviour. If a worm is social, it is attracted to higher levels of pheromone on arcs, mimicking clustering on a food resource.

The value of $\eta_{ij}(t)$ is determined by generating a random variable (r) from the uniform distribution $U(0,1)$. If $r \leq$ RMG, then the worm is deemed to be social and:

$$\eta_{ij} = \frac{1}{d_{ij}}, \quad \text{for all } i = 1,\dots,N, \quad j = 1,\dots,N, \tag{13.3}$$

where d_{ij} is the Euclidean distance from city i to j, and the probabilities of arc selection are calculated using (13.1). Shorter arcs are preferred to longer arcs in the arc selection process.

Conversely, if $r >$ RMG, the worm is deemed to be solitary, it is repelled by pheromone, and selects its next city randomly from all options available. The value of η_{ij} is therefore

$$\eta_{ij} = \frac{1}{N}, \quad \text{for all } i = 1,\dots,N, \quad j = 1,\dots,N \tag{13.4}$$

and (13.1) is altered to

$$P_{ij}^k(t) = \frac{\tau_{ij}(t)^{-\alpha} \cdot \eta_{ij}(t)^{\beta} \cdot \text{ADF}_{ij}}{\sum_{c \in V_i^k} \tau_{ic}(t)^{-\alpha} \cdot \eta_{ic}(t)^{\beta} \cdot \text{ADF}_{ic}}, \quad j \in V_i^k. \tag{13.5}$$

The value selected for RMG will impact critically on the choice of foraging style. In the boundary cases of RMG = 0 and RMG = 1, the worms will only engage in solitary and social behaviours, respectively. The choice of value within the range (0,1) therefore determines the relative balance of social and solitary foraging by the population of worms during the search process. When social foraging is selected, a worm will take advantage of the social information contained in pheromone trails and also the local information available from arc lengths when choosing arcs, emphasising exploitation. In the case of solitary foraging, the choice of outbound arc from a city does not use social or local information, emphasising exploration.

Modelling Toxin Avoidance

When a worm has completed its tour, the quality of the solution is calculated. Based on this, a decision is made as to whether the tour should join the list of bad solutions. A $k \times (N+1)$ array ADFList is created to store up to k bad solutions, with each entry in the array consisting of the ordered list of N cities visited by the worm during the bad tour and the associated cost of that tour (c_{worm}). (The term c_{best} denotes the length of the current best solution.) The parameter k is calculated as $k = \lceil \sqrt{W} \rceil$, where W is the number of active worms in the population. This number can change as the algorithm iterates, as outlined in the discussion below on the Dauer state.

The $N \times N$ array ADF is linked with ADFList so that for every solution stored in the list, its arcs are updated using $ADF_{ij} = \lambda$, where λ is the bad solution factor (a penalty). If an arc is deemed bad because it belongs to a poor solution in ADFList, it is less likely to be selected subsequently by foraging worms.

The pseudocode for modelling ADF, which is applied for every worm, is provided in Algorithm 13.1. The choice of value for λ impacts on the likelihood of an arc being selected. Initially, all arcs are assigned an ADF value of one. As the value selected for λ decreases, then the penalty applied to arcs on bad solutions is increased, and the less likely they are to be selected for traversal by worms.

In the case where $\sum ADF = N \cdot \lambda$, all potential starting arcs for a tour have been added to ADFList. In order to continue the search, a random arc is selected and released.

Algorithm 13.1: ADF calculation

Sort ADFList in ascending order using c_{worm} (ADFList$_{1,N+1}$ contains the worst solution);

Assess current worm's solution (c_{worm});

if $c_{\text{worm}} > $ ADFList$_{1,N+1}$ **then**
  | Add worm's tour to ADFList;
else
  | **if** $c_{\text{worm}} > $ ADFList$_{k,N+1}$ **then**
  | | Replace row k in the list with current worm's tour;
  | **else**
  | | ADFList$_{k,i} := 0$, for $i = 1,\ldots,N+1$ (if generated tour is better than the kth item in ADFList, remove the kth entry in the list);
  | **end**
end
for $l = 1,\ldots,k,\ i = 1,\ldots,N$ **do**
  | **if** ADFList$_{l,i} \neq 0$ **then**
  | | $ADF_{l,i} = \lambda$ (reduce ADF from 1 to λ);
  | **end**
end
for $i = 1,\ldots,N,\ j = 1,\ldots,N$ **do**
  | $\sum ADF = \sum ADF + ADF_{ij}$;
  | **if** $\sum ADF = N \cdot \lambda$ **then**
  | | Choose a random city c (randomly release a starting arc to ensure that search can proceed);
  | | $ADF_{i,c} := 1$;
  | **end**
end

Dwelling and Roaming

A dwelling and roaming mechanism is simulated after each worm generates a tour, and the mechanism acts to stochastically implement a local search around the worm's current tour. A random value (r) is generated from the uniform distribution $U(0, 1)$. If r is less than a threshold value (AIY), then a local search step is initiated.

In the local search step, two cities from the tour are chosen at random and their places in the tour are swapped. If this leads to a better (lower-cost) tour, this replaces the current tour of the worm. If the tour is not better, the original tour is restored and the random swap process is repeated until either a better tour is found or a maximum of t_{AIY} random swaps are undertaken.

The values of AIY and t_{AIY} impact on the degree of local search undertaken in the WO algorithm. As the values of both increase, the degree of local search also increases.

Dauer State

The final simulated mechanism in the WO algorithm is a modelling of a Dauer state. The quality Q of a tour (metaphorically, a proxy for food quality) is assessed after each worm completes its tour using

$$Q = \min\left\{1, \ 1 + \frac{c_{worm} - c_{best}}{c_{best}}\right\}. \tag{13.6}$$

When c_{worm} is larger than c_{best}, the best solution is not improving, and then $Q = 1$.

The concentration of worms in the search space is assessed using the number of tours generated. The number of iterations is updated until it reaches its maximum (t_{max}), which indicates a high concentration of worms. This is then normalised into the range $(0, 1)$ to measure the *worms concentration C* as

$$C = \frac{t}{t_{max}}, \tag{13.7}$$

where t is the current iteration. When $C = 1$, the algorithm has reached the value of t_{max}. This does not terminate the algorithm. After a worm completes a tour, the Dauer status (S_{Dauer}) is assessed using Algorithm 13.2.

The parameter ϕ is the reproductive rate of worms when they are not in a Dauer state. If S_{Dauer} is less than one, the population of worms can reproduce and increase in number up to a maximum of W_{max}. If $S_{Dauer} = 1$, indicating that the best solution has not improved and C is high, then the worm population is decreased. The mechanism acts to increase the population (simulating reproduction) when new best solutions are being found and when the number of iterations of the algorithm is low. Later in the run the population is decreased as no new best solutions are found and as the population converges on the same solutions. The algorithm terminates when the population size of the worms is reduced to zero.

Algorithm 13.2: Assessment of Dauer status

Compute Q and C using (13.6) and (13.7);
Compute $S_{\text{Dauer}} = (Q+C)/2$;
Assess whether worms are in active reproductive or declining stage;
if $S_{\text{Dauer}} < 1$ **then**
$\quad\mid\quad W = \min\{W_{\max}, \lceil W \cdot (1+\phi) \rceil\}$;
end
if $S_{\text{Dauer}} = 1$ **then**
$\quad\mid\quad W = \lfloor W \cdot (1-\phi) \rfloor$;
end
if $W = 0$ **then**
$\quad\mid\quad$ Stop WO algorithm and report c_{best};
end

Algorithm

Algorithm 13.3 brings together the individual elements of the WO algorithm as described above. Here, t is the iteration counter.

Algorithm 13.3: Worm Optimisation Algorithm [17]

Initialise the parameters for the worm optimisation algorithm including τ_{ij};
while $W \neq 0$ **do**
$\quad\mid\quad$ Solve for a tour using (13.1) or (13.5) (RMG, social or solitary foraging);
$\quad\mid\quad$ Find cost of tour (c_{worm});
$\quad\mid\quad$ Implement local search (AIY);
$\quad\mid\quad$ Update pheromone levels if required using (13.2);
$\quad\mid\quad$ Update bad solution list if required using Algorithm 13.1 (ADFList and ADF);
$\quad\mid\quad$ Update number of worms (W) using simulated Dauer mechanism in
$\quad\mid\quad\quad$ Algorithm 13.2;
$\quad\mid\quad$ $t := t + 1$;
end
Output the best solution (best tour) found;

13.3.3 Parameter Setting

Based on a series of experiments examining different parameter settings on four test TSPs ranging in size from 30 to 100 cities, the following parameter settings were suggested in [17]: W (initial number) = 40, $W_{\max} = 7438$, RMG = 0.55, AIY = 0.36, $t_{\text{AIY}} = 38$, $\tau_{ij} = 0.01$, $\phi = 0.3$, $\lambda = 0.01$, $\rho = 0.01$, $\alpha = 1.5$, $\beta = 2.5$ and $t_{\text{best}} = 125$. This study also compared the performance of the WO algorithm at these settings with previously published results from genetic algorithm and simulating annealing heuristics on the same problems and found the performance of WO to be better than that of the other two methods in all four cases.

13.4 Summary

The WO algorithm is broadly inspired by a number of foraging behaviours of the roundworm *C. elegans* and has shown itself to be capable of producing competitive results on a number of TSP problems. Further work is required to assess whether the algorithm can scale to higher-dimensional problems and to determine whether its results are competitive in terms of quality and computational effort required, against those of other heuristics such as ACO. It is also notable that the algorithm bears some similarities with ACO and the BFOA, and a detailed investigation of the mechanisms of the WO algorithm is required in order to better understand the contribution of its distinguishing factors to its performance.

Part V

Nonneuronal Organisms

In the preceding chapters an array of optimisation algorithms derived from foraging behaviours of organisms, including mammals, birds, fish and insects, have been explored. In all of these cases the organisms possess a nervous system, with varying degrees of complexity and plasticity.

In contrast, the foraging activities of most life on earth are not supported by neuronal-based information processing. The vast majority of life forms, including the plant, bacteria, fungi and protist kingdoms of life, do not have a brain or other nervous system hardware [481], with recent work by [29] estimating that more than 99.5% of the earth's biomass is found in nonneuronal organisms such as plants (accounting for 82% of total biomass) and bacteria (13%), with less than 0.45% of total biomass being comprised of animals with neuronal tissue. Despite the absence of neuronal tissue, these organisms live in environments that are no less complex than those faced by organisms with a brain. These *nonneuronal organisms* face the same basic challenges as animals in foraging for food and other resources, and in dealing with competitors, predators and pathogens. This raises the question as to what mechanisms these organisms use for sensing their local environment and internal state, and subsequently undertaking actions that enhance their survival. In single-cell organisms, by necessity, environmental sensing, information processing and effectors operate at the molecular level [578].

It is clear from empirical studies that many nonneuronal organisms are capable of quite complex information processing. For example, plants, fungi and slime moulds sense their environment and search for food resources by growing exploratory structures such as roots, shoots, hyphae or pseudopodia [481]. Typically, growth or movement is enhanced in a direction if a positive gradient for a required resource is detected. Bacteria also explore their environment, moving in the direction of desired resources. *E. coli* bacteria, for example, integrate information on the current strength of a chemical gradient with past information on the strength of that gradient when foraging. If a gradient of an attractant chemical is getting stronger, a bacterium is more likely to continue to move in that direction.

Nonneuronal organisms are also capable in some instances of marking previously explored territory in order to avoid revisiting an already harvested area, creating an external form of memory. For example, slime moulds leave nonliving extracellular slime behind, and if this is subsequently detected during foraging movement, the area is avoided.

Another capability that has been discovered in multiple species of nonneuronal organisms is kin recognition, with examples having been found in plants, amoebae and bacteria [543]. This capability is used to influence the degree of competition for resources that an organism engages in, with greater competition typically being exhibited when nonkin are detected.

Although our understanding of the sensing, information integration and decision-making mechanisms of nonneuronal organisms is still immature, notable strides have been made in the last decade. One positive side effect of these advances has been the development of a range of optimisation algorithms which are inspired by the foraging activities of some of these organisms, specifically bacteria, slime moulds, fungi and plants. In this part of the book, over the next three chapters, we provide an introduction to each of these areas and a range of associated optimisation algorithms.

14

Bacterial and Viral Foraging Algorithms

Bacteria are one of the most populous forms of life on earth, being found in every habitat from the Arctic to deep ocean environments. It is estimated that a single gram of topsoil can contain up to 10^{10} bacterial cells [473]. In addition to being numerous, bacteria are amongst the oldest form of life on the planet, being included (with archaea) in the category of prokaryotes. These organisms have no nucleus and their genetic information is contained in a single loop of DNA.

Despite possessing a relatively simple unicellular structure, typically being some 5–15 times smaller than an animal cell, bacteria are capable of sophisticated inter-actions with their environment and with each other. While an individual bacterium has limited information-processing power, colonies of bacteria can engage in a wide array of social behaviours, encompassing cooperative foraging for resources, cooper-ative defence, construction of protective structures and group-mediated dispersal or reproduction [78]. Bacteria are metaphorically capable of communication with each other (and with other organisms) by means of chemical signals, a process which has parallels with the use of pheromone trails by ants (Sect. 9.2). The chemical signalling mechanisms support cooperative, emergent behaviours, resulting in out-comes which are functionally equivalent to problem solving. These activities are often underpinned by quorum sensing capabilities, whereby bacteria can sense the local density of conspecifics through the detection of chemicals emitted by individ-ual bacteria. The greater the concentration of these chemicals, the greater the local population density, with certain behaviours being triggered at threshold chemical concentrations.

More generally, a correspondence between the foraging activities of nonneuronal organisms such as bacteria and the activities of higher-order animals can be outlined (Fig. 14.1 presents one instance of this). In each case, the organism obtains sensory information concerning its local environment, and this information is processed in conjunction with internal state information to produce a behavioural outcome, such as a locomotory response. As noted in [578], single-cell organisms, by necessity, im-plement sensors, information processing and effectors on the molecular level. Hence, these mechanisms are less 'visible' to us than those of higher animals.

© Springer Nature Switzerland AG 2018
A. Brabazon, S. McGarraghy, *Foraging-Inspired Optimisation Algorithms*,
Natural Computing Series, https://doi.org/10.1007/978-3-319-59156-8_14

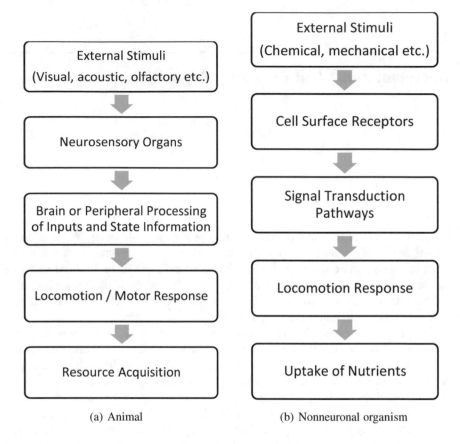

(a) Animal   (b) Nonneuronal organism

Fig. 14.1. Illustration of one instance of the correspondence between stylised foraging in animals with a central nervous system (a), and nonneuronal organisms (b)

The embodiment of learning differs between neuronal and nonneuronal organisms. In the former, learning is embodied via modification of the nervous system. In the case of nonneuronal organisms, there is still considerable debate concerning the mechanisms by which learning is embodied. Proposed mechanisms include epigenetic control systems [210] (i.e. memory being stored as epigenetic modifications of DNA) and modification of chemical signal transduction pathways. It is also notable that we are beginning to uncover numerous examples of electrochemical signalling in multicellular nonneuronal organisms, not unlike the signalling between neurons ([469] provides an example of this in the case of bacterial communities in biofilms). Our understanding of the role of these mechanisms for information processing and communication in nonneuronal organisms is still immature.

In this chapter we describe a number of bacterial foraging behaviours and demonstrate how these can be used to design optimisation algorithms. We also introduce a number of algorithms which are drawn from activities of even simpler biological agents, namely viruses, and illustrate how these agents can metaphorically be considered as foraging for host cells.

In Sects. 14.1–14.3, we need to be able to refer to both a component of an individual vector and the index of a particular vector in a set of vectors. To allow this, we use a subscript to indicate the component, and a superscript to indicate the index of the vector in the set; for example, the set of vectors might be $\{x^1, x^2, \ldots, x^i, \ldots, x^n\}$, and x^i_j would indicate component j of vector i. In Sects. 14.4 and 14.5, we revert to the use of a subscript to indicate the index of a particular vector in a set of vectors.

14.1 Chemotaxis in *E. coli* Bacteria

Many species of bacteria are able to move in response to external stimuli—a phenomenon called *taxis*—and this provides them with a powerful adaptive capability. Examples of such movement include chemotaxis, where bacteria move towards or away from certain chemical concentrations; phototaxis, where they move in response to light; thermotaxis, where they move in response to different temperatures; aerotaxis, where they move in response to different levels of oxygen concentration; and magnetotaxis, where they align themselves in response to magnetic lines of flux. Depending on the environment occupied by a bacterium, movement can be through a medium (for example, through a fluid) or by gliding along a surface.

Escherichia (*E.*) *coli* is a Gram-negative, facultatively anaerobic, rod-shaped bacterium of the genus *Escherichia*. This classification of bacteria includes many individual strains which have significant genotypic and and phenotypic diversity. A simplified diagram of a motile *E. coli* cell is shown in Fig. 14.2. Each *E. coli* cell

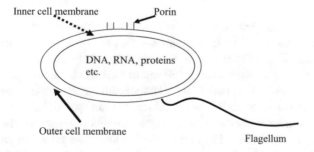

Fig. 14.2. A simplified diagram of a motile *E. coli* cell, showing one flagellum (not to scale) and two porins (protein channels which allow the entry of water soluble nutrients). Typically a cell will have hundreds of porins

is approximately 1 to 3 μm in length, with approximately eight flagella (only one is shown in Fig. 14.2) which allow it to swim through a liquid medium. Each flagellum is approximately 10 μm long, has a rigid left-handed corkscrew shape and forms a propeller. The propeller is powered by a tiny biological electric rotary motor. The motor, and hence the propeller shaft, is capable of turning both clockwise (CW) and counter clockwise (CCW). If the flagella are rotated CCW in tandem, they produce a force against the *E. coli* cell and it moves (swims) forward in its current direction (termed a *run*). If the flagella switch from a CCW to a CW rotation, the cell tends to rotate randomly (termed a *tumble*) to point in a new direction.

E. coli cells are rarely still but instead are engaged in continual run and tumble behaviours [51]. The two modes of movement alternate. Runs tend to last for circa 1 to 2 seconds, whereas tumbles take less time to complete (circa 0.1 to 0.2 seconds). Even this brief time span is sufficient to orientate the bacterium in a random direction, causing successive runs to occur in nearly random directions (there is a slight bias in the tumbling process towards continuing in the same direction as before). Therefore, a bacterium runs, tumbles to face in a new direction and runs again (Fig. 14.3). When a bacterium finds stronger concentrations of a chemical attractant, the mean length of its runs increases. This effect biases its movement towards the climbing of attractive chemical gradients.

Fig. 14.3. A sample movement track for an *E. coli* cell. The track consists of a series of runs punctuated by tumbles (directional changes)

The running and tumbling behaviours of bacteria result from the stimulation of receptors (proteins on the surface membrane that bind to a specific attractant) on the cell for the chemicals of interest to it. If increasing concentrations of chemical attractants are detected, the information is relayed to internal chemical signalling networks that act to increase CCW rotation. Repellant stimuli have the opposite effect. The behavioural response of the cell is temporal rather than spatial, as *E. coli* cells are too short for differences in chemical concentrations between receptors at each end of the cell to be significant [381]. Therefore, a cell cannot directly sense and use (chemical) gradient information from its environment. Instead, it has a short-term working memory based on the strength of stimulation of its receptor proteins [87, 380, 548], lasting up to about two seconds, which allows it to detect a temporal gradient, in other words, whether the concentration of a chemical is changing over time.

The cell is also subject to Brownian motion if it is moving though a liquid medium, as it has very little mass. This causes the cell to tend to drift off course every few seconds, even when it is following a strong chemical trail. *E. coli* bacteria have *molecular noise generators* in their chemotaxis network (i.e. identical chemical inputs do not always produce exactly the same behavioural outputs). This injects a stochastic element into the locomotion of cells, and it has been suggested that this is adaptive as it has been selected for [326].

14.1.1 Chemotaxis as a Search Process

The idea that the chemotactic behaviour of bacteria such as *E. coli* and *Salmonella typhimurium* could be considered as a stochastic search (optimisation) process was originally proposed by Bremermann in 1974 [81]. However, the idea did not attract attention when initially published, and it is only in the last decade that it has been revisited.

14.1.2 Basic Chemotaxis Model

The pseudocode in Algorithm 14.1 describes a search process which simulates a primitive chemotaxis mechanism. A population of S bacteria is created and distributed randomly in the search space, assumed to be a domain in $\mathbb{R}^D$. The position (vector) of each bacterium i is stored as $x^i \in \mathbb{R}^D$. We assume that there is a cost $f^i = f(x^i)$ associated with each location that a bacterium can occupy and that the intention is to find the best location in the search space, i.e. the location with minimum cost. (Note: here we are framing the optimisation problem as a minimisation problem and, hence, the best or fittest location is the one with lowest cost.)

The quality of the location occupied by each bacterium is calculated, after which each bacterium seeks to move. The bacterium tumbles to face in a random direction and then continues to take equal-sized swim steps in this direction while this improves its fitness (i.e. quality of its location), up to a maximum of N_s steps. The best location (with lowest cost) found by any bacterium is stored and is returned at the algorithm's termination.

Although this algorithm produces a search process, it is not particularly effective or efficient as it amounts to a population of bacteria engaging in a series of biased random walks across the environment. Each bacterium searches individually and there is no social communication regarding the location of nutrient-rich or nutrient-poor regions between them.

The underlying chemotactic model is also a notable simplification of real-world chemotactic behaviour. In the next section we outline an algorithm which is based on a more sophisticated model of chemotaxis.

14.2 Bacteria Chemotaxis Algorithm

The *bacteria chemotaxis algorithm* (BCA), proposed by Müller et al. [418] and described further in [419], uses a biological model of the chemotactic process arising from [127] to develop an optimisation algorithm. The underlying model is based on six assumptions, and these are described as follows in [418]:

  i. the path of a bacterium is a sequence of straight-line trajectories (characterised by speed, direction and duration) linked by instantaneous turns;

Algorithm 14.1: Basic chemotactic bacterial foraging algorithm

---

Randomly distribute initial values for the position vectors x^i, $i = 1, 2, \ldots, S$ across the optimisation domain;

Compute the initial cost function value $f^i = f(x^i)$ for each bacterium i;

repeat

    for *each bacterium i* **do**

        Tumble: Apply random tumble to bacterium to face it in a new direction;

        Take a step in this direction;

        Measure fitness of new location;

        while *number of swim steps* $< N_s$ **do**

            if *fitness of new position $>$ fitness of previous position* **then**

                Take another step in current direction and calculate fitness of new location;

            else

                Let number of swim steps $= N_s - 1$;

            end

            Increment number of swim steps;

        end

    end

until *terminating condition*;

Output the best solution found;

---

ii. all trajectories have the same constant speed;

iii. when a bacterium turns, its choice of new direction is determined by a probability distribution which is symmetric about its previous direction;

iv. the angle between two successive trajectories is governed by a probability distribution;

v. the duration of a trajectory is governed by an exponentially decaying probability distribution; and

vi. the probability distributions for the angle and duration are independent of the parameters of the previous trajectory.

In the canonical BCA, each bacterium searches independently using a simulated chemotactic mechanism, with no concept of group communication. Unlike the primitive chemotactic mechanism described in Sect. 14.1.2, there is no assumption in this model that bacteria will continue to run in the direction of an increasing chemoattractant concentration.

14.2.1 Two-Dimensional Case

The algorithm is described as follows in [418] for the two-dimensional case (its extension to the *n*-dimensional case is also outlined below). The bacterium moves around the search space in a sequence of straight steps where the duration (length) and direction of each step are variable, being determined by the workings of the BCA.

Duration of Trajectory

Initially, a bacterium is located randomly in the search space and its velocity (v) is assumed to have a constant value. The duration of the trajectory (τ) for a bacterium is assumed to be a random variable with an exponential probability density function

$$P(X = \tau) = \frac{1}{T}e^{-\tau/T}, \tag{14.1}$$

where the expectation is $\mu = E(X) = T$ and the variance is $\sigma^2 = \text{Var}(X) = T^2$. The time T is given by

$$T = \begin{cases} T_0, & \text{for } f_p/l_p \geq 0, \\ T_0\left(1 + b \cdot |f_p/l_p|\right), & \text{for } f_p/l_p < 0, \end{cases} \tag{14.2}$$

where T_0 is the minimal mean time (user-defined), f_p is the difference between the actual and previous function values (the two-dimensional function $f(x_1, x_2)$ is being minimised in this case), b is a dimensionless parameter (user-defined), and $l_p = \|x_p\|$, where x_p is the vector connecting the previous and actual positions in the search space.

New Direction

The new direction for the bacterium is determined using the probability density distribution of the angle α between its previous direction and the new direction. This is assumed to be Gaussian and, for a right or left turn, respectively (in a two-dimensional environment, the only directional changes possible are either to the left or to the right of the current trajectory), is:

$$P(X = \alpha, v = \mu) = \frac{1}{\sigma\sqrt{2\pi}}e^{-(\alpha-v)^2/(2\sigma^2)},$$

$$P(X = \alpha, v = -\mu) = \frac{1}{\sigma\sqrt{2\pi}}e^{-(\alpha-v)^2/(2\sigma^2)}, \tag{14.3}$$

where the expectation value $\mu = E(X)$ is $62°$ and the standard deviation $\sigma = \sqrt{\text{Var}(X)}$ is $26°$ based on empirical chemotactic measurements on *E. coli* undertaken in [51], and $\alpha \in [0°, 180°]$. The choice of right or left direction, relative to the previous direction, is determined using a uniform probability density distribution, yielding a probability density distribution for the angle α of

$$P(X = \alpha) = \frac{1}{2} \cdot (P(X = \alpha, v = \mu) + P(X = \alpha, v = -\mu)). \tag{14.4}$$

The expectation value and variance are as follows:

$$\mu = 62°(1 - \cos(\alpha)), \tag{14.5}$$

$$\sigma = 26°(1 - \cos(\alpha)), \tag{14.6}$$

with

$$\cos(\alpha) = \begin{cases} 0, & \text{if } f_p/l_p \geq 0, \\ e^{-\tau_c \tau_p}, & \text{if } f_p/l_p < 0, \end{cases} \tag{14.7}$$

where τ_c is the correlation time (user defined) of the angle between successive trajectories, and τ_p is the duration of the previous step. Equation 14.7 is based on the first order approximation of the average value of the cosine of the angle between successive trajectories at times t and $t + \tau_c$ as a function of t. The cosine of the angle (θ) between the trajectories at times t and $t + \tau$ is assumed to be related to the duration of the previous step τ_p with an exponentially decaying function [419].

The expectation and variance are modified to ensure that the new and old directions are correlated if the bacterium is moving towards better solutions and the previous trajectory duration is small.

Once α is calculated, the normalised new displacement vector (n_u) with unit length can be obtained.

Compute New Position

Finally, the bacterium is moved to its new position. The length l of the path is calculated as:

$$l = v\tau. \tag{14.8}$$

The normalised new direction vector n_u, with $\|n_u\| = 1$, is multiplied by l to obtain the displacement vector x:

$$x = l \cdot n_u \tag{14.9}$$

and the new position of the bacterium can then be determined using:

$$x^{new} = x^{old} + x. \tag{14.10}$$

Combination of Steps

The above steps are repeated for each move of the bacterium. As mentioned earlier, the model embeds only a chemotactic mechanism and in the above form, considers only the movement of a single bacterium. There is no populational search or transfer of information between individual bacteria in the environment.

The model contains a number of user-determined parameters, including T_0, b, τ_c and v, and good values of these will be problem-specific. Both [418] and [419] provide a discussion of how these parameters can be adapted dynamically as the algorithm iterates, in order to improve the efficiency of the search process.

14.2.2 Extension to n-Dimensional Case

As described in [418, 419], the above model can be extended to the more general n-dimensional case as follows. The basic features of the BCA remain the same but additional notation is required to define positions and angles for this space.

To define a position $(x_1,\ldots,x_n)$ in polar coordinates in an n-dimensional space, a radius r and $n-1$ angles $(\varphi_1,\ldots,\varphi_{n-1})$ are required. The position in n dimensions is defined as

$$x_1 = r \cdot \prod_{k=1}^{n-1} \cos(\varphi_k),$$

$$x_i = r \cdot \sin(\varphi_{i-1}) \prod_{k=i}^{n-1} \cos(\varphi_k), \quad i = 2,\ldots,n-1, \tag{14.11}$$

$$x_n = r \cdot \sin(\varphi_{n-1}).$$

Using the formulation in (14.11), the two-dimensional version of BCA can be extended to n dimensions as follows.

Duration of Trajectory

The velocity v and duration τ of the trajectory are calculated as before.

New Direction

The probability density distribution of the angle φ_i on the plane defined by the axes (x_i, x_{i+1}), referring to the axis x_i, is Gaussian, and reads as follows for turning left or right:

$$P(X_i = \varphi_i, v_i = \mu_i) = \frac{1}{\sigma_i \sqrt{2\pi}} e^{-(\varphi_i - v_i)^2/(2\sigma_i^2)},$$

$$P(X_i = \varphi_i, v_i = -\mu_i) = \frac{1}{\sigma_i \sqrt{2\pi}} e^{-(\varphi_i - v_i)^2/(2\sigma_i^2)}, \tag{14.12}$$

where $\mu_i = E(X_i) = 62°$, $\sigma_i = \sqrt{\mathrm{Var}(X_i)} = 26°$ and $\varphi_i \in [0°, 180°]$. The right or left direction is determined using a uniform probability density distribution, thereby yielding a probability density distribution of the angle φ_i:

$$P(X_i = \varphi_i) = \frac{1}{2} \cdot (P(X_i = \varphi_i, v_i = \mu_i) + P(X_i = \varphi_i, v_i = -\mu_i)) \tag{14.13}$$

and

$$\mu_i = 62°(1 - \cos(\varphi_i)) \tag{14.14}$$

$$\sigma_i = 26°(1 - \cos(\varphi_i)). \tag{14.15}$$

Once the φ_i are calculated, the normalised new displacement vector n_{u} can be obtained, by adding the new computed angles $(\varphi_1,\ldots,\varphi_n)$ to the old ones and using

(14.11) with r set to one to determine the Cartesian coordinates corresponding to these polar coordinates.

The new position for the bacterium can then be obtained by applying this vector to the old position as for the two-dimensional case.

14.2.3 Discussion

The canonical BCA is a relatively simple algorithm based on a biological model of chemotaxis. In order to strengthen the performance of the canonical algorithm, a series of improvements have been added, including dynamic adaptation of the algorithm's parameters during the optimisation process.

As the performance of the algorithm is dependent on the starting point of a bacterium in the search space, the canonical algorithm can easily be extended to encompass a population of bacteria that engage in individual search from random starting points on the landscape, with the algorithm returning the best location found by any member of the population. Owing to the solitary foraging nature of agents in the algorithm (there is no communication between them), this is equivalent to a multiple restart process.

An extension to the canonical BCA, called the *bacterial colony chemotaxis algorithm*, was introduced in [352]; here, the movement of bacteria is governed both by the individual chemotactic response to attractants and by a social influence factor.

14.3 Bacterial Foraging Optimisation Algorithm

An alternative approach to the design of bacteria-inspired foraging algorithms is to include a richer repertoire of foraging-related behaviours than just chemotaxis. This perspective is taken by Passino [450, 451] in the *bacterial foraging optimisation algorithm* (BFOA). This family of algorithm draws their inspiration from four aspects of *E. coli* behaviour as illustrated in Fig. 14.4.

As before, chemotaxis refers to the tumble-and-run behaviour of an individual bacterium in response to chemical gradients. Swarming refers to the capability of *E. coli* bacteria to emit the chemical attractant aspartate when they uncover nutrient-rich environments. In turn, other *E. coli* bacteria respond by moving (swarming) towards bountiful regions which are marked with attractant. The emission of attractant produces an indirect (chemical) social communication mechanism between bacteria. Under selection for reproduction, the healthiest bacteria are more likely to survive and divide, thereby creating more bacteria which are colocated in nutrient-rich environments. Bacterial elimination and dispersal events, a form of randomisation and movement, occur frequently in the real world. The former occur when a local environmental event kills a bacterial colony, and the latter occur when bacteria are transported to a new location, for example via wind dispersal.

A wide variety of BFOAs can be created from the above highly stylised aspects of bacterial foraging, as the individual characteristics such as reproduction or swarming can be operationalised in many ways. The canonical BFOA is now overviewed (Algorithm 14.2), following the description in [450, 451].

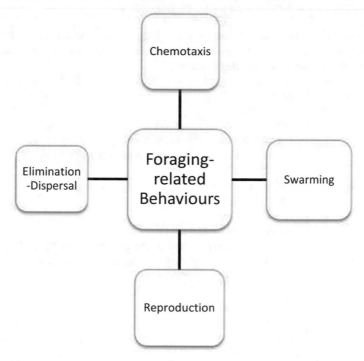

Fig. 14.4. Some stylised foraging-related behaviours of *E. coli* implemented in the BFOA

Initialisation of the Algorithm

As above, each bacterium i is initially randomly located in the search space, with its position being stored as $x^i \in \mathbb{R}^D$. Let $f^i = f(x^i)$ be the cost associated with the bacterium's location. Assume that the intention is to find the location in the search space with minimum cost.

Notation Used

The ordered S-tuple of positions of the entire population of S bacteria at the jth chemotactic step, the kth reproduction step and the lth elimination–dispersal event is denoted by $P(j,k,l) = (x^i(j,k,l) : i = 1,2,\ldots,S)$. As the algorithm executes, $P(j,k,l)$ is updated immediately once any bacterium moves to a different location. The cost associated with location $x^i(j,k,l)$ is denoted by $f^i(j,k,l)$. Each bacterium has a lifetime N_{ch}, measured as the maximum number of chemotactic cycles it can undertake.

Chemotaxis Loop

At the start of each chemotactic loop, the *swarm effect inclusive* (SEI) cost corresponding to each bacterium's current location is calculated as follows:

Algorithm 14.2: BFOA with social communication

Randomly distribute initial values for x^i, $i = 1, 2, \ldots, S$, across the optimisation domain;
Compute the initial cost function value for each bacterium i as $f^i = f(x^i)$, and the
initial total cost with swarming effect as f_{sw}^i;

for $l = 1$ **to** N_{ed} **do**   // Elimination–dispersal loop
    for $k = 1$ **to** N_{re} **do**   // Reproduction loop
        for $j = 1$ **to** N_{ch} **do**   // Chemotaxis loop
            for *each bacterium* $i = 1$ **to** S **do**
                Tumble: Generate a unit length vector $\phi \in \mathbb{R}^D$ in a random direction;
                Move: Let $x^{\text{new}} = x^i + c\phi$ and compute corresponding f^{new};
                Let $f_{\text{sw}}^{\text{new}} = f^{\text{new}} + f_{\text{cc}}(x^{\text{new}}, P)$;
                Swim: Let $m = 0$;
                while $m < N_{\text{s}}$ **do**
                    Let $m = m + 1$;
                    if $f_{\text{sw}}^{\text{new}} < f_{\text{sw}}^i$ **then**
                        Let $x^i = x^{\text{new}}$ and compute corresponding f^i and f_{sw}^i;
                        Let $x^{\text{new}} = x^i + c\phi$ and compute corresponding f^{new};
                        Let $f_{\text{sw}}^{\text{new}} = f^{\text{new}} + f_{\text{cc}}(x^{\text{new}}, P)$;
                  else
                    Let $m = N_{\text{s}}$;
                  end
                end
            end
        end
        Sort bacteria in order of ascending cost f_{sw};
        The $S_{\text{r}} = S/2$ bacteria with the highest f value (the least healthy) die and the
          remaining S_{r} bacteria split;
        Update value of f and f_{sw} accordingly;
    end
    Eliminate and disperse individual bacteria to random locations on the optimisation
      domain with probability p_{ed};
    Update corresponding values for f and f_{sw};
end
Output the best solution found;

$$f_{\text{sei}}^i(j, k, l) = f^i(j, k, l) + f_{\text{cc}}^i(x^i(j, k, l), P(j, k, l)). \qquad (14.16)$$

The SEI cost comprises both the underlying cost of the bacterium's location (as given
by $f^i(j, k, l)$) and a value for the cell-to-cell attraction and repulsion (swarming) term
$f_{\text{cc}}(x(j, k, l), P(j, k, l))$. In the swarming behaviour of the bacteria, each individual is
trying to minimise $f_{\text{sei}}^i(j, k, l)$, so they will try to find low-cost locations and move
closer (but not too close) to other bacteria.

The effect of the cell-to-cell attraction and repulsion term is to create a time-
varying SEI cost function which is used in the chemotaxis loop. As each bacterium

moves about on the landscape, its $f^i_{sei}(j,k,l)$ alters, as it depends not just on the bacterium's own current location but also on the simultaneous locations of all other bacteria in the population. Another way of thinking about this is that the landscape being searched is dynamic and it deforms as the bacteria traverse it.

Once $f^i_{sei}(j,k,l)$ is calculated, it is stored in f^i_{curr} for each bacterium. This value is used later in the chemotaxis loop to determine whether a bacterium's movement is improving its fitness.

The calculation of $f^i_{cc}(x^i(j,k,l),P(j,k,l))$ depends on the proximity of each bacterium to its peers. Each bacterium is attracted to its peers, loosely mimicking the effect of the chemical attractant aspartate. Bacteria are also repelled from one another, mimicking the real-world problem arising from 'too close' location, as the bacteria would then compete for the same nutrients. Each of these mechanisms is included in (14.18):

$$f_{cc}(x, P(j,k,l)) = \sum_{i=1}^{S} f^i_{cc}(x, x^i(j,k,l)) \tag{14.17}$$

$$= \sum_{i=1}^{S} \left[-d_{attract} \exp\left(-w_{attract} \sum_{b=1}^{p} (x_b - x^i_b)^2 \right) \right] \tag{14.18}$$

$$+ \sum_{i=1}^{S} \left[h_{repel} \exp\left(-w_{repel} \sum_{b=1}^{p} (x_b - x^i_b)^2 \right) \right].$$

In (14.18), the parameter p is the dimension of the real vector space $\mathbb{R}^p$ being searched by the bacteria (here $p = 2$ is assumed). The first additive component of (14.18) acts to reduce the SEI cost of each bacterium, as it is restricted to returning a nonpositive value. In the limit, if all bacteria converge to the same location, the exponential terms in this component will tend towards their maximum value of 1 and the SEI costs of all bacteria will be reduced by $-d_{attract} \cdot S$. Metaphorically, the terms represent the depth of the attractant released by a bacterium and the width of the attractant signal, respectively. The second additive component of (14.18) repels the bacteria from one another. If the bacteria swarm to such a degree that they colocate, the SEI costs of all bacteria will be increased by $h_{repel} \cdot S$.

The values of the parameters $d_{attract}$, $w_{attract}$, h_{repel} and $-w_{repel}$ therefore control the strength of the swarming effect and repulsion effect relative to each other, and relative to the impact of the cost function $f^i(j,k,l)$.

In order to move a bacterium in the search space, a tumble followed by run behaviour is simulated. The tumble acts to orientate the bacterium in a random direction, with the bacterium then continuing to move in that direction until either it has taken the maximum number of chemotactic steps (N_s) or its cost function stops declining. In generating a tumble, a vector of unit length and random direction $\phi^i(j)$ are used. The bacterium is then moved a step ($C_i > 0$) in this direction:

$$x^i(j+1,k,l) = x^i(j,k,l) + C_i\phi^i(j). \tag{14.19}$$

The term $\phi^i(j)$ in (14.19) is obtained by generating a vector $\Delta^i \in \mathbb{R}^2$ (assuming the search space is of dimension two), where each component of the vector is randomly drawn from the interval $[-1,1]$, and then normalising Δ^i by dividing it by its own Euclidean norm:

$$\phi^i(j) = \frac{\Delta^i}{\|\Delta^i\|}. \tag{14.20}$$

For example, if $(0.6, 0.3)$ were drawn randomly to be Δ^i, the resulting unit vector would be

$$\left(\frac{0.6}{\sqrt{0.6^2 + 0.3^2}}, \frac{0.3}{\sqrt{0.6^2 + 0.3^2}}\right) = (0.894, 0.447).$$

Once the bacterium has moved in a random direction, its SEI cost is updated:

$$f_{\text{sei}}^i(j+1,k,l) = f^i(j+1,k,l) + f_{\text{cc}}(x^i(j+1,k,l), P(j+1,k,l)). \tag{14.21}$$

Then, if $f_{\text{curr}}^i < f_{\text{sei}}^i(j+1,k,l)$ (the location after the tumble and move has lower SEI cost than the bacterium's location before the tumble), run behaviour is simulated. The bacterium iteratively takes a further step of the same size in the same direction as the tumble, checks whether this has lowered its cost value further and, if so, continues to move in this direction until it has taken its maximum number of chemotactic steps, N_s. Figure 14.5 illustrates the tumbling and swimming processes.

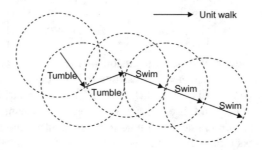

Chemotactic step with tumbling and swimming

Fig. 14.5. Chemotactic step

Reproduction Cycle

After N_{ch} chemotactic steps have been undertaken for the complete population of bacteria, a reproduction cycle is undertaken. In each reproduction cycle, the health of each bacterium is calculated as the sum total of its location costs during its life (over all its chemotactic steps) as follows: $f_{\text{health}}^i = \sum_{j=1}^{N_{\text{ch}}} f_{\text{sei}}^i(j,k,l)$. All the bacteria in the population are then ranked in order of their fitness, with higher costs corresponding to lower health. The $x\%$ (where x is a user-selected parameter) healthiest bacteria split in two (reproduce) at their current location and a corresponding number of less healthy bacteria are eliminated, keeping the total population of bacteria constant. The reproduction cycle is repeated N_{re} times during the algorithm's execution.

Elimination–Dispersal Events

The entire population is subject to a total of N_{ed} elimination–dispersal events. In each of these events, individual bacteria in the population are killed with a probability of p_{ed}. If a bacterium is killed, a new bacterium is generated and randomly located in the search space. If a bacterium is not selected for elimination (this has a probability of $1 - p_{ed}$), it remains intact at its current location.

Parameter Values for the BFOA

The BFOA described above has a number of parameters which the modeller must select. While good choices of parameters will vary depending on the problem to which the algorithm is being applied, Passino [450] used the following in a sample application (see Fig. 14.6): $S = 50$ (number of bacteria), $N_{ch} = 100$ (chemotactic cycles

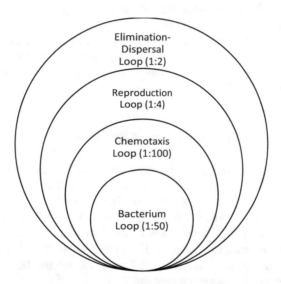

Fig. 14.6. The key loops in the BFOA algorithm

per generation), $N_s = 4$ (number of steps per run), $N_{re} = 4$ (number of generations), $N_{ed} = 2$ (number of elimination–dispersal cycles), $p_{ed} = 0.25$ (dispersal probability per bacterium in an elimination–dispersal cycle) and $C_i = 0.1, i = 1, \ldots, S$ (step size).

In selecting parameter values, a few issues should be borne in mind. As the values of S, N_{ch}, N_{re} etc. increase, the computational cost of the algorithm increases proportionately. The choice of C_i will be domain-specific, depending on the scaling of the solution space that the algorithm is searching. Decreasing the value of C_i during the run will encourage convergence (see [129] for a discussion of a number of schemes for chemotactic step adaptation). The value for N_s determines the maximum number of steps that a bacterium can take in a single run. If $N_s = 0$, the search process becomes a random walk, and as N_s increases, the search process tends towards a gradient descent. Intermediate values trade off the exploration–exploitation balance of the algorithm. The value of N_{re} and the way that bacterial eliminations are implemented impacts on the convergence properties of the algorithm. If a heavy-handed selection process occurs in each reproduction-elimination cycle, the algorithm will tend to converge more quickly but may get trapped in a local minimum. Similarly, if a low value of N_{ch} is chosen, the algorithm will be more prone to getting stuck in a local optimum. The values of N_{ed} and p_{ed} also impact on the exploration–exploitation balance. If both values are large, the algorithm tends towards exploration (random search). The values selected for $w_{attract}$, h_{repel}, w_{repel} and $d_{attract}$ impact directly on the value of $f_{cc}(x, P(j, k, l))$ and therefore define the tendency of the bacteria to swarm. If the attractant width and depth parameters are set to high values, the bacteria will tend to swarm easily, and if extreme values are set for these parameters, the bacteria may prefer to swarm rather than to search for nutrients. On the other hand, if very low values are set for these parameters, the bacteria will search individually and will ignore communications from other bacteria.

A substantial literature on the BFOA has been produced since it was introduced, addressing some issues with the canonical algorithm such as limited scalability and convergence performance. Many papers describing applications of the BFOA have also been published.

14.4 Bacterial Colony Foraging Algorithm

The *bacterial colony foraging* (BCF) algorithm was introduced by Chen et al. [109] and extends the BFOA by adding a cell-to-cell communication process, loosely based on the particle swarm optimisation algorithm (PSO) [314]. The BCF algorithm also embeds a self-adaptive foraging strategy mimicking area-restricted search (Sect. 2.5.2), in which local search behaviour in an area is intensified after the prior detection of food resources in that area. The implementation of this mechanism in the BCF algorithm allows the algorithm to self-adapt the relative balance of exploration and exploitation during the search process.

The foraging process is a function of both social and personal information, the latter arising from current sensory inputs (chemotactic information concerning the quality of the current location) and from a memory of recent foraging successes. Both the social communication and memory mechanisms are implemented quite differently from those in the BFOA.

Algorithm

In the BCF algorithm, a population of simulated bacteria are initially dispersed randomly in the search space and all bacteria are assigned a run length ($C_{initial}$). Next, an iterative loop commences in which each bacterium has the opportunity to move from its current position.

A new direction for this step is calculated, based on the last direction that the bacterium has moved in, modified by the direction between its current position and that of the best position it has ever located, and also modified by the direction between its current position and the global best location ever found by any member of the bacterial colony. Therefore, each bacterium adjusts its tumble angle based on its own experience and based on social information drawn from the whole colony as to the global best position.

The bacterium continues to swim in this new direction using its current chemotactic step size, for up to N_s swim steps, as long as its fitness continues to improve.

The chemotactic step size C_i for each bacterium i is adapted as the algorithm progresses, depending on the search experience of the bacterium. The step size becomes smaller if the bacterium is uncovering fitness improvements, and gets larger if no fitness improvements have been discovered for some time.

The pseudocode for the BCF algorithm (assuming a minimisation problem) is provided in Algorithm 14.3, followed by a discussion of how the key elements of the algorithm are operationalised.

Direction

In each iteration of the algorithm, a bacterium determines the direction in which it will move by blending information from its previous direction, simulating a notion of directional persistence (see Sect. 2.5.2), information concerning its current location relative to the best location it has ever found and information concerning its current location relative to the best location ever found by any member of the bacterial colony. The new direction is calculated as

$$D_i(t+1) = kD_i(t) + \phi_1 R_1(x_i^{best} - x_i(t)) + \phi_2 R_2(x^{best} - x_i(t)), \qquad (14.22)$$

where $D_i(t)$ is the direction component for bacterium i at time t, k is the weight given to the previous direction of the bacterium, ϕ_1 and ϕ_2 are learning coefficients which influence the role of previous experience and social experience in determining the next swim direction, and R_1 and R_2 are randomly drawn from a uniform distribution $U(0,1)$ to ensure that the weights attached to previous and social experience in the direction update are stochastic. The terms x_i^{best} and x^{best} refer to the best location ever found by bacterium i and by the entire bacterial colony, respectively. Thus $x_i^{best} - x_i(t)$ (respectively, $x^{best} - x_i(t)$) is a unit vector indicating the direction between the current location of bacterium i and its previous best (respectively, the global best).

Self-Adaptive Chemotactic Step

Initially, all bacteria have the same chemotactic step length ($C_{initial}$). This can be altered over time for each bacterium based on its search performance. Two other parameters are defined at the start of the algorithm: λ, which is used to reduce the size of the step when local search is being emphasised; and $\varepsilon_{initial}$, which is the initial precision goal. As the algorithm runs, the step length and precision goal parameters can

---

Algorithm 14.3: Bacterial colony foraging algorithm

Set values for C_{initial}, $\varepsilon_{\text{initial}}$, N_s, K_u, ϕ_1 and ϕ_2;
Randomly locate each bacterium $i = 1, \ldots, s$ at a location x_i in the search space;
Set iteration counter $t = 0$;
Determine location of best bacterium (x^{best}) and set personal best location (x_i^{best}) for
  each bacterium to its current location (x_i);
while $t < t_{\max}$ **do**
    for *each bacterium $i = 1$* **to** *s* **do**
        Calculate new direction $D_i(t + 1)$ for each bacterium in turn using (14.22);
        Move bacterium to its new location $x_i(t + 1)$ using (14.23) (simulating tumble
          and run);
        Set $m = 1$ (counter for swim steps);
        while $m < N_s$ *(maximum number of swim steps)* **do**
            if *objective function value at location $x_i(t + 1)$ < objective function value*
            at location $x_i(t)$ **then**
                Take another swim step from current location using (14.23);
                Update the personal best location (x_i^{best}) for that bacterium;
                Update the global best location (x^{best}) if necessary;
                $F_i := 0$;
                $m := m + 1$;
            else
                $F_i := F_i + 1$;
                Terminate the **while** $m < N_s$ loop;
            end
        end
        if $F_i > K_u$ **then**
            Reset $C_i(t + 1)$ and $\varepsilon_i(t + 1)$ to C_{initial} and $\varepsilon_{\text{initial}}$ as per Algorithm 14.4;
        end
        if *improvement in fitness after move step > $\varepsilon_i(t)$ or $F_i \leq K_u$* **then**
            Keep $C_i(t + 1)$ and $\varepsilon_i(t + 1)$ at same values as $C_i(t)$ and $\varepsilon_i(t)$ as per
              Algorithm 14.4;
        else
            Reduce values of $C_i(t + 1)$ and $\varepsilon_i(t + 1)$ to $C_i(t)/\lambda$ and $\varepsilon_i(t)/\lambda$ as per
              Algorithm 14.4;
        end
        Set $i := i + 1$;
    end
    Set $t := t + 1$;
end
Output the best solution found;

alter. These are denoted by $C_i(t)$ and $\varepsilon_i(t)$ for bacterium i at time t. The pseudocode for the adaptation of the step size is outlined in Algorithm 14.4.

Algorithm 14.4: Self-adaptive chemotactic step for BCF algorithm

---

for *each bacterium i* **do**
    if *criterion 1 (exploitation)* **then**
        $C_i(t+1) = C_i(t)/\lambda$;
        $\varepsilon_i(t+1) = \varepsilon_i(t)/\lambda$;
    else
        if *criterion 2 (exploration)* **then**
            $C_i(t+1) = C_{\text{initial}}$;
            $\varepsilon_i(t+1) = \varepsilon_{\text{initial}}$;
        else
            $C_i(t+1) = C_i(t)$;
            $\varepsilon_i(t+1) = \varepsilon_i(t)$;
        end
    end
end

Under criterion 1 in Algorithm 14.4, the bacterium uses a smaller step size to more intensively examine the search regions around its current position. This mode is activated if the bacterium has improved its fitness since the last iteration of the algorithm by more than a threshold amount (determined by the required precision level of ε_i).

If the fitness of the bacterium has not improved for more than K_u consecutive generations (a user-defined parameter), criterion 2 is activated. The value of ε_i is restored to its initial value of $\varepsilon_{\text{initial}}$, thereby lengthening the step size and promoting exploration of the search space.

Movement

The movement of each bacterium at each step is determined by its previous position, its step size and its direction from the last iteration of the algorithm. Hence, its new position ($x_i(t)$) at time t is calculated as

$$x_i(t+1) = x_i(t) + C_i(t)D_i(t). \tag{14.23}$$

Discussion

The BCF algorithm is a relatively compact algorithm which adopts a different operationalisation of both social communication and memory from the canonical BFOA.

Unlike the BFOA the BCF algorithm does not include reproduction or elimination/dispersal steps. However, [109] notes that the self-adaptive mechanism for determining the chemotactic step size for each bacterium in the BCF algorithm plays a similar role in maintaining diversity and avoiding premature convergence of the search process, mirroring the role of the elimination/dispersal mechanism in the BFOA. The cell-to-cell communication mechanism in the BCF algorithm which broadcasts information on the location of the global best location is also considered to mirror the reproductive process in the BFOA.

The process of selection of the direction in the BCF algorithm bears similarities with the PSO algorithm, in that both algorithms use the concepts of a personal best and global best [109]. Therefore, the BCF algorithm can be considered as a hybrid algorithm, combining PSO and area-restricted search mechanisms. In terms of strict biological plausibility, the implicit assumption that a bacterium has awareness of the global best location is questionable, given the limited sensory range of bacteria.

14.5 Bacterial Colony Foraging Optimisation Algorithm

The canonical BCF algorithm was extended by Chen et al. [110] to include a *life cycle* component, with the authors naming the resulting algorithm as the *bacterial colony foraging optimisation* (BCFO) *algorithm*.

In this version of the BCF algorithm, the cell-to-cell communication mechanism and the self-adaptive chemotactic step of the BCF algorithm remain as before, being supplemented by the inclusion of *birth*, *reproduction*, *death* and *migration* processes. The BCFO algorithm takes a stronger inspiration, therefore, from the biological life cycle of bacteria than is the case with the BCF algorithm.

In the BCFO algorithm, each bacterium has an associated nutrient level and energy level. If a better location is found during the chemotactic step, the nutrient level of a bacterium is increased by 1; otherwise, it is decreased by 1, as follows:

$$N_i(t+1) = \begin{cases} N_i(t)+1, & \text{if } f(x_i(t+1)) < f(x_i(t)), \\ N_i(t)-1, & \text{otherwise,} \end{cases} \tag{14.24}$$

where $f(x_i(t))$ is the fitness of the ith bacterium at time t (assuming a minimisation problem, a lower fitness is better) and $N_i(t)$ is the nutrient level of bacterium i at time t. Initially, all bacteria have a nutrient level of zero and the level only increases when they move to a better location, simulating a gain in energy from moving to a new location which has food resources.

The energy level of each bacterium ($E_i(t)$), which determines whether it reproduces, dies or migrates, is calculated as

$$E_i(t) = \eta \frac{H_i(t)}{\sum_{j=1}^{S(t)} H_j(t)} + (1-\eta) \frac{N_i(t)}{\sum_{j=1}^{S(t)} N_j(t)}, \quad \eta \in [0,1], \tag{14.25}$$

where

$$H_i(t) = \frac{f(x_i(t)) - f_{\text{worst}}(t)}{f_{\text{best}}(t) - f_{\text{worst}}(t)} \qquad (14.26)$$

with $f_{\text{worst}}(t)$ and $f_{\text{best}}(t)$ being the current (at time t) worst and best fitnesses, respectively, of the bacterial colony.

If bacterium i accumulates sufficient energy during its foraging process, it reproduces by splitting into two bacteria, located in the same position. In this case, the overall population of the colony increases by one. The reproductive step occurs when the energy level exceeds a threshold, calculated using

$$E_i(t) > \max\left\{ E_{\text{split}}, E_{\text{split}} + \frac{S(t) - S}{E_{\text{adapt}}} \right\}. \qquad (14.27)$$

The values of the parameters E_{split} and E_{adapt} are user-defined, and $S(t)$ is the current colony size. As the value of E_{split} increases, a bacterium requires more energy before reproduction can take place. As the population level increases, the value of $S(t)$ also increases. Therefore, the value of $(S(t) - S)/E_{\text{adapt}}$ increases, again making it harder for reproduction to take place. This simulates a crowding mechanism, whereby reproduction rates decline as the density of the population increases.

If the energy level of a bacterium drops below a threshold, calculated as

$$E_i(t) < \min\left\{ 0, \frac{S(t) - S}{E_{\text{adapt}}} \right\}, \qquad (14.28)$$

the bacterium dies and is removed from the colony (population). Unlike the BCF algorithm, the population size in the BCFO algorithm can alter over time.

A migration mechanism can also be implemented, whereby a bacterium either is migrated to a new random position (with a probability of p_{mig}) and has its nutrient value reset to zero if it fails the test in (14.28), or is killed (with a probability of $1 - p_{\text{mig}}$). The parameter p_{mig} is a user defined *migration* probability.

Embedding these steps into Algorithm 14.3, after each bacterium has completed its chemotactic movement (in which its nutrient level is altered) and self-adaptation phase, its energy level is assessed to determine whether it reproduces. If there is no reproductive step, the bacterium's energy level is assessed to see whether a death or migration event is triggered. The population size adapts dynamically following birth or death processes. A series of benchmark experiments in [110] used a variety of parameter settings, with the following values being typical: $E_{\text{split}} = 0.1$, $E_{\text{adapt}} = 50$, $\eta = 0.5$, $p_{\text{mig}} = 0.2$, $\varepsilon_{\text{initial}} = 100$, $K_u = 20$, $\lambda = 10$, $k = 0.6$, $\phi_1 = \phi_2 = 2$ and $N_s = 4$. A full description of the algorithm is provided in [110].

14.6 Viral Algorithms

While bacteria are relatively simple unicellular organisms, viruses are an even more basic biological agent. Viruses are found in every environment on earth that contains cellular organisms, with the total number of distinct types of virus being very large, numbering in the millions [80]. Most of the genetic diversity on earth resides in viral genomes and only a very small portion of these have ever been studied [381].

Although it is known that viruses date from ancient times, we do not have a clear picture of how they initially developed. Theories concerning their origin include the *regressive theory*, which suggests they stemmed from free-living cells that evolved to parasitise other cells, eventually losing genes not required for this task. The *cellular origin theory* suggests that viruses may have evolved from strands of DNA or RNA that escaped from cells. An alternative theory is that viruses may have arisen as self-replicating entities *before* the last universal common ancestor came into existence (i.e. before the origin of all cellular life) [266].

A virus is usually composed of an RNA genome (in the case of an *RNA virus*) or a DNA genome (in the case of a *DNA virus*) which carries genetic information, encapsulated in a protective protein coat called a *capsid*. In some viruses, typically viruses that infect animal cells, the protein coat is in turn surrounded by a lipid (fat) envelope. Unlike bacteria, viruses cannot directly reproduce by processes of cell division, and must *infect* a host cell in order to reproduce. Once a virus takes over a suitable host cell, it utilises the cell's resources, both energy and cellular machinery, to manufacture and assemble many more viral particles. These eventually leave the host cell and go on to infect other cells.

14.6.1 Viral Life Cycle

Although there are multiple distinct types of virus each with their own life cycle, we can consider the viral life cycle in stylised form as consisting of five stages (Fig. 14.7) [381].

In the initial stage of infecting a host cell, proteins on the external surface of the viral particle bind with specific molecules (receptors) on the surface of the host cell. Viral particles are quite specific in terms of the host cells that they will bind to and cannot infect a cell unless they attach to it. Once a viral particle has attached to a target host cell, it penetrates the cell membrane, with either the entire viral particle or just its genome (in the case of bacteriophages) entering the host cell. During viral replication, the cellular machinery of the host cell is hijacked in order to produce viral nucleic acid and proteins, which are then assembled into viral particles. Finally, the viral particles are released from the host cell, either through the bursting of the membrane surrounding the host cell (a process called *cell lysis*) or by *budding*, a process by which virus particles are slowly released from the surface of the host cell with the host cell remaining alive (i.e. a persistent infection). Cell lysis can result in the release of 100 or more new viral particles.

Normally, the end point of viral infection is the death of the host cell. However, in some cases, the virus genome can become incorporated into the host cell's own chromosome (a process called *lysogeny*). Typically, the virus then becomes dormant, with its genome being replicated as the host cell itself replicates. The dormant virus can later become reactivated by external stimuli such as exposure to chemicals or radiation.

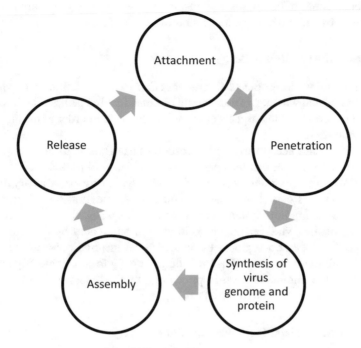

Fig. 14.7. Virus life cycle

Viruses can infect all cellular life, including animals, plants and bacteria. Viruses that infect bacteria and archaea are known as *bacteriophages* (or *phages* for short). Cellular organisms have evolved a number of defence mechanisms against viral infection, including innate and adaptive immune defences in the case of animals. The interplay of immune mechanisms and viruses has produced a coevolutionary arms race, with viruses evolving in order to overcome immune defences and immune defences also evolving in order to reduce the chance of viral infection.

Are Viruses Alive?

There is a long-standing debate as to whether viruses should be considered as living organisms. Although they possess genes and can evolve via processes of natural selection, viral particles are obligate intracellular parasites as they are not capable of independent metabolism and require a host cell to reproduce. Some scientists consider viruses to be organic structures that interact with living organisms rather than being a form of life in their own right, with the phrase 'organisms at the edge of life' [501] sometimes being applied. The difficulty in unambiguously defining 'life' contributes to the difficulty in classifying viruses.

Viruses are not the only agents that require host cells to reproduce, with some bacteria, including Rickettsia and Chlamydia, also being host cell parasites. Many other organisms, including some plants, insects, worms and fungi, are also parasitic, requiring hosts in order to complete their life cycle.

Viruses and Algorithmic Design

Notwithstanding philosophical discussions concerning the definition of what it means to be 'living' and whether viruses should be considered as being 'alive', we can draw inspiration from varying aspects of the viral life cycle in order to design algorithms for optimisation.

Viral particles need to appropriate resources from host cells in order to reproduce and, consequently, we can metaphorically consider viral particles as 'foraging' for host cells. Other important aspects of viral propagation are evolutionary mechanisms by which viruses can change their genome over time, and the interplay between viruses and the immune system in the organisms they attack.

Many distinct viral-inspired algorithms can be created by drawing on various elements from the above processes. Indeed, viral-inspired optimisation algorithms overlap with other families of natural computing algorithms including evolutionary algorithms and immune-system-inspired algorithms. Readers requiring background on either of these areas are referred to [77].

14.6.2 Viral-Inspired Optimisation Algorithms

The literature on viral-inspired optimisation algorithms can be divided into a number of strands. One strand, typified by the *virus evolutionary genetic algorithm* [340], simulates a population of host cells, each representing a potential solution to the problem of interest (for example, a host cell could be represented by a binary- or real-valued vector, where the components of this vector are parameters of a model that we are seeking to optimally fit to some test data), which are under attack from a population of viral particles. During the optimisation process both the host cells and the viral particles coevolve, with the viral replication mechanisms of reverse transcription and transduction helping to generate diversity in the population of host cells and also in the population of viral particles. The aim of the simulated coevolutionary process is to generate a host cell which is a good solution for the problem of interest.

In the second strand of these algorithms, the *virus optimisation algorithm* [292, 355, 373], each viral particle represents a potential solution to the problem of interest and the host cell represents the entire search space. A viral replication process is simulated in which stronger viral particles, corresponding to better solutions, are more likely to survive and replicate. The newly created particles are generated at local random locations around stronger viral particles, corresponding to an exploitation mechanism. If a new particle is generated from a poorer-quality viral particle, the range in which it can be generated is larger, corresponding to a greater degree of exploration. The range within which the new viral particles are randomly generated decreases as the algorithm runs, leading to finer-grained search. Initially, when the algorithm commences, there are a small number of viral particles. As the number of these particles increases owing to the replication step, a basic immune system concept is implemented which kills off poorer-quality viral particles in order to keep the viral population to a maximum (user-defined) size.

A number of other optimisation algorithms inspired by aspects of the virus life cycle have been implemented, including (amongst others) the virus colony search algorithm [351] and the virulence optimisation algorithm [414]. While it is beyond the scope of this book to provide a detailed discussion of each of the above viral-inspired optimisation algorithms, we introduce the oldest such algorithm (the virus evolutionary genetic algorithm) for illustrative purposes. As will be seen, whilst the algorithm embeds a foraging metaphor it also draws on other metaphors, including an evolutionary metaphor.

Virus Evolutionary Genetic Algorithm

The *virus evolutionary genetic algorithm* (VEGA) was first proposed by Kubota et al. in 1996 [340]. In the VEGA, a canonical genetic algorithm is hybridised by adopting some aspects of viral replication, specifically *reverse transcription*, whereby part or all of a viral genome is inserted into the genome of its host cell, and *transduction*, whereby part of the genome of a host cell is extracted by a virus and later potentially carried by the virus into the genome of another host cell.

In the VEGA there are two populations, a population of fixed-length bitstrings (i.e. host cells) representing potential solutions to the problem of interest, and a population of shorter bitstrings representing virus particles. Just like the canonical genetic algorithm, the VEGA implements selection, crossover and mutation on the population of potential solutions (hosts). A steady-state genetic algorithm is implemented, with two parents being replaced by the two children resulting from the selection, crossover and mutation steps. The evolutionary process is then supplemented by a virus infection step as in Algorithm 14.5. In the description of the infection step below, we outline a slightly modified version of the original VEGA, drawing on the version described by Fountas et al. [186].

Algorithm 14.5: Virus evolutionary genetic algorithm

Initialise the population of fixed-length bitstrings representing potential solutions (i.e. host cells);
Initialise the population of bitstrings representing virus particles;
repeat
    Select host cells (parents) for breeding from the current population;
    Generate new cells (offspring) from these parents using crossover and mutation;
    Implement virus infection step;
until *terminating condition*;
Output the best solution found;

As mentioned above, the virus infection step consists of two components, reverse transcription and transduction. In the reverse transcription component, new (variant) individuals are created from the host population using the viral particles, i.e. *infected* host cells are created. While this can be operationalised in many ways, including random selection of hosts for infection, we adopt the following approach.

For each host in turn, the host is presented with each virus and virus i stochastically infects the host if a randomly drawn variable $r \leq r_i^{inf}$, where r has a uniform distribution on $(0, 1)$ and r_i^{inf} is the *infection rate* of virus i. A virus with a high value of r_i^{inf} is therefore likely to infect more hosts. If a host is to be infected, the reverse transcription step is operationalised by copying the bitstring of the virus i into the host's genome. After all hosts have been subject to potential infection, the fitness of any infected host is (re)calculated and if it is higher than that of the preinfected host, the infected host survives into the next generation, replacing its preinfected version.

In addition to an infection rate, each virus i also has an associated fitness or *infection strength*, denoted as f_i^{virus}. Assume that f_{j*}^{host} is the fitness of host j after infection by virus i, and that f_j^{host} is the fitness of the host prior to infection. The value of f_{ij}^{virus} is given by

$$f_{ij}^{virus} = f_{j*}^{host} - f_j^{host}.$$ (14.29)

The value of f_i^{virus} is then calculated (14.30) as the overall sum of fitness changes it produces in each host (j), summed over the set S of all hosts which are infected by virus i:

$$f_i^{virus} = \sum_{j \in S} f_{ij}^{virus}.$$ (14.30)

If this value is positive, it implies that the virus has a positive effect overall on its hosts. If a virus has a positive fitness, its infection rate r_i^{inf} (i.e. the chance it infects a given host) is increased by a constant a (14.31), otherwise it is decreased by a, with [340] suggesting an initial infection rate of 0.05 and a maximum infection rate of 0.10 based on a virus population size of 10. Obviously, $0 \leq r_i^{inf} \leq 1.0$ indicates the viable boundary values for r_i^{inf}:

$$r_i^{inf}(t+1) = \begin{cases} (1+a)r_i^{inf}(t), & f_i^{virus} > 0, \\ (1-a)r_i^{inf}(t), & f_i^{virus} \leq 0. \end{cases}$$ (14.31)

The second element of the infection step consists of the application of a simulated transduction operator. This is operationalised using a *life force* parameter l_i for each virus. The value of this life force alters during the algorithm according to

$$l_i(t+1) = r \cdot l_i(t) + f_i^{virus}$$ (14.32)

where t is the iteration number and r is the *life reduction rate* (i.e. a decay parameter) set at 0.9 in [340].

If $l_i(t+1)$ takes a negative value this implies that the virus is not a positive influence on its hosts and, as a result, the viral individual i transduces a new substring (i.e. overwrites its current bitstring, thereby creating a new viral genome) by copying some of the bitstring from a randomly selected host. In other words, if a virus is not helping to generate better solutions, it is replaced. Otherwise, the virus transduces a partially new substring from one of the infected host individuals.

As the algorithm iterates, the string length of viral individuals gradually extends over time. This allows the generation of more complex virus particles, which can capture more widely separated epistatic relations between potential solution components on a viral particle.

In summary, the VEGA simulates the coevolution of a population of host cells and viral particles. The algorithm uses ideas from the replication process of the viral life cycle in order to generate diversity in the population of candidate solutions and the population of virus particles. Readers requiring more detail on the implementation of the VEGA, including suggested parameter settings, are referred to [340, 338, 339] and related work [185, 186].

Related Algorithms

A number of studies have applied the above concept of viral infection of a host cell to hybridise other optimisation algorithms, including PSO. In *virus evolutionary particle swarm optimisation* (VEPSO), there are two populations or swarms. One population consists of the candidate solutions (i.e. hosts) and the second population consists of simulated viral particles. As was the case for the VEGA, these viral particles are substrings of the candidate solutions. In each iteration of the VEPSO algorithm, the velocity and position update steps of canonical PSO are applied to the population of candidate solutions. This is then followed by a simulated virus infection step. This works in a similar manner to that of the VEGA, with candidate solutions whose fitness improves as a result of their infection being updated and surviving into the next iteration. Further details on the implementation of VEPSO can be found in [205].

14.6.3 Communication Among Viruses

A common theme running though many foraging settings is the social transmission of information between foragers. In group settings, quorum-based decision making can enable a group of even simple organisms to come to a consensus as to what to do next (see Sects. 8.1.2, 9.11 and 10.4).

There is emerging evidence that viruses, specifically temperate phage viruses, can metaphorically communicate with phages in other infected bacterial cells and effectively engage in quorum-like decision making concerning their development in individual host cells [170].

After infecting a host cell, temperate phage viruses have two life cycle options, either to enter the lytic development pathway, which produces many new viral particles and kills the host cell, or to follow a lysogenic pathway, wherein the phage genome is incorporated into the genome of the host cell. Erez et al. [170] found that during infection, the phage (phi3T in their study) caused the host bacterium to produce and secrete a small-molecule peptide. When this molecule was absorbed by another bacterium infected by the same phage, the probability that the virus in that cell would follow a lysogenic pathway was increased. At high concentrations of the peptide, the chances of cell lysis were substantially reduced and most infected cells followed a lysogenic pathway.

In essence, this process provides functional intercellular communication between the phages in the infected bacterial cells resulting from chemicals that the phages caused their infected host cells to emit. The mechanism is adaptive for the viral population as cell lysis would dramatically, and rapidly, increase the density of virus particles in a local region. As potential host cells were killed off, there would be fewer remaining potential hosts available. Increasing the propensity of phage infections to turn dormant would allow for the potential dispersion of the infected host cells, with viral production resuming when environmental conditions were more favourable.

Similar communication systems using differing peptides have been found in more than 100 other phages, indicating that this capability is quite widespread [170]. The specificity of the peptide used in each case reduces the chance of inadvertent crosstalk between different types of phage. As commented by Davidson [130, p. 467] this allows the '... phages to "speak" different molecular languages and so convey messages only to their own kind'. As yet, such direct viral communication capabilities have not been incorporated into viral-inspired optimisation algorithms.

14.7 Summary

This chapter has described four optimisation algorithms which are drawn from the foraging processes of bacteria and viruses. The primary underlying principle is the sensing of a local chemical gradient by a bacterium and the stochastic following of this to the source of an attractant. Of course, there are limits to a simple sensory-gradient-climbing foraging strategy such as bacterial chemotaxis. While the strategy can work well in finding a strong source in a nonturbulent environment, it will perform less well in an environment characterised by turbulent flows, where the sensory signal is discontinuous and patchy [308].

Recognising this, variants on the canonical chemotaxis algorithm include additional features such as a simulated bacterial communication mechanism and/or a more holistic model of the bacterial life cycle. The algorithms included in this chapter have been selected to illustrate examples of each of these categories, with evident scope to design a wide array of related algorithms with slightly differing operationalisations of each mechanism. The chapter also provides a short introduction to the subfield of viral-inspired optimisation algorithms, and notes that, metaphorically, the process of finding and infecting a host cell can loosely be considered as foraging.

As with many other organisms mentioned in this book, our understanding of the foraging and communication processes of bacteria (and virus particles) is still maturing. One aspect of bacterial behaviour not yet incorporated into optimisation algorithms is that individual bacterial colonies live in a rich ecology with other microorganisms, including other bacterial colonies. These microorganisms can engage in complex interactions (including competition and cooperation), leading to the transmission of information concerning the environment via cues and perhaps signals. It is also known that some species of bacteria can identify kin. Kin recognition can support the development of cooperative behaviour, as the cooperation can be directed to genetic relatives. Thus far, only a few instances of kin recognition have been researched in microbial species, with [543] finding evidence of this capability in *Bacillus subtilis*, a soil bacterium in which nonkin bacteria were found to exhibit clear boundaries between groups whilst swarming.

Apart from bacteria, many other organisms also use chemical information to guide their foraging process. As mentioned above, local chemical gradient cues pointing directly to the source of a food resource are not always available, as chemical signals may be diffused via a flowing medium such as air or water. Therefore, signals from a target item may only be detectable at intermittent intervals during the search process. In these conditions, if the chemical signal is lost, many organisms, including birds and flying insects, engage in a zigzag search pattern across the direction of flow of the medium in order to reacquire the signal, at which point they again begin to move towards the source. This search process is known as *casting* and has been used as the inspiration for the design of a search algorithm [587], with subsequent development of this algorithm in [308].

In addition to bacterial communication via chemical signalling, it has recently been discovered that bacteria may also be capable of communication and recruitment via electrical signalling [469, 271]. Previously it had been thought that only multicellular organisms were capable of this. The study [469] found that messages can travel via ion channels (proteins on a cell's surface which control the flow of charged particles) into and out of a cell, with these messages changing the state of neighbouring bacteria, inducing them to release such particles and thereby pass on the message to their neighbours in turn. Coordinated behaviour in large groupings of bacteria was found to result, with the authors speculating that these signalling mechanisms could play a role in bacterial foraging. Although further work is required to better understand these processes which bear some similarity to neuronal signalling, these mechanisms could provide possible inspiration for the design of new bacterial foraging algorithms.

Slime Mould and Fungal Foraging Algorithms

Slime mould is a shorthand name for a number of eukaryotic organisms which have similarities in their life cycles, with their primary reproductive stage entailing the formation of a fruiting body and the release of spores. The organisms can live freely as single amoebae, aggregating together for spore production and dispersal. The phrase *slime mould* refers to the appearance of these organisms during the aggregative phase of their life cycle, in which they look like a gelatinous slime.

While slime moulds do not appear at first glance to be capable of particularly sophisticated behaviour, on closer examination their problem-solving capabilities are quite remarkable despite their lack of a brain or nervous system.

Research on slime mould computing over the past decade has taken two main tracks. The first adopts physical slime mould as a biological (or living) computer, examining the capability of slime mould to directly solve computational problems. The second track, which we concentrate on in this book, seeks to develop computational algorithms based on simple models of slime mould behaviours.

In this chapter, we initially introduce the main groupings of slime moulds and outline some of their foraging behaviours. Relevant behaviours include how a slime mould moves in order to access and harvest food resources efficiently, and the capability of a slime mould to assess food quality, to create a memory of already-visited areas and to trade off risk versus return when making foraging decisions. We then discuss the topologies of foraging networks that slime moulds can create when harvesting multiple food sources simultaneously, and describe how slime mould has been used to solve a number of well-known network and maze problems (Sect. 2.6 provides a short introduction to networks and related terminology). We introduce a number of optimisation algorithms which have been inspired by behaviours of slime moulds.

Fungi are another type of organism which can form complex structural networks while foraging, in response to local environmental conditions [250]. Many species of fungi explore their environment in order to discover resources such as nutrients and water by growing thin tube-like extensions called hyphae. These hyphae branch as they grow, creating a tree-like structure. Like slime moulds, fungi are nonneuronal and their morphological self-organisation results from emergent phenomena rather

© Springer Nature Switzerland AG 2018
A. Brabazon, S. McGarraghy, *Foraging-Inspired Optimisation Algorithms*,
Natural Computing Series, https://doi.org/10.1007/978-3-319-59156-8_15

than from 'top-down' control. As yet, there have been very few studies which have attempted to design computational algorithms drawing inspiration from the foraging behaviours of fungi. In this chapter we provide a brief introduction to the kingdom of fungi and explore aspects of the search and foraging behaviours of some species of fungi. We introduce some previous studies which have attempted to mathematically model fungal foraging behaviours, and, finally, some suggestions for future work are provided.

15.1 Groupings of Slime Moulds

Figure 15.1 outlines the three main groupings of slime moulds. Myxomycetes or

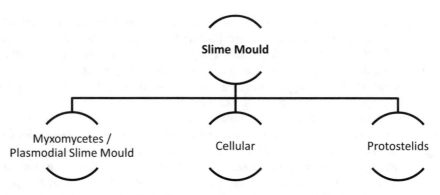

Fig. 15.1. Three groupings of slime mould

plasmodial slime moulds have a relatively complicated life cycle, characterised by two different trophic (feeding) stages, followed by a spore-producing reproductive stage [547]. In the first trophic stage, individual amoeboid cells resulting from germinated spores reproduce by binary fission and build up a population of unicellular microorganisms. In the second trophic stage, two compatible haploid amoeboid cells fuse, creating a diploid zygote, which feeds and undergoes repeated mitotic nuclear division, giving rise to a plasmodium. In time, the plasmodium produces fruiting bodies containing spores [547].

The plasmodium consists of a bag of cytoplasm encased within a thin membrane which acts a single organism. Unusually, the organism in this state consists of a large single cell with thousands of nuclei (i.e. it is a unicellular, multinucleated eukaryote [66]). About 875 species of plasmodial slime moulds are known to exist [547] and they can grow to a considerable size, extending up to $930\,\mathrm{cm}^2$ in the case of *Physarum polycephalum* [347]. During the motile phase of their life cycle, plasmodia are capable of movement by means of cytoplasmic streaming (the internal movement of cell contents) at rates of up to 5 cm/h [347].

The cellular slime moulds have a different life cycle from the plasmodial slime moulds. Taking the case of the *Dictyostelida* cellular slime moulds, they spend most of their lives as separate single-celled amoebae. When food runs out and the amoebae begin to starve, they signal to each other by releasing chemoattractants such as cyclic adenosine monophosphate (cAMP). They then coalesce by chemotaxis to form an aggregate that becomes surrounded by a slime sheath. The aggregate forms a fruiting body which eventually releases many spores. These disperse, allowing the life cycle to recommence.

The smallest grouping of slime moulds, the *protostelids*, has only 35 described species, although it is thought that there are probably as many undescribed species [547]. The protostelids consist of amoebae that are capable of making simple fruiting bodies consisting of a cellular stalk topped by one or a few spores. These organisms are very small and can typically only be seen with the aid of a microscope.

15.2 Foraging Behaviours of Slime Moulds

As with all organisms, a key behaviour of slime moulds is foraging for food resources. Species of slime mould feed on a variety of resources, including dead plant material, living bacteria, yeasts and fungi, by engulfing their prey and digesting the item (*phagocytosis*). Unlike fungi, the digestion process is internal to the organism.

To forage effectively, these organisms need to be able to process sensory information and integrate this with internal state information in order to decide for how long to exploit already-discovered food resources and when and where to move to next in the environment. Perhaps the best-studied plasmodial slime mould is *Physarum polycephalum*. Despite not having neuronal tissue, this organism has been found to exhibit complex foraging behaviours.

In the following subsections we consider a number of characteristics of *Physarum polycephalum*, including its ability to move, to assess food quality and to engage in simple learning. We also discuss the ability of the slime mould to assess risk, and its decision-making biases. All of these factors impact on its foraging capabilities.

15.2.1 Mobility

Under idealised conditions of plentiful resources, *Physarum polycephalum* plasmodia are sedentary and grow steadily. If they are located on non-nutrient-bearing substrates they search for food resources by migrating at a rate of up to a few centimetres per hour directed by external stimuli such as gradients of sugar and proteins [166].

When a slime mould chemotactically senses attractants such as food via the binding of chemicals to receptor molecules presented on its outer membrane surface, cytoplasm flows towards the attractant, thereby inducing movement of the organism in the direction of the food [482]. More technically, when a plasmodium comes in contact with food, the oscillation frequency of the region of the plasmodium closest to the food increases. In turn, this promotes an increase in the oscillation frequency of nearby biomass in the plasmodium, creating a positive feedback loop, and these oscillations cause biomass to flow towards the attractant [348].

When a slime mould comes into contact with a food item, it fully or partly engulfs it by covering it with biomass, later resuming exploration by extending pseudopodia into the surrounding environment while remaining in physical contact with the initial food source [347].

15.2.2 Food Quality and Foraging

Physarum polycephalum is able to assess the quality of current food resources and use this information to determine how much time to allocate for exploration of the environment for new resources, and to determine what search strategy to use during exploration [347]. Lower-quality food sources produce earlier resumption of search and earlier abandonment of the food source, even in some cases before the food source is exhausted. This behaviour is consistent with the marginal value theorem (Sect. 2.1.2). The quality of a food source influences subsequent search behaviour, with higher quality food sources promoting intensification of local search around that food source, and lower quality food sources encouraging explorative search of the environment elsewhere [347].

Another common problem faced by foraging organisms is the optimal diet problem (Sect. 2.1.2), wherein an organism needs to ingest multiple items of food in order to obtain all necessary nutrients. A study [166] investigated whether *Physarum polycephalum* is capable of solving this problem by providing an environment consisting of food patches with differing nutritional composition. The results indicated that the organism was able to grow to contact patches of different nutrient quality in the precise portions necessary to compose an optimal diet. The nutrients absorbed by different parts of the slime mould are circulated internally, along with various chemical signals, by means of cytoplasm streams, ensuring that all parts of the organism obtain their required nutrition.

15.2.3 Spatial Memory

Given the nature of the food resources consumed by a slime mould, it would take some time for a previously harvested area to regenerate new food resources [480]. As *Physarum polycephalum* moves in its environment, it deposits nonliving extracellular slime. It has been found that *Physarum polycephalum* preferentially avoids areas if they are found to be so marked [482]. This response is adaptive, as such regions are likely to have been previously harvested of resources and therefore represent a poor choice of foraging location.

If extracellular slime is detected during foraging search, it could have been deposited by that specific slime mould. In this case, the ability to detect that an area has already been visited creates a form of external memory, somewhat akin to the pheromone deposits of trail-marking ants. The use of an external spatial memory frees the forager from having to internally store information on which areas have already been searched. It has been speculated that external memory processes may be

a functional precursor to the development of internal memory systems, allowing biological organisms with primitive information-processing systems to solve complex
tasks requiring spatial memory [482].

Slime moulds coexist with many other species of amoebae, all engulfing and digesting similar prey of yeast, bacteria and other microorganisms. However, the food
preferences of each species vary somewhat and it would be useful for a slime mould
to be able to distinguish between extracellular slime deposited by a conspecfic and
that deposited by a heterospecific. Investigation of this issue found evidence that
Physarum polycephalum can indeed distinguish between the two categories, with
the avoidance response being less pronounced when the slime had been deposited
by other species [480]. This is plausible, as heterospecifics will not overlap completely in terms of food preference with *Physarum polycephalum*. The mechanism
also indicates that *Physarum polycephalum*'s behavioural response to the presence
of extracellular slime is flexible and not just a simple stimulus–response pathway
[480].

More generally, there is evidence that slime moulds are capable of simple learning, specifically habituation (Sect. 4.1.2). A study by Boisseau et al. [66] found that
Physarum polycephalum could become habituated to repeated presentations of non-
toxic substances, including caffeine and quinine, while foraging, despite displaying
initial aversive behaviour to those substances. The exact mechanisms behind these
learning capabilities are not yet understood but it is speculated that the experience
could have resulted in the epigenetic modification of the slime mould's nuclei (i.e. a
physical embodiment of the learning event). A follow-on study by Vogel and Dussutour [595] found that this learning could be shared with another slime mould when an
informed slime mould fused with one which had not been habituated to the stimulus
(i.e. the fused slime mould, consisting of both an informed and a naïve constituent,
exhibited habituation behaviour).

15.2.4 Foraging and Risk

An important capability when foraging is the ability to make trade-offs between the
potential returns (and risks) of specific foraging strategies. For example, some territories may contain plentiful food resources but may be subject to higher predation
risk.

Physarum polycephalum is photophobic, as exposure to strong light can damage
the organism. In experiments where *Physarum polycephalum* was presented with
food sources at different locations, with varying strengths of light field illuminating
the space between the locations, the tubes connecting the slime mould's biomass
on each food source were found to connect the food sources not by the shortest
path between them but according to the relative strength of the light fields in the
environment, indicating an ability to trade off the risk of damage to the organism
from light against the cost of creating a longer path between the food sources [425].
The increased biomass allocated to create longer paths reduces the food absorption
capacity at each food source (as less biomass remains for placement on the food) and
has a cost in terms of lower nutrient absorption capability per time period.

15.2.5 Decision-Making Biases

In traditional models of rational economic decision making, the information input into a decision can be characterised as a vector of attributes. The decision-making process then consists of decision makers assigning an absolute value to each attribute and the calculation of a weighted sum of these values. The highest-scoring decision alternative is then selected.

However, it is known that many animals, including humans, frequently use a *comparative valuation model* rather than an *absolute valuation model* in decision making, and this can lead to irrational decision making in that a poorer decision alternative ends up being chosen. Under a comparative valuation model, a ranking of each of the attributes between the decision alternatives being considered is undertaken, and the sum of these rankings across each attribute is used to decide the final choice. In this process the ranking of an attribute, and therefore the decision, depends on what is in the decision choice set and is not based on an absolute value. Although this approach can result in economically irrational decisions (i.e. adding a poorer alternative to a decision set can change the decision made), it has the advantage that it is less computationally intensive as it only requires relative assessments of attributes.

Physarum polycephalum has also been shown to be subject to similar decision-making biases, using comparative rather than absolute values, and therefore shares this bias with other organisms [348]. This was the first case of the discovery of comparative valuation rules in a unicellular organism, and it is speculated that use of such rules may be widespread amongst biological decision makers [348].

15.2.6 Foraging Networks in *Physarum polycephalum*

Foraging networks were introduced in Sect. 2.6.2.

If a number of small sources of food are presented at various positions on a surface to a starved plasmodium of *Physarum polycephalum*, it endeavours to reach them all. In nutrient-poor environments, the plasmodium grows by extending pseudopodia towards nearby chemoattractant sources, producing a parallel search process guided by the chemoattractant gradients in the local environment. In nutrient-rich conditions, the plasmodium grows radially in all directions.

The organism attempts to optimise the shape of its biomass in order to facilitate the harvesting of food resources. As the plasmodium has a limited mass, it has to decide how much of this mass to place over each food source in order to absorb nutrients (the more biomass placed over a food source, the faster it can be absorbed) and how much to allocate to the construction of tubes between the food sources in order to maintain connectivity and intracellular communication throughout the organism [426].

15.2.7 Slime Mould Foraging Behaviours: a Summary

In summary, *Physarum polycephalum* manifests sophisticated foraging capabilities including adaptive movement in its environment, an ability to assess the quality of food sources, a primitive memory of already-harvested locations, and a capability to trade off risk and return when foraging. Slime moulds also appear to exhibit similar decision-making biases to those found in more complex organisms. As slime moulds do not possess a nervous system, they rely on intracellular information processing and protoplasmic flow to integrate local sensory information and to generate a response to stimuli [579].

15.3 Slime Mould Computation

A slime mould can be used to undertake computation in two ways, either as a biological computer (biocomputer) or by using elements of its behaviour to inspire the design of computational algorithms. Each of these is discussed below.

15.3.1 Slime Mould as a Biocomputer

Although slime moulds are relatively simple organisms, they are capable of both sensory and locomotive behaviours, and are also capable of adapting the morphology of their protoplasmic tube network in response to environmental conditions. These capabilities have resulted in slime moulds such as *Physarum polycephalum* being used as a living computational material, which can be programmed by appropriate placement of external stimuli such as food or repellents such as chemicals or light [291], with the result of the program being the structure of the final slime mould network.

Graph and network design problems are a natural fit for slime mould computing applications, as the growth and adaptation of slime mould naturally forms edges (protoplasmic tubes of slime mould) between discrete nodes (which can be located on nutrients such as food flakes) [291]. This has led to multiple studies which have applied slime mould as a biocomputer to solve these problems.

In [426] three separate food sources were presented to *Physarum polycephalum*, located at the vertices of a triangle. In repeated experiments, the resulting tubular networks connecting the food sources were often similar in design to a Steiner minimum tree approximating the mathematically shortest path connecting the vertices. Further investigation of the ability of the slime mould to construct smart networks indicated that it can simultaneously meet multiple requirements of a smart network, trading off total length of connections against tolerance of accidental disconnection of the tubes [427, 479].

Solving Maze Problems

Physarum polycephalum has also demonstrated a capability to solve maze problems, which are a standard test of intelligence in animal psychology [258]. In an experiment [428], a large piece of plasmodium was cut into a number of smaller pieces which were then distributed at intervals within a physical, two-dimensional maze structure. Initially, the pieces coalesced to form a single organism that filled the entire maze but when nutrient blocks of oat flakes were placed at two locations in the maze, the pseudopodia reaching dead ends shrank, resulting in a single thick pseudopodium spanning the shortest path between the nutrient blocks, indicating the capability of the slime mould to solve the maze by adapting its physical structure (Fig. 15.2).

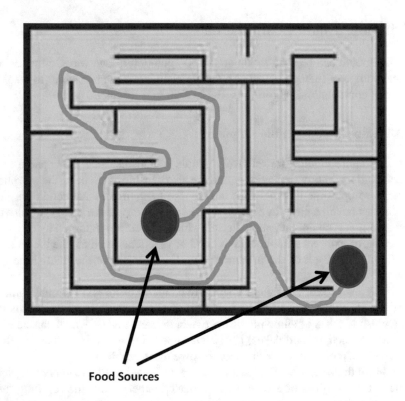

Food Sources

Fig. 15.2. Maze with two food sources. A slime mould places mass on each food source with each source being connected via protoplasmic tubes of slime mould approximating the shortest path in the maze between the sources

Reproducing Human-Designed Transport Networks

An interesting variant on the above studies investigated the capability of *Physarum polycephalum* to design transport networks, with the results being benchmarked against existing human-designed transport networks. In [567], food sources were placed on a flat, wet dish at locations corresponding to 36 major cities in the greater Tokyo area and bright light was shone at appropriate places in the environment in order to simulate the challenges posed to network designers by geographic features such as mountains and waterways. The slime mould initially filled the entire space with plasmodium and then thinned to describe a network linking the food sources. In repeated experiments, many of the networks bore notable similarity to the actual rail system linking Tokyo and the outlying cities. Analysis of the networks produced indicated that they had characteristics similar to those of the rail network in terms of cost, transport efficiency and fault tolerance [567].

The above study used a two-dimensional representation of the transport nodes, and an open question was whether the results would carry over to a more complex three-dimensional representation of the terrain between transport nodes. This issue was investigated [2] using plastic three-dimensional terrain models of both the US and Europe and the results obtained illustrated *Physarum polycephalum*'s capability to approximate Route 20, the longest road in the United States (running from Boston, Massachusetts to Newport, Oregon, this road is 5415 km in length), and Autobahn 7 in Germany (running from Flensburg to Füssen, this road is 963 km in length), the longest national motorway in Europe. In these experiments, a plasmodium of *Physarum polycephalum* was placed at one end of the road and nutrient sources were placed at the other end point, and the plasmodium was allowed to explore the three-dimensional terrain, solving the problem in each instance.

The capability of slime mould to solve hard network problems is noteworthy, as this is achieved without any neural tissue and without global knowledge of its environment or indeed of its own topology. The adaptive process is driven by local responses to physical forces, resulting in an emergent problem-solving capability.

Robot Control Using *Physarum polycephalum*

A practical problem in creating robot controllers (Sect. 18.3) is to ensure that the controller is able to respond sensibly to changes in the environment. A top-down approach, wherein every possible environmental scenario is associated with a pre-programmed response, is infeasible for all but the simplest environments.

A study by Tsuda et al. [578], followed up in [579], investigated the use of *Physarum polycephalum* for the control of movement of an omnidirectional, six-legged, walking robot. As noted by the authors, the study was one in a line of work concerned with the integration of living cells into robotic devices in order to import the plasticity and adaptability of a living organism into a device's architecture [578].

In the experiment, a light sensor was placed on each of the robot's six legs and the robot was placed in a flat floor environment which was lit from above using several lights. The level of light could be varied over time, corresponding to changing environmental conditions. The plasmodium was contained in a six-pointed, star-shaped, circuit. Initially, the plasmodium was grown so that it covered the entire star shape.

Sensory (light) signals from the macrophysical environment of the robot were detected via the sensors on each of its legs. These were processed using a computer, resulting in a light pattern being shone on the star-shaped circuit containing the plasmodium. As *Physarum polycephalum* is sensitive to white light, the plasmodium reacted via shuttle streaming, with plasmodial mass being reduced in the lighted areas. This impact was detected using a camera and was fed back to alter the motion of the robot. In essence, the sensory signals from the macroscopic physical environment of the robot were transduced to a cellular scale and processed using the information-processing capabilities of *Physarum polycephalum*. The result from this cellular computation was then amplified to yield an output action in the robot's environment [579]. The end result was a walking robot which sought to avoid bright light, where the robot's behaviour was controlled by the phototaxis response of the slime mould.

15.3.2 Slime Mould Algorithms

In addition to using slime mould as a physical computing substrate (computing *in vivo*), a number of researchers have sought to mathematically model aspects of the behaviour of slime mould. Although the underlying molecular mechanisms which produce problem-solving behaviour in slime mould are not completely understood [389], these simplified models have proven to be high-quality problem solvers in their own right. The resulting algorithms provide an example of computing *in silico*.

Tero Model

One of the most commonly used models is based on the work of Tero et al. [566]. This model of the evolution of transport networks within *Physarum polycephalum* is based on an assumption that protoplasmic flux through network veins produces an autocatalytic effect whereby increases in the flux in a protoplasmic tube cause it to change structure (becoming wider), in turn facilitating even greater levels of flux.

The starting point for the Tero model, as applied to shortest-path-type problems, is a complete network of simulated protoplasmic tubes which connect nodes (metaphorically, nutrient sources), which act as sources and sinks for flux. An adaptation process, as outlined in the model, is simulated, producing an optimal (shortest) path through the network. In Sect. 15.4 we outline the detail of the model's implementation and describe two algorithms for combinatorial optimisation which are derived from this.

While these algorithms have shown good performance, the underlying model includes only a subset of *Physarum polycephalum* behaviours [291]. Specifically, the model does not consider the processes of initial network formation, plasmodium growth or adaptation to a changing nutrient environment. Therefore, there is scope to develop alternative models of *Physarum polycephalum* behaviours, which in turn could produce novel problem-solving algorithms.

Other approaches to deriving inspiration from slime mould behaviours to design optimisation algorithms include the cellular automaton (CA) model of Gunji et al. [234], which considers the processes of plasmodial growth and amoeboid movement, developing an algorithm for application to maze solving and shortest path problems based on these processes. A multiagent approach to reproducing behaviours of slime moulds is outlined in [291]. This study also provides a useful review of other recent work on slime mould computing.

15.4 Graph Optimisation Using Slime Mould

The majority of published work on slime-mould-inspired algorithms concerns graph optimisation applications, particularly the shortest path problem and close variants of it. In this section, two sample algorithms from this work are introduced. Recall that in the *shortest path problem* or SPP (Sect. 2.6), given a start vertex and an end vertex, the objective is to find a path (a sequence of vertices and their connecting arcs, with no repeated vertex) of least possible distance or time or other cost, which goes from the start to the end vertex.

The first algorithm, the improved *Physarum polycephalum* algorithm, employs the Tero model but modifies this through the addition of an energy parameter. The resulting algorithm is used for a series of graph optimisation problems. The second algorithm, the *Physarum*-based ant colony system, illustrates how the model of Tero et al. [566] can be used to help modify the pheromone matrix in an ant colony system application.

15.4.1 Improved *Physarum polycephalum* Algorithm

As described in earlier sections of this chapter, plasmodia of *Physarum polycephalum* in a starved state can find the shortest path between two points in a maze, where each end of the maze is marked with food. In this scenario, as the explorative (foraging) phase progresses, tubular pseudopodia that are not on the shortest path will shrink and eventually disappear, whereas pseudopodia on the shortest path between discovered food sources will be reinforced and become thicker. In essence, *Physarum polycephalum* self-adapts, reconfiguring its biomass morphology to the resource availability in its environment.

In this section we introduce the improved *Physarum polycephalum* algorithm (IPPA) developed by Zhang et al. [662]. In this study, the authors note that basic *Physarum*-inspired algorithms can suffer from a low rate of convergence, resulting in slow performance when applied to shortest path problems. The IPPA adapts the canonical *Physarum*-inspired shortest path algorithm based on the Tero model [566] by adding a new energy parameter in order to improve the speed at which good solutions are found. The authors argue that the addition of an energy component to the model is plausible from a biological perspective as energy is consumed during the process of tube expansion and during slime mould movement, whereas, at the same time, these processes can facilitate the capture of new resources. Therefore, dynamic adaptation of the morphology of a slime mould involves a trade-off between the consumption and absorption of energy.

Initially, the canonical *Physarum* model for determination of the shortest path across a network is described, and this is followed by a description of the modifications made to this in the IPPA.

Canonical *Physarum* Algorithm for Shortest Path

The canonical *Physarum polycephalum* algorithm for calculation of the shortest path in an undirected network is based on [566]. In this model it is assumed that the network formed by the *Physarum* is represented by a graph, in which a plasmodial tube is an edge and the junction between two tubes is a node. A key feature in the model is a positive feedback phenomenon, based on hydrostatic pressure, between flux and tube thickness. As noted in [566], hydrodynamic theory implies that short, thick tubes are the most effective for internal transportation. If a slime mould can form short, thick tubes, it can enhance its survival as this enables it to place most biomass over discovered food, and to transport nutrients and chemical signals effectively across its structure.

Tubes in the network become thicker in a given direction when streaming of protoplasm persists in that direction for a period of time, and thicker tubes enable greater conductance as the resistance to flow is reduced as the tube widens. In turn, greater conductance enables a greater flux, leading to a further thickening of the relevant tubes. Therefore, tubes with a large flux tend to grow, while those with a small flux tend to disappear.

Below we describe the model following the description provided in [566, 662].

Let N_1 and N_2 be the source (starting) and sink (ending) nodes, respectively, of the graph, let the edge between nodes N_i and N_j be expressed as ij, and let Q_{ij} be the flux (or flow) in tube (edge) ij. If the flow in this tube is approximately a Poiseuille flow (a laminar flow of an incompressible fluid induced by a constant positive pressure difference or pressure drop in a pipe), then the flux Q_{ij} can be calculated as

$$Q_{ij} = \frac{D_{ij}}{L_{ij}}(p_i - p_j), \tag{15.1}$$

where p_i is the pressure at node N_i, D_{ij} is the conductivity of tube ij and L_{ij} is its length. As the inflow and outflow at each node must be balanced (the principle of *conservation of flow*), we have

$$\sum_{j \neq 1,2} Q_{ij} = 0. \tag{15.2}$$

For the source node (N_1) and the sink node (N_2), the following hold:

$$\sum_i Q_{1i} + I_0 = 0,$$

$$\sum_i Q_{i2} - I_0 = 0, \tag{15.3}$$

where I_0 is the total flux flowing out of the source node (a fixed value; equivalently, I_0 is the total flux flowing into the sink node). It is assumed that the conductivity D_{ij} changes over time as the flux Q_{ij} changes, so the evolution of $D_{ij}(t)$ can be expressed as

$$\frac{d}{dt}D_{ij} = f(|Q_{ij}|) - rD_{ij}, \tag{15.4}$$

where r is a decay parameter for a tube. The conductivity of a tube will vanish over time if there is no flux along it.

From (15.1)–(15.3), the network equation for the pressure can be deduced [662]:

$$\sum_i \frac{D_{ij}}{L_{ij}}(p_i - p_j) = \begin{cases} +1, & \text{for } j = 1, \\ -1, & \text{for } j = 2, \\ 0, & \text{otherwise.} \end{cases} \tag{15.5}$$

By setting $p_2 = 0$ as a basic pressure level, all p_i can be determined from (15.5) and, in turn, Q_{ij} can be calculated.

Following the assumption in [566] that $f(Q) = |Q|$, and with the flux calculated, the conductivity can be calculated using the following equation:

$$\frac{D_{ij}^{n+1} - D_{ij}^n}{\Delta t} = |Q| - D_{ij}^{n+1}. \tag{15.6}$$

A more detailed description of the above is provided in [566]. In addition to its application to shortest path problems, a variant on this model has been applied to solve the constrained shortest path problem [610].

Description of Improved *Physarum polycephalum* Algorithm

In the IPPA, the above mathematical model is modified to incorporate an energy constraint. Under this modification, it is assumed that tubes need to consume energy in order to be maintained, and that this energy is obtained from nutrients in the flow through the tubes. If the net energy gain is positive, the tubes grow and the conductivity increases, otherwise the tubes wither and vanish over time. As the tubes alter their physical state, the flux in each tube also changes. Over time, the tubes in the network tend to converge to a steady state. The IPPA is described as follows in [662].

The energy E, the flux Q, and the conductivity D are defined as below:

$$E_1 = f(Q), \tag{15.7a}$$
$$E_2 = g(D), \tag{15.7b}$$
$$\Delta D = h(E_3), \tag{15.7c}$$

where E_1 is the energy that can be provided by the tube when its flux reaches Q, E_2 is the energy consumed by the tube when its conductivity is equal to D and ΔD shows how the conductivity changes when the remaining energy (i.e. the energy provided by the flux less the energy consumed by the tube) is E_3. Therefore, (15.6) becomes

$$\Delta D_{ij} = h(f(Q_{ij} - g(D_{ij})))\Delta t, \tag{15.8}$$

or, taking the derivative,

$$\frac{d\Delta D_{ij}}{dt} = h(f(Q_{ij} - g(D_{ij}))). \tag{15.9}$$

Next, the functions f, g and h are modified to render them operational. In order to preserve the law of conservation of energy (as noted in [662], the formulation in (15.6) does not obey this law), it is assumed that the total energy provided by the flux from the start to the end node is constant, and is therefore independent of path. Therefore, f can be defined as

$$f(Q_{ij}) = \frac{Q_{ij}(p_i - p_j)}{p_s - p_e},\tag{15.10}$$

where s and e are the starting and ending nodes, respectively, and p_i and p_j are the pressure at nodes i and j respectively.

It is assumed that the energy required to maintain a tube is a function of the conductivity and length of the underlying tube, and g is defined as

$$g(D_{ij}) = D_{ij}L_{ij}.\tag{15.11}$$

The function h which relates the level of net energy in a tube to the change in conductivity is defined so as to take tube length into account:

$$h(E_3) = \frac{E_3}{L_{ij}}.\tag{15.12}$$

The longer a tube is, the more energy it requires for its maintenance.

Combining (15.10)–(15.12), the following can be obtained:

$$\frac{dD_{ij}}{dt} = \frac{Q_{ij}(p_i - p_j)}{L_{ij}(p_s - p_e)} - D_{ij}.\tag{15.13}$$

The pseudocode for the IPPA is presented in Algorithm 15.1. As illustrated in [662], the IPPA can produce competitive results when compared with ant colony optimisation for a shortest path problem.

Algorithm 15.1: Improved *Physarum polycephalum* algorithm

Let L be an $n \times n$ matrix, where L_{ij} is the length between nodes i and j. Let V denote the set of arcs;
Let s and e be the start and end nodes;
Let $D_{ij} \in (0, 1]$ (for all $i, j = 1, 2, \ldots, N$);
Let $Q_{ij} := 0$ (for all $i, j = 1, 2, \ldots, N$);
Let $p_i := 0$ (for all $i = 1, 2, \ldots, N$);
Set iteration counter $t = 0$;
while *termination criteria not met* **do**
    Let $p_e := 0$ (pressure at ending node e) ;
    Calculate the pressure of every node in network solving (15.5);
    Let $Q_{ij} := D_{ij}(p_i - p_j)/L_{ij}$ using (15.1);
    Let $D_{ij} := \frac{1}{2}(Q_{ij}(p_i - p_j)/(L_{ij}(p_s - p_e)) + D_{ij})$ using (15.6);
    Set $t := t + 1$;
end
Output the best solution (shortest path) found;

15.4.2 *Physarum*-Based Ant Colony System

In this section we highlight work of [374, 663], which combined the Tero model with an ant colony system (ACS) algorithm to develop a hybrid *Physarum*-based ant colony system (denoted as PM-ACS).

Canonical versions of ant colony optimisation (ACO) algorithms can be prone to premature convergence, and a multitude of variant algorithms have been developed to overcome this issue, for example by limiting the rate of pheromone build-up on arcs. These algorithms attempt to maintain continual exploration of novel solutions over time and therefore avoid undue exploitation of already discovered tours. A key suggestion in [374, 663] is that a hybrid PMM-ACO algorithm, where the updating strategy for the pheromone matrix is partially based on a PMM (*Physarum*-inspired mathematical model), can improve the efficiency and robustness of ACO algorithms.

The PMM used in this algorithm focuses on feedback regulation of the thickness of each tubular pseudopodium (*tube* in the network) arising from changes in the internal protoplasmic flow. Higher rates of protoplasmic flow stimulate an increase in the tube diameter (i.e. tubes get bigger as internal streaming rates increase) and low flow rates lead to a reduction and eventual disappearance of tubes (Figs. 15.3 and 15.4).

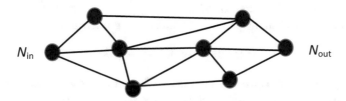

Fig. 15.3. PMM: initial network. N_{in} and N_{out} are the inlet and outlet nodes

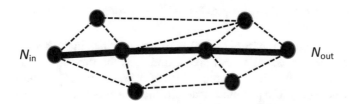

Fig. 15.4. PMM: final network after shortest path has been reinforced. N_{in} and N_{out} are the inlet and outlet nodes

Below, we outline the workings of the algorithm, drawing on the description provided in [374]. To complement this description, readers are referred to Sect. 9.4

for a description of the application of ACO algorithms for discrete optimisation and to Sect. 9.6 for a description of the ACS algorithm.

Algorithm

The PM-ACS algorithm assumes that there is a *Physarum* network with simulated pheromone flowing along the tubes in the network. When the model is applied to the TSP (Sect. 9.5), nodes represent cities and the tubes of the *Physarum* network are the paths connecting the cities.

As the algorithm iterates, the quantity of pheromone in each tube of the *Physarum* network dynamically changes. The update of the global pheromone matrix (Sect. 9.4) considers both the pheromone released by ants as they construct tours and the simulated pheromone flows in the *Physarum* network which also models these cities and their connections. The PMM acts to modify the *Physarum* network over time, reinforcing (thickening) the arcs on the shorter tours of the cities. The classical formulation of the TSP is adopted and it is assumed that the object is to construct the shortest tour of n cities, such that each city is visited exactly once, and the tour ends in the city in which it started.

As in the canonical ACS algorithm, the ants construct tours using the information in the pheromone matrix to guide their choice of exit path at each node (city) as they build their tour. In canonical ACS, the deposition and evaporation steps which determine the updates in the pheromone matrix are governed by

$$\tau_{ij}(t+1) = \tau_{ij}(t)(1-p) + p\Delta\tau_{ij}^*, \tag{15.14}$$

where only the arcs traversed by the best-so-far ant (on tour T^*) participate in the pheromone deposition/evaporation process. The term $\Delta\tau_{ij}^*(t)$ is equal to $1/L^*$, where L^* is the length of the best-so-far tour.

Under PM-ACS, the global pheromone matrix update rule of ACS is modified by adding an additional term which considers the quantity of pheromone in the *Physarum* network on all the arcs in the best-so-far solution, as follows:

$$\tau_{ij}(t+1) = \tau_{ij}(t)(1-p) + p\Delta\tau_{ij}^* + \epsilon\frac{Q_{ij}M}{I_0}, \quad \text{for all arcs } ij \text{ on } T^*, \tag{15.15}$$

where M is the number of tubes (arcs) between the cities in the TSP, Q_{ij} is the flux through a tube connecting cities i and j, and I_0 is the flux (assumed to be a fixed quantity) between the inlet and outlet nodes of the network (i.e. it represents the flow across the network). The parameter ϵ determines the effect of the flowing pheromone in the *Physarum* network on the final update in the pheromone matrix. The value of ϵ is calculated using

$$\epsilon = 1 - \frac{1}{1 + \lambda^{t_{\text{PMM}}/2 - (t+1)}}. \tag{15.16}$$

In (15.16), t_{PMM} is the total number of steps of iteration affected by the PMM process, t is the current iteration number (time step) and $\lambda \in (1, 1.2)$. As t gets large, the value of ϵ becomes smaller, therefore, the impact of flows in the *Physarum* network reduces as the PM-ACS algorithm iterates and as the ants converge on a good solution.

The relationship between conductivity and flux for each tube is modelled as:

$$Q_{ij} = \frac{1}{M} \sum_{m=1}^{M} \left| \frac{D_{ij}}{L_{ij}} (p_i^m - p_j^m) \right|, \tag{15.17}$$

where Q_{ij} represents the flux through a tube connecting node i and j, L_{ij} is the length of tube connecting nodes i and j, D_{ij} is a measure of the conductivity of the tube connecting nodes i and j, and the pressure at node i is p_i^m. The flux through a tube is related to its conductivity, its length, and the pressure at the node on each end of the tube.

The conductivity can be considered as the flow capacity of a tube and is related to the tube's thickness (diameter). All tubes have an initial assigned value for D_{ij}, and if a tube subsequently becomes thicker, its conductivity will be enhanced. In turn, as conductivity increases, all other things being equal, so does the rate of flux.

In an iteration of the PMM, each pair of nodes connected by a tube can be selected as inlet/outlet nodes. The flux input to a node must equal the flux output from that node under an assumption of conservation of flow. As above, I_0 is a fixed quantity, being the flux between the inlet node and the outlet node of the network. When two nodes a and b, connected by the mth tube are selected as inlet and outlet nodes, respectively, the pressure on each node p_i^m is calculated using Kirchhoff's Law as follows:

$$\sum_i \frac{D_{ij}}{L_{ij}} (p_i^m - p_j^m) = \begin{cases} -I_0, & \text{for } j = a, \\ I_0, & \text{for } j = b, \\ 0, & \text{otherwise.} \end{cases} \tag{15.18}$$

The above process iterates until all pairs of nodes in each tube have been selected as inlet or outlet nodes. The flux Q_{ij} is calculated using (15.17). The conductivity of a tube adapts according to the flux based on

$$\frac{dD_{ij}}{dt} = \frac{|Q_{ij}|}{1 + |Q_{ij}|} - D_{ij} \tag{15.19}$$

The conductivities at the next iteration step are fed back to (15.18), and the flux is updated using (15.17). Based on the positive feedback mechanism between conductivity and flux, the shorter tubes (called *critical tubes*) become wider and are maintained as connections, while other tubes become narrower and eventually disappear.

The pseudocode for the PM-ACS algorithm is outlined in Algorithm 15.2. Aspects of the ACS algorithm which remain unchanged under PM-ACS, such as the processes by which ants construct a new tour and the local pheromone matrix update step, are not discussed here, and readers are referred to Sect. 9.6.

Algorithm 15.2: PM-ACS algorithm

Set values for $\alpha, \beta, p, s, q_0, \lambda, I_0, t_{PMM}, t_{max}$;
Set initial values for pheromone levels on each arc $\tau_{ij}(0)$ and conductivity of each tube D_{ij};
Set iteration counter $t = 0$;
while $t < t_{max}$ **do**
    for $k = 1$ **to** s *(all s ants)* **do**
        Construct a tour by ant k using approach in canonical ACS algorithm;
        Update the local pheromone matrix;
    end
    Let $k_{best} :=$ global best ant (shortest tour found);
    Let $S_{min} :=$ length of tour generated by ant k_{best};
    Calculate flowing pheromone in the *Physarum* network using (15.17)–(15.18) and
      update the conductivity of each tube using (15.19);
    Update the global pheromone matrix using (15.15);
    Let $t := t + 1$;
end
Output the best solution (best tour) found;

Parameters

The application of the PM-ACS algorithm requires the setting of several parameters. The parameter values used in [374] are outlined in Table 15.1. A sensitivity analysis

Table 15.1. Parameter values used in [374]

Parameter	Description	Value
α	Relative importance of pheromone trail	1
β	Relative importance of heuristic information	2
p	Pheromone evaporation rate	0.8
s	Number of ants	Number of cities
q_0	Parameter in range $[0, 1]$	0.1
λ	Parameter impacting ϵ	1.05
I_0	Fixed flux flowing in *Physarum* network	20
t_{PMM}	Total steps of iteration affected by PMM	300
t_{max}	Maximum number of iterations of algorithm	300
$\tau_{ij}(0)$	Initial pheromone levels on each connection	1
D_{ij}	Initial conductivity of each tube	1

concerning the impact on the results of different choices of parameter settings is provided in [663].

Discussion

The TSP is representative of a wide array of network problems, and algorithms for efficient solution of TSP problems are of general interest. A feature of slime moulds when foraging is their capability to allocate biomass efficiently in order to capture resources. One aspect of this is their ability to stream protoplasm to pseudopodia which have encountered resources and away from locations without resources.

In the PMM-ACS algorithm, a model of this streaming process is used to contribute to the updating of the pheromone matrix which is used by the ants in constructing their solutions, tours in the case of the TSP. This approach can also be applied to other ant colony algorithms with relatively little modification, with [663] illustrating similar applications to a canonical ACO algorithm (PM-ACO) and to an MMAS algorithm (PMM-MAS). A formulation of the heuristics for application to multiple-objective versions of the TSP is also outlined in [663].

15.5 Real-Valued Optimisation Using Slime Mould

Although most applications of slime mould algorithms for optimisation purposes have concerned combinatorial problems, a paper by Monismith and Mayfield (2008) [409] introduced the *slime mould optimisation algorithm* (SMOA). This provides an illustration of how a real-valued, single-objective optimisation algorithm could be designed based on slime mould behaviours.

In contrast to the algorithms discussed earlier in this chapter, this algorithm draws inspiration from several behaviours over the life cycle of the cellular slime mould *Dictyostelium discoideum*, a common soil-living amoeba which feeds on bacteria.

15.5.1 Life Cycle of *Dictyostelium discoideum*

Figure 15.5 provides a stylised life cycle diagram for *Dictyostelium discoideum*. Initially, spores are released from a mature fruiting body, and under the right environmental conditions these dispersed spores hatch and give rise to individual amoebae. During their vegetative stage, the amoebae divide by mitosis as they feed. When the surrounding food supply becomes exhausted, the cells starve and begin to enter the aggregation stage. The cells become sensitive to cAMP and are attracted to the strongest cAMP gradient in their locality (the cell emitting the strongest cAMP signal is termed the *pacemaker* cell). As *Dictyostelium discoideum* cells move towards the pacemaker cell they also deposit cAMP, trails which in turn can also encourage other cells to chemotactically move towards the pacemaker.

As the cells come into close proximity, they aggregate into an intermediate structure known as a *mound*. This is a protective structure wherein the amoebae coalesce and form a slime sheath around themselves, producing a motile pseudoplasmodium (or *slug*).

The slug (i.e. slime mould) is mobile and will seek a suitable location for spore production. Once the slug has found a suitable environment, the anterior end of the

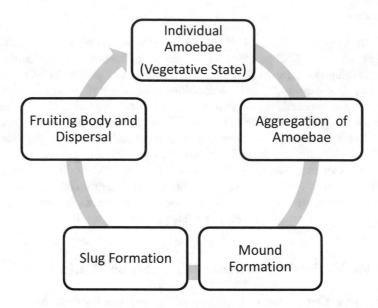

Fig. 15.5. Stylised life cycle diagram for *Dictyostelium discoideum*

slug forms the stalk of the fruiting body and the posterior end forms the spores of the fruiting body. When the fruiting body is mature, the spores are released and are dispersed by abiotic (wind, water or gravity) or biotic (insect or animal) dispersal mechanisms (see Sect. 16.3.4 for a discussion of related plant seed dispersal processes). The life cycle recommences when these spores form new amoebae.

15.5.2 Overview of Algorithm

Drawing a parallel with the life cycle of the amoeba, in the algorithm each search agent (or simulated amoeba) can take on a number of states. Initially, all amoebae begin in the vegetative state and are assigned to random locations in the search space. In the vegetative state, each amoeba searches locally around its current location in isolation, via a simulation of the ability of amoebae to extend pseudopodia in multiple directions. A parameter k determines the number of simulated pseudopodia extended in this phase, corresponding to the number of search probes undertaken by each individual.

Next, it is assumed that the population enters a starved state, controlled by a parameter $t_{unimproved}$ which keeps track of the number of iterations of the algorithm that have passed without an amoeba uncovering a better location. The counter $t_{lifetime}$ keeps track of the number of iterations since the last dispersal event (i.e. it is a measure of the lifetime of an amoeba). The longer since an amoeba has uncovered a better location, the greater the likelihood that it begins to starve.

As the density of starving amoebae in a location increases above a preset threshold $A_{\min}$, the probability of a mound forming increases. Both mounds and slug formations engage in a search process in their region.

Finally, once the search process in the slug stage stagnates, a dispersal process is simulated and amoebae are relocated to random individual locations. Therefore, the search process consists of a blend of stochastic local search and periodic random reinitialisation with restart.

The life cycle metaphor is supplemented in the algorithm by a *cellular automaton* mechanism where, a mesh is formed using the initial locations of the amoebae and the approximate nearest-neighbour algorithm (the ϵ-ANN algorithm, where ϵ is the number of nearest neighbours being considered). Cellular automata are idealised models of systems in which space and time are discrete, where elements of the system can take only a finite set of values and where the elements interact locally (i.e. the state of a cell is influenced by the states of neighouring cells). The global behaviour of the system emerges from the collective activity and interaction of simple components. Figure 15.6 illustrates a simple two-dimensional cellular automaton (CA). More generally, cellular automata can be multidimensional and neighbourhoods can be defined in various ways.

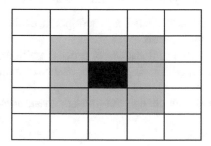

Fig. 15.6. Two-dimensional cellular automaton where the state of the middle (black) cell is determined by its current state and the state of the surrounding grey cells. This example illustrates a Moore neighbourhood, where all directly adjacent cells contribute to determination of the state of the centre cell

15.5.3 Algorithm

The SMOA is described as follows in [409]. The population consists of n amoebae, each of which represents a potential solution $x_i = (x_{i1}, x_{i2}, \ldots, x_{id})$ and has a corresponding value based on the value of the objective function at that location.

Individual amoebae store a number of items of information, including: a state (vegetative, aggregative, mound, slug or dispersal); a current location; a current objective function value; a personal best location (location of best objective function value they have uncovered); a personal best objective function value; the amount of time since their personal best location was improved; and the length of time they have existed since the last dispersal event. The behaviour of an amoeba depends on its state and we next outline the relevant behaviour in each of these states.

Vegetative State

In this state an individual amoeba is assumed to engage in random, local search by extending its pseudopodia in multiple directions. The search process is simulated using

$$x_i = crE(d_{\text{neighbour}}),\tag{15.20}$$

where c is a constant (< 1), r is a random variable drawn from the uniform distribution $U(0,1)$, x_i is the location of the ith amoeba and $E(d_{\text{neighbour}})$ is the average distance to a neighbouring ϵ-ANN mesh point. A total of k searches are simulated.

With a probability of p, the amoeba moves towards the location of the end of one of the k pseudopodia, chosen at random. Alternatively, with a probability of $1 - p$, the amoeba performs a roulette wheel choice (weighted using the objective function values at the k points) to determine the pseudopod to which it will move. The amoeba also continues to track the location of its closest lattice point in the search space.

Hence, the process of vegetative movement is as follows:

i. Using (15.20), the amoeba extends k pseudopodia, evaluating the objective function at each end point location.
ii. Movement is based on a roulette wheel process (with a probability of p), with better-located pseudopodia having a higher chance of being chosen; alternatively (with a probability of $1 - p$), a random pseudopod location is chosen.
iii. If an amoeba is starving, change the state of the amoeba based on (15.21) to aggregative.

Aggregation

It is assumed during the vegetative state that amoebae are fed by values from the solution space. As each amoeba finds a better solution (feeds), it updates its personal best. The determination as to whether an amoeba is starving is made probabilistically using

$$P\left(x < \frac{t_{\text{unimproved}}}{t_{\text{lifetime}}}\right),\tag{15.21}$$

where x is a random variable with a uniform probability distribution $U(0,1)$, $t_{\text{unimproved}}$ is the amount of time since the amoeba last improved its personal best (i.e. found a new best location) and t_{lifetime} is the number of time steps since the last dispersal of that amoeba (i.e. how many time steps it has lived for).

If an amoeba is determined to be starving, it emits cAMP. The mechanics of cAMP emission and the subsequent process of aggregation movement make use of a cellular automaton, supplemented by a sparse mesh structure, where the mesh is constructed using the starting points of the algorithm and a simplified nearest-neighbour structure (here, the ϵ-approximate nearest-neighbour structure, denoted by ϵ-ANN). The mesh structure is used in order to reduce the infeasible computational load that would arise if a traditional grid structure was used to represent the search space. The detailed operationalisation of the CA is somewhat involved, and readers requiring full details for implementation are referred to [409] and references therein. Below, using part of the description in [409], we provide a short synopsis of a number of the key steps in the aggregation process.

Each aggregating cell may be considered as part of a CA. In [409], this is represented as a mesh formed using the ϵ-ANN algorithm. Every point in the mesh can assume a set of values in the range $0, 1, \ldots, k$, where k is reasonably large (i.e. > 100).

An amoeba i deposits an amount of cAMP at its current location in proportion to the norm (absolute value) of its current objective function value relative to the value of the overall best objective function value. The rate of deposition depends on how an amoeba ranks relative to its peers:

$$\sigma_i(t+1) = \begin{cases} \sigma_i(t) + c \left| \dfrac{f(x_{\text{current}})}{f(x_{\text{best}})} \right|, & \sigma_i(t+1) < k, \\ k, & \sigma_i(t+1) \geq k. \end{cases} \tag{15.22}$$

In the deposition rule as outlined in (15.22), c is a user-defined parameter which controls the rate of deposition in the current time step and k is the maximum level of cAMP that can be contained on a node in the lattice.

Apart from the deposition of cAMP, the CA model provides for an evaporation process after all updates to positions have been made. The evaporation rule (15.23) removes a single unit of cAMP from every lattice point on the mesh (subject to a lower bound of zero cAMP in any location) after all updates have been completed for a given time step:

$$\sigma_i(t+1) = \begin{cases} \sigma_i(t) - 1, & \sigma_i(t) > 0, \\ 0, & \sigma_i(t) = 0. \end{cases} \tag{15.23}$$

A neighbourhood update (which functions as a smoothing operator) is implemented as in (15.24). This process simulates the diffusion of cAMP from a given node in the lattice to adjoining nodes:

$$\sigma_q(t+1) = \begin{cases} \frac{1}{N+1} \sum_i \sigma_i(t), & \sigma_q(t+1) < k, \\ k, & \sigma_q(t+1) \geq k, \end{cases}$$
$$q \in \{1, 2, \ldots, N\} ,$$
$$i \in \{\text{the neighbours of } \sigma_q(t) \text{ and } \sigma_q(t) \text{ itself}\} . \tag{15.24}$$

The three rules above allow starving amoebae to aggregate about a single point and also model the attraction of other starving amoebae to this point. The pseudocode for the aggregation process is:

i. If a pacemaker amoeba exists, move using (15.25); otherwise, undertake vegetative movement.

ii. Drop cAMP at closest mesh point using (15.22).
iii. If no pacemaker exists, this amoeba becomes the pacemaker and emits cAMP from its location.
iv. If (15.26) is satisfied, at the next time step, convert all amoebae within a radius r of the pacemaker's location to the mound state.

Movement of an amoeba towards a pacemaker amoeba during the aggregation stage is undertaken using formulae similar to those for particle swarm optimisation:

$$v_i(t+1) = c_1 v_i(t) + c_2(x^{\text{best}}_{\text{ANNNeighbour}}(t) - x(t))$$
$$+ c_3(x_{\text{pacemaker}}(t) - x(t)) + c_4 E(d_{\text{neighbour}})N(0,1), \qquad (15.25)$$
$$x_i(t+1) = x_i(t) + v_i(t+1) \ .$$

In (15.25), x_i is the location of amoeba i, v_i is its velocity, $x^{\text{best}}_{\text{ANNNeighbour}}$ is the nearest mesh location with the largest amount of cAMP (excluding the amoeba's current mesh location) and $x_{\text{pacemaker}}$ is the location of the pacemaker amoeba. The parameters $c_1, \ldots, c_4$ are randomly drawn from a uniform distribution in the range $[0, 2]$. $E(d_{\text{neighbour}})$ is the mean distance to the amoeba's nearest neighbour in the mesh.

Under this formulation, each amoeba tends to move stochastically in the direction of the strongest nearby cAMP sources and takes a random step scaled by the distance to its nearest neighbours. Therefore, the communication mechanism in the model is partly stigmergic (see Sect. 9.1), as a change in the cAMP environment by one individual impacts on the activities of other cells in future time steps.

Mound Formation

The propensity to form a mound is determined by the density of starving amoebae at a lattice point. As this density increases beyond the predetermined minimum threshold required for mound formation, the probability of mound formation also increases. Therefore, if

$$x > \frac{A_i - A_{\min}}{A_{\max}}, \qquad (15.26)$$

a mound is formed. In this formulation, x is a random variable with a uniform probability distribution $U(0, 1)$, A_i is the number of amoebae at position i in the grid, $A_{\min}$ is the minimum threshold needed to form a mound and $A_{\max}$ is the maximum threshold needed to form a mound.

Once a mound is determined to be at a particular location, a dense lattice is formed over the points located in the mound using each of their locations and making use of a nearest-neighbour algorithm.

The steps in the mound formation process are:

i. Add an amoeba to its pacemaker's ANN mesh based on its current location.
ii. At the end of this time step, create the entire mesh for this mound with locations for cAMP.
iii. Convert all amoebae within the mesh to the slug state.

After mound formation, a local search is undertaken in the area of the mound, and this is discussed next.

Slug Movement

During the slug stage, updates are performed using the rules for the sparse lattice on the dense lattice for the slug. Updates are performed only on the locations within the denser lattice. As necessary, the location of the dense lattice is updated to better reflect the location of the slug. A dense-lattice reset is performed after the centroid $C_{current}$ of the amoeba locations has moved more than a threshold distance from the original centroid C_{old} of the slug:

$$P\left(x < \frac{d(C_{current}, C_{old})}{std\{old\ lattice\ points\}}\right) \tag{15.27}$$

where x is a random variable with uniform probability distribution $U(0,1)$.

The process for slug movement is therefore:

i. Perform local search using (15.28).
ii. Update the function value and personal best if necessary.
iii. If, after performing all movements for this mound (15.27) is satisfied, reset the mesh for the slug.
iv. If (15.29) is satisfied, convert this amoeba to the dispersal state.

Movement of each amoeba within the slug is carried out using formulae similar to (15.25):

$$v_i(t+1) = c_1 v_i(t) + c_2(x^{best}_{neighbour}(t) - x(t))$$
$$+ c_3(x^{best}_{slug}(t) - x(t)) + c_4 E(d_{neighbour})N(0,1), \tag{15.28}$$
$$x_i(t+1) = x_i(t) + v_i(t+1) \ ,$$

where x_i is the location of amoeba i, v_i is its velocity, $x^{best}_{neighbour}$ is the location of its nearest neighbour with the best personal best and x^{best}_{slug} is the location of the best value found within the slug. As before, the parameters $c_1, \ldots, c_4$ are randomly drawn from a uniform distribution in the range $[0,2]$. $E(d_{neighbour})$ is the mean distance to the amoeba's nearest neighbour in the slug's mesh.

The action of (15.28) is to stochastically move each amoeba in a direction which is determined by the location of a neighbour that has found a good-quality location during its search, the location of the best value found within the slug and a factor which is determined by the closeness of an amoeba to its neighbours (if they are tightly clustered, the last term in (15.28) will be small).

Fruiting Body and Dispersal

Once there are too few personal updates for a slug (i.e. the search process within the slug is stagnating), it is dispersed in order to restart the search process. The following formula is used to determine the probability that dispersal will occur:

$$P\left(x > \frac{\text{Number of updates}}{\text{Update threshold}}\right), \tag{15.29}$$

where x is a random variable with a uniform probability distribution $U(0, 1)$. Therefore, dispersal will occur with higher frequency when there are fewer updates to personal bests of amoebae within the slug. The dispersal process consists of moving an amoeba to a new random location in the search space, clearing its personal best memory and resetting the state of that amoeba to vegetative.

Pseudocode for the SMOA is provided in Algorithm 15.3.

Algorithm 15.3: Slime mould optimisation algorithm

Set parameters n, k, ϵ, A;
Generate amoeba population (for all x_i) at random locations in the search space;
Evaluate fitness of population;
Store objective function values and locations and set these as personal best values and
    locations for each amoeba;
Set state of all amoebae to Vegetative;
Store best objective function value and associated location;
Input location of each amoeba to ϵ-ANN;
Create a mesh based on results of ϵ-ANN;
for *each amoeba i* **do**
    Determine amoeba state;
    if *state=Vegetative* **then**
        | Undertake a Vegetative movement;
    end
    if *state=Aggregative* **then**
        | Undertake Aggregation;
    end
    if *state=Mound* **then**
        | Undertake Mound formation;
    end
    if *state=Slug* **then**
        | Undertake Slug movement;
    end
    if *state=Dispersal* **then**
        | Undertake Dispersal;
    end
end
Output the best solution found;

15.5.4 Discussion

In the SMOA, the life cycle of *Dictyostelium discoideum* is used to inspire the design of the algorithm. Initially, each amoeba in the population is assigned to a random location. Over the life cycle of the amoebae, the search strategy varies in terms of exploitation and exploration, ranging from local random search in the vegetative state to intensive searching of a local region by mounds and slugs, with the restarting of the search process anew at random locations during the dispersal stage of the life cycle. This life-cycle-inspired search process is operationalised using elements drawn from cellular automata (including the ϵ-ANN approximate nearest-neighbour algorithm) and elements from the particle swarm optimisation algorithm.

The SMOA is a relatively complex algorithm, consisting of a number of embedded search processes. It is not trivial to determine, from the algorithm's description alone, which of its elements are most significant in determining its overall performance. However, the algorithm does offer an interesting, alternative example of how metaphorical inspiration can be taken from aspects of the life cycle of a slime mould in order to design an optimisation algorithm.

15.6 Fungi

Fungi, along with plants and animals, are one of the three main kingdoms of eukaryotic life. The kingdom of fungi contains a huge array of species, encompassing yeasts, rusts, smuts, mildews, moulds, and mushrooms. Taken together, these species have a worldwide distribution across all environmental settings, including soil, the air, aquatic settings, and are also distributed on and within plants and animals. Fungi have widely varying lifestyles, preferred habitats and morphologies, ranging from unicellular organisms to vast fungal networks. While the exact number of species which comprise the kingdom of fungi remains unknown, estimates range from 1.5 million to 5 million separate species [58, 441]. Only about 5% of these have been studied.

The vegetative body of all but the simplest fungi consists of a mass of thread-like, branched, tubular filaments called *hyphae*, filled with cytoplasm and various organelles [547]. Biomass can flow within these tubular structures, facilitating the movement of nutrients, water and signalling molecules throughout the organism. Fungal hyphae, like the roots of vascular plants, grow primarily at the tip, elongating and branching repeatedly, creating a radially expanding tree like network called the *mycelium*.

Despite the surface similarities between fungi and plant root networks, fungi are not plants, as they lack chlorophyll and therefore cannot undertake photosynthesis [547]. They must forage for resources. The mycelium grows by absorbing nutrients from the environment. On reaching a certain stage of growth, it begins to produce spores (i.e. seeds of the fungus) either directly on the somatic hyphae or, more often, on special sporiferous (spore-producing) hyphae, which may be loosely arranged or grouped into intricate structures called fruiting bodies, or sporophores. The spores are dispersed and, upon reaching a suitable substrate, the spores germinate and develop hyphae that grow and become the mycelium of a new individual.

Fungi and Slime Moulds

The foraging behaviours of fungi and slime moulds bear some similarities, and both have served as inspiration for the design of computational algorithms. In each case, the organism faces the same problem, finding food in a patchy resource environment, and then transporting nutrients from this resource to other parts of the organism. Hyphae can branch apically or subapically (i.e. before the apex or tip of the hypha) to form a tree-like mycelium which bears substantial similarity to the network structure of slime moulds. In each, the structure allows the organism to efficiently explore its environment and to dynamically alter its morphology in response to resource location by concentrating biomass and search efforts in areas of high promise, and withdrawing biomass from areas which are devoid of resources [250]. The process governing exploration of the environment is driven by local information rather than by a centralised control mechanism. It is also notable that both types of organism have exhibited complex behaviours, including maze solving.

However, there are important differences between these organisms. Slime moulds have a number of different life stages. During the trophic (feeding) stage in their life cycle, the organism lacks a cell wall, facilitating the engulfment and internal digestion of food. In contrast, fungi engage in external digestion, secreting enzymes on food resources and absorbing the resulting soluble breakdown products through their cell wall.

Of course, both types of organisms must forage for food and the behaviour of each can inspire the design of computational algorithms. While there is an emerging literature concerning slime mould computing and slime mould algorithms, as yet, little inspiration has been drawn from the behaviours of fungi for algorithm design. In the following sections we introduce aspects of the foraging activities of some species of fungi which undertake an active search for new resources by forming hyphal networks to explore their environment. We also briefly refer to a number of algorithms which have been designed based on inspiration from fungal foraging behaviours.

15.6.1 Fungal Foraging Behaviours

Food resources are often patchily and sparsely distributed. To survive, fungi must be able to relocate to new food resources before their current food base is exhausted [62]. The majority of saprotrophs (fungi which forage on decomposing organic matter such as leaf litter or tree stumps and branches) are *resource-unit-restricted* in that they decompose solid, organic resources and are confined to those resources, i.e. they relocate to new resources by creating and dispersing spores in the environment [62].

In contrast, some fungi, including those that colonise patches of leaf litter on the forest floor, and cord and rhizomorph formers, which typically extend between spatially discrete woody resources, are able both to produce spores and to grow hyphae out of their current food base in search of new resources. These fungi are termed *non-resource-unit-restricted* [62]. These fungi divert energy into hyphal growth in order to explore the environment and to acquire new food resources, rather than re-lying solely on spore dispersal. This can be viewed as an alternative way of investing their biomass, as the production of spores, many of which never germinate, is costly.

Growth Behaviour

The rate of growth can vary markedly between fungi which develop mycelial net-works, depending on the nature of their foraging strategy.

At one extreme, some species engage mainly in *sit-and-wait* foraging, capitalis-ing on any new resources which land on their mycelial network (e.g. leaf fall) [63]. In contrast, other species engage in more active search, growing their mycelial net-work in order to find and colonise new resources [63]. The precise architecture of the resulting network depends on the species of fungus involved [571].

The rate of growth varies by species. Some carry out a slow but intensive explo-ration of the environment by a diffuse colony margin of fine hyphae. Other species generate more open networks with rapidly growing cords that are better suited to discovery of large, sparsely distributed resources.

The former case corresponds to area-concentrated search, and is likely to be suc-cessful in exploiting abundant, homogeneously, distributed resources. The latter is more akin to *near–far* search. These approaches are not mutually exclusive, as the nature of growth is generally adaptive in response to foraging feedback.

Although growth from an initial food resource is typically reasonably symmetri-cal, as this resource is depleted the colony typically switches to faster, less symmetri-cal growth in a limited number of directions. As new food resources are encountered, the degree of local growth in the region of the new resource increases markedly, at the expense of growth elsewhere in the mycelial network. A typical adaptive forag-ing strategy switches to area-restricted growth on the discovery of a significant new food resource, intensifying subsequent foraging efforts around that location.

Within these general foraging strategies, the precise pattern of growth of a mycelial network is also dependent on a number of factors, including the nutritional state of the fungus, the quality and quantity of current resource(s) being exploited, the nature of the new resource, the order in which past resources have been encountered, the degree of competition from other fungi at resource locations and microclimate factors [571].

Despite the lack of a nervous system, changes in foraging behaviour, resource flows and growth are integrated across hyphal networks. For example, if a significant new food resource is discovered, global coordinated reorganisation of the whole net-work is triggered [63]. There is a strengthening of the cords linking to that resource (they become larger and thicker), facilitating enhanced transport of nutrients away from that location to other parts of the organism or to resource storage sites, with a re-duction or cessation of growth elsewhere in the network. The regression of mycelium from areas with no resources allows the redistribution of biomass via translocation within the mycelium to support hyphal exploration elsewhere [62, 174, 613]. Hence, activity in one part of network can impact elsewhere, even though the local conditions at the latter location are unchanged. Each part of the network can influence or be influenced by other parts [250].

15.7 Fungal Algorithms

As is the case for many other organisms, the foraging processes of fungi can provide metaphorical inspiration for the design of computational algorithms. Fungi can explore their environment when looking for new food resources, by generating interconnected hyphal networks which adapt over time in response to feedback from foraging efforts. Growth is promoted in regions in which food resources are found with the hyphal network regressing from less promising areas. As with plasmodial slime moulds, these networks are constructed by local iterative development processes rather than via centralised control, with growth involving overproduction of links and nodes, followed by selective pruning of some links and reinforcement of others [39, 250]. There are also similarities between these mechanisms, those in ant colony optimisation, and those implemented in a number of plant algorithms (Chap. 16).

Most research work on mathematically modelling the foraging and growth activity of hyphal networks has sought to enhance our understanding of the critical mechanisms underlying these behaviours. Most commonly, a simulation approach ([71, 400], to name but two such studies) is taken and the *in silico* fungal networks which result are examined to determine how realistic their growth patterns are compared with those of real-life networks. These studies can provide insight into the minimal set of mechanisms which are required to generate life-like growth behaviours and can be used to inspire the subsequent design of computational algorithms for application to real-world problem domains. As yet, this latter area is quite unexplored.

15.7.1 Innate Growth Behaviour in Three-dimensional Environments

Although the structures of hyphal networks are known to be sensitive to the location of food resources and local environmental conditions, the impact of structural features in the environment such as obstacles on fungal growth is much less well understood. Traditionally, many laboratory experiments concerning fungal growth have limited their attention to growth in Petri dishes, i.e. in two dimensions, owing to the difficulty of observing fungal growth in natural environments. More recently, with improvements in manufacturing technologies such as additive manufacturing and other technologies, the examination of the physical growth of fungi in three-dimensional microfluidic structures has become possible.

The growth behaviour and optimality of the search behaviours of basidiomycetous fungi in microfluidic mazes and networks was examined in [238], with the study finding that growth was strongly modulated by the geometry of the structure being traversed. The key issues faced in exploring the environment include in what direction hyphae should grow, when and where should they branch, and which hyphae should be maintained over time. The growth/search strategy of the mycelium was found to consist of two main elements:

i. long-range directional memory of individual hyphae; and
ii. inducement of branching by physical obstructions (*collision-induced branching*).

Hyphae were found to grow in a relatively constant direction. If they had to alter direction temporarily to get around an obstruction, growth would continue along the wall of the maze (a three-dimensional structure in this case) and then revert to the original direction once the obstruction or corner was circumvented. Accordingly, it appears that each hypha is 'programmed' to explore an arc of the environment on branching from the mycelium, with memory of this direction not being lost even as the hypha grows away from the branch point [238]. However, if an obstacle is sufficiently large a hypha may find it difficult to circumvent it, becoming trapped behind it (there is a limit to the degree of turning a growing hypha will undertake). This problem is partially overcome by the second element, that the branching rate of hyphae is impacted by the rate of collisions with barriers, with more branching occurring as the rate of collisions increases. This facilitates the escape of a hypha from a barrier, as, after collision with it, either the tip branches or subapical branches form further back along the hypha (depending on the species of fungus), and these could potentially branch around the obstacle. Therefore, we can think of a hypha and its 'offspring' as exploring a region of space.

Simulations using these two mechanisms, where each was parameterised using data drawn from empirical examination of real fungi, indicated that this simple growth strategy maximised both survival and biomass homogeneity in microconfined networks and mazes, even in cases where the networks or mazes contained no nutritional resources.

It is critical to note that these behaviours are independent of the resource distribution in the environment, and appear to be innate growth behaviours. Although real-world growth behaviours are influenced by the distribution of nutrients and the avoidance of inhibitory substances, growth is not simply a matter of responses to positive and negative feedback. Even without chemotactic cues, fungi are capable of implementing an intelligent search strategy, allowing them to solve complex geometric problems in their environment [238]. In contrast, most exemplars of slime mould intelligence (such as maze solving) rely on nutrient cues.

The performance of the above fungal-inspired search strategy was benchmarked in [19, 20] in a variety of simulated two-dimensional maze scenarios (where the maze is represented by a network structure) against the performance of a number of standard discrete-space-searching algorithms, including depth-first search and A*. In the simulations, the size of the maze was varied, as well as the locations of the maze entrances and exits. The results indicated competitive performance by the fungal-inspired algorithm against several standard algorithms, but underperformance against the more powerful 'informed' algorithms such as A*. The degree of underperformance did not increase as the size of the problem was scaled upwards.

The results are encouraging, indicating that even a simple algorithm extracted from simple fungal search behaviours can perform reasonably well on these maze problems. As noted in [238], the underlying model considers only two search mechanisms and, consequently, does not consider all aspects of real-world search behaviour by these fungi. It remains an open question as to whether the incorporation of some additional mechanisms would strengthen the results further.

15.7.2 Biomass Recycling

An issue to consider when developing algorithms based on fungal foraging behaviour is how (simulated) biomass is recycled in response to feedback from the environment. Unlike animals, fungi are indeterminate organisms which can exhibit considerable phenotypic plasticity, enhancing their survival in complex environments where the location of required resources is ephemeral [174]. A key element of the adaptiveness of mycelial networks is their ability to recycle biomass from regions of less promise to regions with greater resource potential.

Falconer et al. [174] examined this process, drawing a distinction between *immobile biomass* such as that incorporated in structural or storage elements of the mycelium (e.g. the external structure of hyphae and cords) and *mobile biomass* which circulates within the mycelium. The movement of mobile biomass facilitates the transport of nutrients and other requirements around the network. In recycling, local immobile biomass is broken down and converted into mobile biomass which can be relocated, either to provide energy for hyphal growth or for storage.

In [174], the mycelium is characterised as consisting of both immobile and mobile biomass. Immobile biomass is considered as being either noninsulated biomass, meaning it can take up resources from the external environment (e.g. at the hyphal tip), or insulated biomass, which does not take up resources from the environment. In the mathematical model developed, the external resources absorbed by the mycelium are redistributed, along with remobilised elements, to growing tips (sinks) via the movement of internal mobile biomass. The model was used as the basis for a simulation study of growth processes. On examining the resulting growth forms, Falconer et al. found that simulated colonies displayed typical empirically observed colony-scale features.

The key finding from the simulations is that apparently complex growth behaviours, and associated phenotypic morphology, can result from fairly simple interactions between localised processes governing the recycling of biomass and macroscopic processes associated with colony-scale transport. Hence, growth and network form can be explained as an emergent phenomenon arising from the interplay of local processes.

FUNnet

The mathematical model of [174] was operationalised in [237] to create FUNnet, an algorithm applied for the purpose of developing a routing protocol for a data communications network. In the study, the topology of the network was fixed, the nodes being sources and sinks for data packets, the hyphae being connections between the nodes, and the mobile biomass being the data traffic (a full mapping between the problem domain and the fungal setting is provided in [237]). The objective was to develop a protocol for routing data across the network.

In order for data to be routed through the network from origin to destination, a routing decision must be taken each time data arrives at a node in order to identify the next node to which the data should be sent. The choice of the next node is probabilistic depending on the capacity of each node and the levels of mobile and immobile biomass. When there is a demand to send a data packet, an amount of mobile biomass is generated, corresponding to the data. As traffic moves across the network, flows of biomass occur (initial biomass input diffuses over the route the data follows), with mobile biomass transforming into immobile biomass. In turn, this impacts on the structure of a link. If more data flows along a link, its capacity is increased. Therefore, the network structure is adaptive to the data flowing across it.

15.7.3 Fungal Search Algorithm

A different approach to the design of a fungal-inspired search algorithm is taken in [321], which metaphorically draws on the life cycle of fungi rather than focusing on the detailed modelling of search behaviours. The algorithm, named the *fungal search algorithm*, was applied in a proof-of-concept study of maze exploration using randomly generated 100-cell mazes. The authors note that the algorithm can be used as a general-purpose pattern search algorithm.

Initially, a fixed number of artificial spores (AS) are scattered in the search space. A spore germinates if one or more artificial nutrients (AN)—in other words, a location with some desired characteristic depending on the nature of the problem being addressed—can be located by a search process which takes place within a fixed radius r_{max} around that AS. If no AN can be found within that region, the AS effectively remains dormant, simulating a nongerminating spore. On the other hand, if an AN is discovered by an AS, an artificial hypha (AH) is constructed from the location of that AS to the newly discovered AN, with subsequent search for further ANs also taking place from this AN. This loosely simulates a hyphal growth process, with search proceeding onward from resources after they are discovered. When no more sources of AN can be located by a growing network of hyphae, a sporulation process is simulated, with new AS being distributed, allowing the search process to recommence in new regions. As the algorithm runs, multiple AH will be formed, and some of these may link up at commonly discovered ANs to form an increasingly large artificial mycelium (AM). In the application to a maze problem, each AN represents a node (a location in the search space), each AH represents an arc and the entire AM is a graph. Further studies are planned by the authors to test and benchmark the performance of the methodology.

15.8 Similarities with Social Insects?

There are striking similarities between the group decision-making processes of organisms such as social bacteria, slime moulds and fungi, and those of social insects. In each case the colony-level decision is based on information gathered by individual organisms (or cells), in a bottom-up, emergent process. Feedback loops play an important role, as the actions of individual organisms (or cells) are crucially driven by both private and public information.

It is speculated that similar mechanisms may govern the decision-making processes in higher animals. Marshall et al. [387] compare the decision-making processes of social insect colonies and the brain and note that both individual ants and individual neurons are relatively simple information processors. Sensory information may be ambiguous and time varying, and therefore individual information processors have local rather than global information concerning the environment. In insect societies, slime moulds, fungi, bacterial colonies and brains, decisions are not made by individual information processors but rather as a result of an emergent process.

15.9 Summary

All organisms are faced with a multitude of decisions when foraging, including how to search for resources, the optimal diet problem, efficient allocation of resources between exploitation of current food resources and exploration for new resources, anticipation of periodic environmental events, and making risk–return trade-offs when foraging. While complex behaviours are usually associated with animals, the capacity to integrate sensory and internal state information as a coordinated individual is a ubiquitous but poorly understood feature of nonneuronal organisms such as slime moulds and fungi [6]. This suggests that brains and neuronal networks are not prerequisites for decision making [481]. Although the mechanisms for coordination in slime moulds and fungi are not clearly understood, it is proposed in [6] that it may involve a simple feedback between a signalling molecule and a propagating contraction front. Under this theory, a localised nutrient stimulus releases a signalling molecule which triggers an increase in the contraction amplitude initiated at the stimulus site, with this front self-propagating across the organism. Further research is required to uncover the exact workings of the underlying coordinating mechanisms.

In this chapter we have provided a brief introduction to some of the foraging behaviours of slime moulds and mycelial fungi. Growth in each organism is indeterminate, as their precise morphology is crucially determined by environmental influences, including resource distribution. We also described a number of algorithms which have drawn inspiration from these behaviours in order to solve graph and real-valued optimisation problems. All of these algorithms are relatively new, and further work is required to fully assess their efficiency and effectiveness.

16

Plant Foraging Algorithms

Most plants are autotrophic and, therefore, unlike animals, are capable of making their own food from inorganic matter. The success of plants in colonising the planet suggests that they have evolved robust resource capture and reproductive strategies. In spite of the preponderance of plant life, relatively little inspiration has been drawn from plant foraging activities for the design of optimisation algorithms.

In this chapter we initially provide some background on plant behaviours, highlighting some of the key distinctions between plants and animals. Then we outline the comprehensive sensory capabilities of plants which help provide information to inform their foraging activities, and follow this with a description of plant *above-ground* and *below-ground* foraging behaviours. Finally, we discuss a number of algorithms which have been inspired by plant foraging behaviours.

16.1 Plants and Animals

Despite the differences between plants and animals in how they earn a living, there are many commonalities in the challenges they face. Plants, like animals and, indeed, simpler organisms, need to undertake a variety of complex tasks in order to acquire necessary resources, defend themselves against predators and pathogens, and reproduce in the dynamic environment which they inhabit. As with all living creatures they need to be able to sense their surroundings and internal state, make decisions based on this information, and adapt their activity accordingly. Plants, like animals, also need to be able to make trade-offs among allocation of energy to resource acquisition, defence and growth [78] in order to optimise their reproductive potential to pass on their genes. In the following subsections we contrast animals and plants along three dimensions, namely, mobility, learning and growth.

© Springer Nature Switzerland AG 2018
A. Brabazon, S. McGarraghy, *Foraging-Inspired Optimisation Algorithms*,
Natural Computing Series, https://doi.org/10.1007/978-3-319-59156-8_16

16.1.1 Mobility

A key distinction between plants and animals is that plants are *sessile*, being fixed in a location owing to their root network. Plants primarily exploit their two environments (above and below ground) by growth of their shoots and roots [577].

Apart from growth, plants are capable of some limited movement, and examples include *tropic* or directional movement of a plant or part of a plant in response to a stimulus. This response may be positive or negative, and common examples include plants moving their leaves and stems in response to light, and root movements in response to gravity, chemicals or water. Other examples of plant movement include *nastic movement*, where a plant responds to an external stimulus but the movement is independent of the direction of the stimulus, for example, the opening of flowers in daytime [514]. The most significant long-term movement of plants occurs during their reproductive phase when they disperse seeds, or asexual propagules such as tubers, runners or clonal plants.

While a small number of animals, including barnacles, corals, sea anemones and sponges, are also sessile for some of their life cycle, most animals possess a locomotive capability.

16.1.2 Learning

A characteristic of animal behaviour is that it is influenced by prior experience and other forms of learning. In animals, memory is maintained in differentiated neural cells and in the cells of their immune system [577]. An open question concerning plants is whether, given their lack of neuronal tissue, they are capable of displaying behavioural responses as a result of lifetime learning.

Several instances of plant behaviour appear to indicate that plants can anticipate and adapt to future environmental conditions [299]. Examples include trees shedding leaves in autumn as if they are anticipating winter conditions, which would have a high probability of damaging leaves and branches if shedding did not occur. On closer examination, many of these cases are more plausibly explained as being instances of plants innately responding to cues that are good correlates with future conditions such as a shortening daily photoperiod as the year enters autumn, rather than being examples of true cognition.

A more tenable example of learning, involving habituation, is illustrated in a study of the *Mimosa pudica* plant (also known as the *sensitive* plant) which displays a leaf folding response when its leaves are touched. While it is long known that this plant displays habituation to stimuli such as drops of water (see [13] for a review of this work), in [194] the plants were exposed to repeated treatments (a 15 cm fall or drop) that caused the leaves to close, in order to test whether the plant could learn that these stimuli should be ignored. The experiments were undertaken under both low-light and good-light conditions in order to see whether plants in the low-light environment would be faster learners, as leaf closure is more costly in these environments owing to the shortage of light for photosynthesis. The results indicated that habituation was more pronounced and persistent for plants growing in light-deprived environments. It was also found that plants could display the response even when left undisturbed in a more favourable environment for a month (i.e. indicating a long-lasting behavioural change as a result of previous experience).

While examples of associative learning concerning animals are plentiful, no instances of this had been documented in plants. A recent paper in [197], based on experimental evidence concerning the garden pea *Pisum sativum* on a Y maze task, suggested that plants are capable of associative learning.

However, as the actual mechanism(s) of learning in both of the above cases are not described, many biologists hesitate to describe these results as providing conclusive evidence that plants are capable of learning.

How Could Plants Encode Learning?

While nonneuronal organisms such as plants clearly do respond to sensory inputs, as noted in [210] these responses can only be defined as resulting from learning if they are coupled to memory mechanisms, and only if partial memory traces which facilitate a future response to the recurring input occur. Our understanding of how plants integrate signals from their environment and produce behavioural responses is very incomplete, as is our understanding of how plants could potentially encode lifetime learning.

One view is that learning could occur as a result of epigenetic changes or, alternatively, as a result of structural changes in signal transduction pathways. Under the latter proposal, learning by plants would entail three steps as illustrated in Fig. 16.1. Trewavas [577] suggests that when the relevant signal stops, a chemical or physical trace of the activated transduction pathway remains, and this acts as a memory of the initial stimulus and its pathway. As a result, a threshold of response is reduced, so that subsequently, either the signal required is lower than before, or the response to a signal is quicker or stronger.

Another intriguing speculation is that plants may embed information-processing structures akin to those in multilayer perceptrons (neural networks) [509]. Under this proposal, inputs are converted into a signal by a plant by means of a chemical reaction, with protein molecules and gene promoters forming processing units (akin to hidden-layer neurons in a multilayer perceptron) that are connected through biochemical connections (edges), in turn resulting in phenotypic outputs. Although this is an interesting analogy, with profound implications if it were true, Scheres and van der Putten [509] point out that there are important differences between biological networks and their electronic counterparts. Biological networks are connected in more intricate ways, inputs are not processed synchronously, intermediate layers are modified by both environmental and developmental signals that operate over different timescales, and there are multiple levels of feedback in these networks. They also note that although feedback exists, there is as yet no evidence to show that the molecular networks involved in plant responses to the environment learn by other mechanisms apart from the random adjustment of weights through mutation and selection that shapes their evolution.

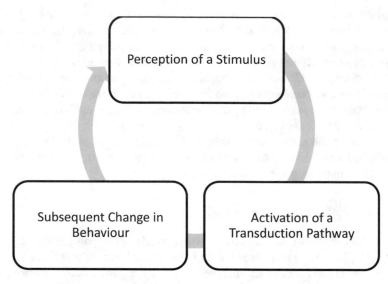

Fig. 16.1. Proposed mechanism for encoding of plant learning [577]

In summary, in spite of the ubiquity of plant life, we have limited insight as to how these organisms can process information and the degree of lifetime learning of which they are capable.

16.1.3 Growth

Another distinction between plants and animals is that the size of animals is determined genetically. In contrast, the size of a plant is not fixed a priori. Subject to resource limitations, they can continue to grow throughout their lives [514].

Indeed, the lifetimes of some perennial plants can be very long by human standards. While the longest-living vertebrate is thought to be the Greenland shark, potentially living for about 400 years [435], the longest-living perennial plants (woody-stemmed plants such as trees) can survive in exceptional cases for thousands of years. The oldest living nonclonal tree in North America (a Great Basin Bristlecone Pine found in the White Mountains of eastern California) is thought to be over 5000 years old [89]. Some plants can form interconnected root systems known as colonies with one of the most famous long-lived examples of this being a colony in Utah called Pando, which is thought to be 80,000 years old [313]. This clonal colony of 43,000 trees originally arose from a single male quaking aspen, with this tree sending out horizontal stems or roots which travelled some distance, taking root and subsequently growing into new connected trees. Hence, all the trees share the same root system and possess the same genes. However, individual trees in the colony die over time and no individual tree dates back to the origin of the colony. There are also examples of very long-lived sea plants, including a clonal colony of *Posidonia oceanica* (seagrass) located south of the island of Ibiza in the Mediterranean Sea which is speculated to have existed for over 10,000 years [18].

Plants have a modular construction and are less well integrated than animals. The modular nature of plants arises from iterative active meristems that have the ability to grow into organs of undetermined characteristics, varying in size, shape and number. As a result, plants can transform in radically different ways in response to different environmental cues [299]. Although modular construction may appear more basic than the integrated design of animals, one advantage of modularity in plants is that it fits well with a sessile lifestyle. If plants had critical functions in specialised tissues (e.g., distinct organs as in animals) then even limited predation on a plant could prove fatal [577]. Therefore, a sessile lifestyle, with its associated risks of predation, may have promoted an evolutionary path towards modular construction.

16.2 Sensory Modalities of Plants

In order to facilitate adaptive capabilities, plants need to be able to capture information about their environment and take action based on this. As plants are fixed in place, their survival depends on capturing and responding to detailed information on their immediate environment. Plants are sensitive to a wide list of stimuli including light, temperature, mechanical vibrations, water, gravity, electrical signals, and chemicals.

The perceptual world of every organism has evolved out of the demands of the environment it inhabits. These perceptual worlds can vary markedly, with species having differing sensory modalities and degrees of acuity for each sense. As commented by Wynne and Udell [631], it is not reasonable to expect animals to have the same sensory, communication or decision-making capabilities as humans, since the precise challenges each species faces depend on its ecological niche. The same is true of plants in that they earn their living in a very different way to humans. Therefore, their sensory and decision-making requirements will be distinctly different. In this section we overview some of the key sensory modalities of plants.

16.2.1 Light Reception

As the perception of light signals and the capture of energy from sunlight are of crucial importance to the survival of plants, unsurprisingly, they have strong photoreception capabilities. As with all living organisms which are responsive to light, plants detect light when receptor molecules absorb electromagnetic radiation at the specific frequency needed to push the receptor molecule into a higher oscillatory state. Once the energy is absorbed, it causes a cascade of other changes in photosynthetically active plant tissue or in the nerve cells of animals [118, 300].

One of the most important ways that plants can sense light is through a number of groupings of photochromic protein photoreceptors whose origins have been traced to photosynthetic prokaryotes [533]. The various groupings of photoreceptors (phytochromes, cryptochromes and phototropin) respond to different wavelengths of light by switching between inactive and active forms in response to light containing those wavelengths. In essence, each photoreceptor acts a sensor for a particular range of wavelengths of light.

The combination of signals from these plays a key role in the regulation of plant development, providing crucial information which is relevant for seed germination, for the timing of flowering, for the response to neighbour competition, and for the allocation of resources to root, stem, leaf, reproductive or storage structures [533].

Light signals allow the plant to determine where it is in space and time during a day [533] or during the year (season) [514]. The latter can be determined from changes in the length of the day–night cycle, with plants being particularly sensitive to night length. Seasonal information is crucial in order to ensure flowering at the optimal time of the year.

Shade Avoidance Behaviour

Sensory signals can be considered as the modulations of a specific carrier for which the sensory system has sensitivity [620]. In the case of light, all organisms generate visual signals passively as light reflects off them or is altered in wavelength by passing through them. One example of a behavioural response to light signals is the *shade avoidance* behaviour exhibited by many plants.

Daylight usually contains roughly equal portions of red and (longer wavelength) far red light. If a plant is shaded by another plant, this ratio is altered as light falling on the shaded plant will have a lower proportion of red light (and therefore a higher proportion of far red light), as this will have been absorbed by photosynthetic pigments in the leaves of the vegetation which is causing the shade [533].

On absorption of red photons, biologically inactive phytochromes are converted to their active version, and, conversely, the active form is converted back into its inactive form if a far red photon is absorbed. The ratio of active to inactive forms is therefore a function of the characteristics of the light falling on the plant, and this ratio provides information on the degree of local competition for light or, in other words, the local community density. This information is used by plants to modify their growth in order to avoid being shaded, by enhancing elongation growth, thereby projecting leaves into unshaded daylight [533]. A critical characteristic of shade avoidance behaviour is that it is initiated before any notable shortage of light resource takes place and therefore anticipates increased future competition for this resource before it actually materialises.

16.2.2 Chemoreception

As with light, chemoreception in plants occurs via proteins that act as receptors. In order to allow chemical cues from the external environment to enter the plant, one end of the protein resides outside the exterior cell wall and the other end of the protein resides within the cell. A cascade of chemical reactions are triggered when the exterior portion of the protein binds with a specific chemical cue [300].

Plants engage in a multiplicity of interactions with their environment, such as the acquisition of necessary chemical nutrients from the soil. They possess a variety of chemoreceptors to facilitate these processes. Plants are also capable of detecting airborne chemicals and can generally respond to three volatile compounds, namely [514]:

i. ethylene (a plant hormone, employed as a signal for some development pathways);
ii. jasmonic acid (a signal pathway for some plant defence mechanisms); and
iii. nitrous oxide (released in some cases following pathogen infection of a plant).

Ethylene serves as a communication signal between plants or parts of an individual plant as to their respective development state. The chemical is released during the ripening of fruits and there are receptors for ethylene on the surface of plants [514]. Detection of this chemical will speed up the ripening of a plant's own fruit.

Chemoreception capabilities play a key role in allowing plants to respond to attack by pathogens or herbivores. Plants can recognise *microbe-associated molecular patterns* (MAMPs) of pathogens and initiate defence mechanisms in response. Plants are also capable of detecting chemicals emitted by herbivores (*herbivore-associated molecular patterns*, or HAMPs) and initiate defences such as thickening the cuticle of their leaves.

Plants can emit chemicals in response to attack by herbivores or disease. One of the best-known examples of this is the emission of the volatile chemical methyljasmonate as an alarm signal by tomato plants when under attack from herbivores. Other tomato plants in the neighbourhood can detect this chemical and prepare for the attack by, in turn, producing chemicals that defend against insects. Other instances of plants defending themselves from insect attack, by emitting volatiles to attract predators of the attacking insects, have also been identified.

Another example of chemical signalling is the drought alarm signal of the garden pea *Pisum sativum*. As well as perceiving and responding to the chemical stress cues emitted by drought-stressed neighbours by closing their stomata, unstressed plants signal information about the impending drought conditions and in turn, elicit stress responses in plants located further away [192]. It is speculated that this could be a form of communication rather than a mere response to a cue, as the process plausibly benefits both the receiver and the signaller [192]. In the first case, receipt of the signal allows early response to changing conditions, and in the latter case, transmission of the warning may benefit the signaller as it could reduce the attraction of the stand of plants for herbivores, as the initiation of a stress response leads to a decline in leaf size and stem growth.

Plants also emit chemical signals such as floral scents to attract mutualists such as seed dispersers and pollinators.

16.2.3 Touch

Many plants display a response to touch. For example, tendrils in climbing plants, such as vines or bindweed, will tend to coil around a supporting object, with differential cellular elongation taking place on the side opposite the supporting object, leading to the coiling behaviour.

16.2.4 Sound

Sound, or more generally a vibration, can produce effects in plants, with [284] documenting the existence of sound-responsive genes in plants. These genes are also known to be light-responsive and this suggests that sound could represent an alternative to light as a gene regulator. The roots of corn are responsive to vibration, displaying directional root growth towards the source of soil-borne vibration [114]. Corn roots can also emit structured spike like acoustic emissions but the production mechanisms and adaptive value of these acoustic emissions are not understood.

More generally, plants can emit sound energy, particularly at the lower end of the audio range (10–240 Hz), as well as ultrasonic acoustic emissions ranging from 20 to 300 kHz. The acoustic emissions are generally interpreted as arising from the abrupt release of tension in the water transport system of plants following cavitation as water is pulled by transpiration from the roots though the xylem (see Sect. 16.3.1) to the leaves, cavitation occurring when dissolved air within the water expands in the xylem conduits, eventually generating air bubbles, causing an embolism which blocks the conduits [192].

Do Plants Use Sound Energy to Communicate?

While many plant biologists argue that these acoustic signals are simply emitted as an incidental by-product of the biomechanical process of cavitation, others speculate that plants may use sound vibrations as a means of communication, noting that the number of sound emissions makes it unlikely that all are due to cavitation alone [192, 196].

Acoustic communication could offer advantages to plants over chemical communication in that sound propagates faster and over longer distances than chemical signals, and allows the transmission of more complex information [192, 564]. The utility of sound as a transmission channel for short-range signalling, possibly modulating the swarm behaviour of growing roots, is noted in [193].

A communication capability would require that plants have the ability to sense sound energy. While plant responses to sound energy vibration have been observed, no specific receptor mechanisms for sound energy have yet been identified in plants [564]. It is speculated that hair-like structures such as root fibres could serve as a sound receiver and detect vibrations in the ground [193].

Vibration

Closely related to the question as to whether plants can sense sound energy is whether plants can detect and respond to vibrations. The evidence suggests that plants are sensitive to vibration and respond to it in a variety of ways. One of the best-known examples of vibration-responsive behaviour is exhibited by the *Mimosa pudica* plant whose leaves fold inward and droop if touched. The response is thought to be a protective behaviour against herbivory.

There is also evidence that plants can detect and respond to the specific vibrations caused by insect herbivores when they attack leaves. Chewing herbivores produce characteristic high-amplitude vibrations that can travel rapidly to other parts of the plant. A study by Appel and Cocroft [12] reports that *Arabidopsis thaliana* plants exposed to (simulated) chewing vibrations produced greater amounts of chemical defences in response to subsequent herbivory. In other words, the simulated vibrations primed chemical defence reactions in plants and therefore plants were able to respond to herbivore-generated vibrations in a selective and meaningful way. The study also found that plants can discriminate between these vibrations and those caused by wind and other factors, with the latter producing no defence reaction.

16.2.5 Magnetoreception

A variety of animals and bacteria display a capability to sense magnetic fields and therefore have a *magnetoreception* sensory modality. The question as to whether magnetic fields can affect plants, and whether plants can detect these fields, has attracted research for decades and a good review of this work is provided in [202]. Empirical studies on the effects of magnetic fields on plants have produced a wide variety of results. For example, some species of plant exhibit faster germination and growth if orientated parallel to the earth's magnetic field. Other species exhibit these features when orientated towards a specific magnetic pole. The strength and direction of magnetic fields can also affect the orientation of the root system of some species.

While it is known that magnetic fields can impact on plants, why this is so remains a poorly understood issue. Many studies examining the effect of magnetic fields on plants used fields much stronger than those typically experienced by plants in nature. The earth's geomagnetic field is fairly weak and relatively homogeneous (being approximately 35 µT near the equator and 70 µT near the magnetic poles).

Plants, like most living organisms, emit weak magnetic fields. A study in [195] tested the ability of young chilli plants to sense their neighbours when all known plant signalling pathways, including light, touch and chemical, were blocked. The results indicated that the presence of a neighbouring plant had a significant influence on seed germination even when all known sources of communication signals were blocked. Seedlings allocated energy to their stem and root system differently depending on the identity of their neighbours. The authors speculated that the plants may have been able to detect the magnetic field generated by their neighbours. While this is a controversial claim, the question as to whether plants can perceive magnetic fields remains open. No specific magnetoreceptors have yet been identified in plants. In animals, magnetoreceptors are typically composed of magnetites; however, few instances of magnetite have been documented in plants.

16.2.6 Summary

From the above discussion, it is evident that plants have significant sensory capabilities and are capable of capturing fine-grained information about their environment. In turn, this information provides a basis for adaptive behavioural responses to changes in their environment on the part of plants. It is also evident that despite a plethora research studies, our understanding of the sensory capabilities of plants is incomplete.

In the next section we provide some background on how plants forage. We then follow this with an introduction to a range of algorithms inspired by various foraging behaviours.

16.3 Plant Foraging Behaviours

Plants require a multiplicity of resources to grow and reproduce successfully. Metaphorically, we can consider plants as foraging for light energy and carbon dioxide above ground, and foraging for nutrients/water below ground. Plants search their environment by growing exploratory structures, roots and shoots, in an effort to capture resources [481]. In this section we provide background on some of these processes and on the complex ecology that plants inhabit. We also consider the processes of seed dispersal and pollination, each of which can also be considered within a foraging metaphor, that of foraging for fertile conditions for seed germination and that of foraging for mates, respectively.

16.3.1 Photosynthesis

Most plants and algae (nonflowering aquatic plants that do not have vascular tissue or roots/stems) use photosynthesis to convert carbon dioxide and water into a variety of complex organic compounds (particularly carbohydrates) using energy from sunlight to drive the chemical process, releasing oxygen as a waste product. Plants create carbohydrates mainly in the form of sucrose and starch. Sucrose is the soluble form in which carbohydrate is transported around the plant, whereas starch is an insoluble carbohydrate which is made and stored in the chloroplasts. Plants make sucrose as the dominant carbohydrate early in the day, gradually switching to the manufacture of starch as the day progresses. At night, the starch is metabolised into sucrose to provide energy for the plant [514].

The basic equation for photosynthesis is

$$CO_2 + H_2O + \text{sunlight} \rightarrow (CH_2O)_n + O_2 \qquad (16.1)$$

with nitrogen playing an important role to help catalyse this process. Despite the plentiful supply of nitrogen (N_2) in the atmosphere, plants cannot directly use nitrogen in this form, and rely on absorption of nitrogen compounds from the soil through their root network for their required supply.

Virtually all of a plant's water requirement is obtained through its root network and passes upwards from the roots to the leaves, where it evaporates through openings in the leaf surface called *stomata*. If a plant is enduring a water drought, its stomata will close in order to avoid water loss, at the expense of carbohydrate production.

The movement of water and other nutrients from the soil upwards through the plant is known as *transpiration* (Fig. 16.2). Plants have two primary internal trans-

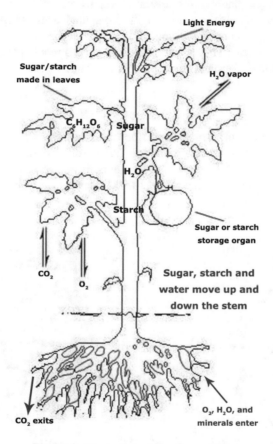

Fig. 16.2. Illustration of plant transpiration

port systems. The *phloem transport system* moves carbohydrate and amino acids around the plant. In order to facilitate this, each photosynthetic cell in a leaf is typically within four cells of vascular tissue [514]. The *xylem* transports water and other nutrients from the roots to the stem and leaves.

As photosynthesis is the critical means of generating food resources for the plant, we might assume that the process has evolved to be quite efficient at converting light energy into carbohydrate. Curiously, even under optimal conditions, the theoretical efficiency of photosynthesis is only around 35% and, in practice, the realised efficiency is typically far lower, often approaching single digits [514]. While this appears very low, there are often other more important restrictions on plant growth than efficient energy conversion [514].

16.3.2 Root Networks and Foraging

The root network of plants fulfils several functions in addition to foraging for resources, including holding the plant upright so that the leaves can be properly placed for light harvesting, anchoring the plant and acting in some cases as a storage system for surplus carbohydrate. The root structure also plays a role in plant propagation, as it impacts on the above-ground development of a plant.

Root Structure

There are two common plant root network architectures. In a *tap root system*, the root consists of a main (or tap) root and multiple lateral (or secondary) roots growing off this. Tap root systems are features of dicot plants. In a *fibrous root system*, as found in monocotyledonous plants, the root network consists of a web of roots of similar diameter (Fig. 16.3).

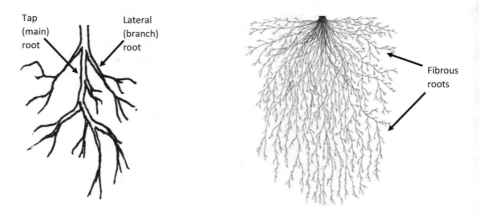

Fig. 16.3. Illustration of a tap root system (left) and a fibrous root system (right)

In all root systems, towards the end of each root, there are many fine root hairs and these are the main interface between the plant and the surrounding soil. The hairs vastly increase the surface area of the root system and the volume of soil that can be exploited. The majority of water and mineral absorption takes place via root hairs, as these have no cuticle and therefore present no barrier to diffusion of water and solutes from the soil [514].

Soils present a complex, dynamic environment. Topsoil (by volume) typically consists of minerals (45%), organic matter (5%), water (25%) and air-filled gaps. It presents an environment through which plant roots can propagate. There are many soil characteristics that plant roots can assess, including soil density, water content, mineral gradients, pH, the presence or absence of other plant roots, the presence of volatile gases and the presence of light [577]. Roots may be assessing up to 20 signals at any time, and our understanding as to how all this information is integrated into root growth decisions is incomplete [577].

In general, if a root encounters a higher concentration of a chemical that is required by the plant, the root system responds by initiating lateral root growth in that area, increasing the root biomass in nutrient-rich areas [300, 514]. Under *Liebig's law of the minimum* [577, p. 124], plant growth rates are dependent on the least available resource and, consequently, root foraging decisions are likely to focus on the constraining resource rather than foraging equally for all items.

Required Resources

Plants need to access several critical minerals in the soil, including nitrogen (for photosynthesis), phosphorus (needed for growth) and potassium (to support metabolic processes in the plant). Most nitrogen in the soil is not directly accessible, and to overcome this problem, plants can enter into symbiotic associations with bacteria to fix nitrogen and therefore make it accessible to the plant (for example, by converting atmospheric nitrogen (N_2) into ammonia (NH_3) which plants can absorb and use). In these cases, the plants house N-fixing (nitrogen-fixing) bacteria in root nodules and receive nitrogen from the bacteria in return for providing a supply of photosynthe-ate to the bacteria. There is evidence that plants can sanction bacteria that perform nitrogen-fixation poorly in a nodule by withholding oxygen supply or by varying carbohydrate delivery depending on the amount of nitrogen fixed [577].

Plants also need to obtain phosphorus, but most forms of this in soil are insoluble and are not directly accessible by plants. In contrast, many species of fungi (Sect. 15.6) in soil form extensive networks of thread-like cells called *hyphae* and are very effective at accessing large soil volumes and mobilising insoluble phosphates. The growth-limiting factor for most fungi is the scarce availability of free carbohydrates in soil. Therefore, a symbiotic relationship, encompassing over 90% of land plant species, has evolved, where plants provide carbohydrates via their root networks to attached fungi in return for minerals such as phosphorus [547]. The two major fungus groupings which form these symbiotic relationships are *endophytic fungi* and *mycorrhizal fungi*. In the latter case, mycorrhizal fungi are *obligate biotrophs* (i.e. they enter into a long-term feeding relationship with a host without killing it) and cannot survive without a plant host [46]. This relationship is known as *mycorrhizal association* [514]. In phosphorus-deficient soils, roots secrete strigolactone, which attracts mycorrhizal fungi, and when the fungus reaches the plant root, there is chemical communication allowing the fungus to enter the root without initiating a defence mechanism by the plant [577].

The symbiotic relationships between plants and other organisms in the soil bear direct parallel with the importance of microbial communities in the digestive tracts of humans and animals. In each case, the symbiont supports nutrient acquisition by the host organism [583].

Root Network Ecology

Mycorrhizal association can therefore be seen as creating an extension of the root network of a plant. It allows the plant to access resources in a far larger volume of soil than that covered by its direct root network. Fungi can grow faster than plant roots and this can facilitate faster adaptation by plants to changes in mineral availability in the soil.

Hyphal systems of neighbouring fungi of the same species can also link together, and multiple plants can be linked to each hyphal system, potentially forming a large network of interlinked organisms. As many plants, of differing species, can share the same mycorrhizal network, this provides a potential pathway for communication and exchange of materials between plants [148, 577]. Plants in these networks can bidirectionally exchange carbon, water and other nutrients [530]. Plants can also communicate defence signals by means of a mycorrhizal network. Song et al. [538] illustrates an example of such communication, with a pathogen attack on one plant attached to a mycorrhizal network resulting in defence responses in other healthy plants attached to the same network. The existence of these networks suggests that many plants are part of a complex, interconnected underground ecosystem rather than being isolated organisms [529].

While our understanding of the nature of chemical communication between plants and species of fungi or bacteria remains limited, it is known that many of the key signalling molecules such as strigolactones play an important role in the control of the development of plants and/or their bacterial and fungal symbiotic partners. It is speculated that over evolutionary time these molecules have acquired an additional function, whereby host plants and their symbiotic partners have learnt to interpret them as symbiotic messengers, in turn impacting on each other's metabolism and development [68].

Competition Between Root Networks

Other than for reproductive purposes, the dominant social interaction between plants is competition for light, water and nutrients [162]. Access to resources from the soil is crucial to a plant's survival and consequently we would expect to see competition between plants for these resources.

Self Versus Nonself Discrimination in Root Networks

Plants have shown a capability to distinguish between self and nonself roots. The capability to distinguish between self and nonself has been documented for many multicellular organisms [301], and it is therefore not surprising that this capability would be found in plants. An ability to distinguish between self and nonself is plausible from an evolutionary standpoint, as competition for resources between the roots of an individual plant would be wasteful of energy.

In the case of self root networks, behaviours that reduce competition among roots of the same individual have been reported for several species, with plants developing fewer and shorter roots in the presence of other roots from the same plant [301].

The picture concerning behaviours when nonself roots are detected is more complex. Some studies of root behaviour illustrate enhanced lateral root development in the vicinity of root networks of neighbouring plants, presumably in an effort to capture soil volume for itself [175]. Other studies suggest that some plant species exhibit avoidance of neighbouring root networks, segregating the soil volume spatially into territories. In both cases, an ability to detect neighbouring root networks is implied.

The current understanding is that plant responses to neighbours are heavily governed by the availability of nutrient resources in the soil [53], with plant root networks avoiding one another in conditions of homogeneous soil nutrient availability and expressing a preference to colonise areas of high resource availability. It would also appear that multiple strategies can be employed by the same species of plant depending on soil conditions [481]. Despite the observed behavioural responses to the presence of nonself roots, the means of detection of these roots is not understood, as avoidance behaviour has been observed even in cases where root networks have not physically touched [175].

Kin Recognition in Root Networks

Kin selection theory recognises that individuals increase their inclusive fitness by behaviour that increases the fitness of related individuals [162]. Kin selection requires kin recognition. This is widespread in social animals and has also been identified in some species of microbes [53]. The question has been posed as to whether the root networks of plants can recognise root networks of genetically related plants and, if so, what impact this has on the interactions of those root networks.

Current evidence suggests that plants are capable of kin recognition in at least some cases, but the mechanism underlying this is not understood as there is no strong evidence for kin recognition entailing root secretions [53, 422].

One of the first studies to suggest kin recognition in plants was that of [162], which found increased allocation to growth of root mass when groups of nonkin plants shared a pot with a constrained soil volume but not when groups of siblings shared a pot. This result was interpreted by the authors as evidence that plants can identify kin. A subsequent study [422] also suggested that plants were capable of kin recognition but only in the presence of another plant's roots. No kin recognition was displayed in response to above-ground competition for light from a kin plant. Only a small number of studies have yet looked at the issue of kin recognition in plants and it is still unknown whether this capability is widespread [422].

16.3.3 Parasitic Plants

The ability of plants to store energy makes them attractive targets for pests and parasites looking to exploit their resources such as carbohydrates, nutrients and water [251]. One significant grouping of parasites is the parasitic plants, of which there are some 4000 species [514]. These prey on other plants in order to obtain some or all of their required resources rather than support themselves through photosynthesis. The aim of these plants is to establish a link between their tissue and the vascular tissue of the host. The attachment may be via the stem or the root network of the target plant. Obligate parasites, which obtain all of their resources from other plants, can take on unusual forms as they are freed from the usual constraints on structure imposed by photosynthesis.

16.3.4 Seed Dispersal

The key imperative of a plant's life is to maximise its number of viable offspring [324]. Many species of plants reproduce by producing seeds and then dispersing them in the landscape. The seed dispersal process can therefore be considered as a foraging process, the search being for fertile conditions in which seeds can germinate. If the seeds are placed in a suitable location, they germinate and in turn reproduce themselves. Therefore, dispersal mechanisms which perform this task well in a given ecological niche will be selected for by evolution.

Dispersal Mechanisms

Plants make use use of multiple dispersal mechanisms, as illustrated in Fig. 16.4. Dispersal mechanisms can be classed as *abiotic* (nonbiological dispersal mechanisms such as wind, water or gravity) or *biotic* (biological dispersal mechanisms such as insects or animals). Many plants use more than one dispersal mechanism.

In the case of wind dispersal, seeds which have characteristics such as small size, wings, hairs etc., fall more slowly, essentially by lowering their wing loading (ratio of mass to surface area), and this promotes wider seed dispersal. Species with these adaptations are very common, comprising some 10–30% of all plants, and up to 70% of the flora in temperate plant communities [430]. Wind-dispersed plants are common in dry habitants such as deserts [269].

Adaptations for animal dispersal include the offering of rewards for dispersion, such as fleshy nutritious fruits which attract the attention of frugivores (fruit eaters) which consume the fruit. The seeds contained in the fruit pass through the digestive tract of the animal and are eventually excreted back into the environment. This means of seed dispersal is common, with some 50–75% of tree species in tropical forests producing fleshy fruits adapted for animal consumption [269]. Other (non-reward) adaptations for animal dispersal include clinging structures such as hooks or resin, whereby seeds stick to the fur or feathers of animals and are accordingly dispersed as the animal moves around the environment. Many types of animals are seed dispersers including various species of mammals, birds, bees, fish and reptiles

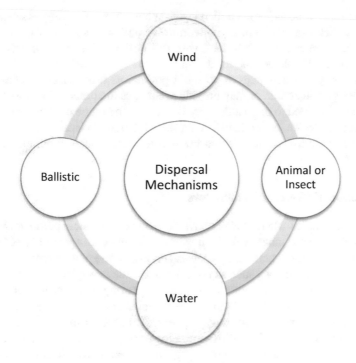

Fig. 16.4. Four primary dispersal mechanisms for plant seeds

[285, 606]. Animals and insects can also play a role as secondary dispersers. For example, ants and dung beetles can transport seeds which have fallen from plants.

Apart from wind and animal dispersal, seeds can also be dispersed by water, for example via buoyant coconuts. Some plant species have evolved ballistic fruits that open explosively and can toss seeds several metres from the parent plant.

16.3.5 Pollination

Pollination plays a key role in the reproductive process of flowering plants, and the search for mates can be considered as a foraging process. In flowering plants, male pollen is transferred to the female reproductive organs in the flower, enabling fertilisation and reproduction. Pollen may be dispersed by abiotic vectors such as the wind (this is common in species of grass), but in most cases (approximately 90% of the time) the pollinating vector is an insect or other animal.

Pollination can occur via *cross-pollination* or *self-pollination*. In the case of self-pollination, fertilisation occurs via pollen from the same flower or plant. In cross-pollination, the pollen comes from another plant. Some plants predominately utilise one form of pollination but many plants use a combination of both.

As with seed dispersal, plants may provide a reward to encourage potential pollinators to visit in order to collect pollen and transport it. Typically, these rewards are in the form of nectar or the pollen itself. When potential pollinators visit a flower to collect nectar (a source of carbohydrate) or pollen (a source of protein), pollen will commonly stick to their bodies, with this pollen being subsequently deposited by them on another flower and fertilising it.

Plants may advertise cues to potential pollinators using visual displays such as flower design and colour, or may provide scent cues to attract pollinators [300]. The visual cues provided to pollinators may lie outside the range of visual signals which humans can perceive, with many nectar guides (elaborate visual signals which a pollinator can learn to associate with a reward) to insects being within the ultraviolet range.

Coevolution of Plants and Pollinators

There are examples where flowers have coevolved with their pollinators such that each depends on the other, with the flower depending on a specific insect or animal to pollinate it, and the pollinating species relying on the plant's reward for vital resources. Some pollinators, such as honeybees, can develop *flower constancy*, where they preferentially visit certain flower species, ignoring other species which also offer similar rewards. Flower constancy can be beneficial for plants as it increases the chance that pollen from that plant will be deposited on a conspecific, thereby increasing the chance of successful fertilisation. Flower constancy can also provide benefits to the pollinators as they need not engage in trial-and-error learning by sampling the rewards offered by diverse plants, nor need they remember the relative rewards provided by a range of plants.

16.4 A Taxonomy of Plant Foraging Algorithms

Until recently, relatively little attention was paid to the potential utility of plant metaphors for the design of computational algorithms. However, the last few years have seen the development of a literature in this area, inspired by various plant foraging behaviours. The majority of these algorithms fall into three categories, namely those inspired by:

  i. plant propagation behaviours;
 ii. light-foraging behaviours; and
iii. the nutrient- and water-foraging behaviours of root networks.

Figure 16.5 outlines the algorithms discussed in the following sections based on these categories.

16.5 Plant Propagation Algorithms

Unlike animals, which can forage in new areas by moving around a landscape, plants access new foraging territories by growth and through propagation behaviours. Plants

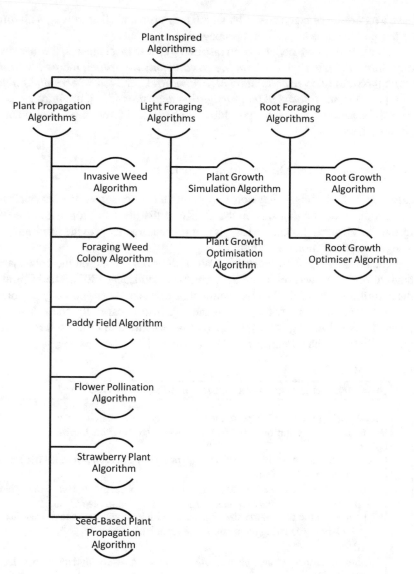

Fig. 16.5. Taxonomy of plant-inspired foraging algorithms discussed in this chapter

have a repertoire of mechanisms by which they propagate themselves, including seed and/or pollen dispersal, and root propagation.

Six algorithms which have been inspired by these mechanisms, the *invasive weed algorithm* [397], the related *foraging weed colony algorithm*, the *paddy field algorithm* [468], the *flower pollination algorithm* [641], the *strawberry plant algorithm* [502] and the *seed-based plant propagation algorithm* [557], are introduced below. As will be seen, each of the algorithms emphasises different aspects of plant propagation behaviours.

16.5.1 Invasive Weed Optimisation Algorithm

Effective seed dispersal plays an important role in ensuring the survival of plant species. In turn, this depends on the ability of the plant to propagate its seeds into resource-rich areas. Processes for this can metaphorically provide inspiration for the design of optimisation algorithms.

The *invasive weed optimisation* (IWO) *algorithm*, based on the colonisation behaviour of weeds, was proposed by Mehrabian and Lucas [397]. The inspiration for the algorithm arose from the observation that weeds or more generally any plant, can effectively colonise a territory unless their growth is carefully controlled. Two aspects of this colonising behaviour are that weeds thrive in fertile soil and reproduce more effectively than their peers in less fertile soil.

Algorithm 16.1: Invasive weed optimisation algorithm [397]

Generate $p_{initial}$ seeds and disperse them randomly in the search space;
Determine the best solution in the current colony and store this location;
repeat
    Each plant in the population of p plants produces a quantity of seeds depending on the quality of its location;
    Disperse these new seeds spatially in the search space, giving rise to new plants;
    if $p > p_{max}$ *(the maximum number of plants has been exceeded)* **then**
        Reduce the population size to p_{max} by eliminating the weakest (least fit) plants, simulating competition for resources;
    end
    Assess the fitness of new plant locations and, if necessary, update the best location found so far;
until *terminating condition*;
Output the best solution found;

Pseudocode for the IWO is provided in Algorithm 16.1. The three key components of the algorithm are:

  i. seeding (reproduction),
 ii. seed dispersal, and

iii. competition between plants.

Mehrabian and Lucas operationalised these mechanisms in the following way in the IWO algorithm.

Seed Production

Each plant produces multiple seeds, based on its fitness relative to that of the other plants in the current colony of weeds. A linear scaling system is used whereby all plants are guaranteed to produce a minimum number $N_{\min}^{\text{seeds}}$ of seeds and no plant can produce more than a maximum number $N_{\max}^{\text{seeds}}$ of seeds. The number of seeds produced by an individual plant x in the current generation is calculated as

$$N_x^{\text{seeds}} = N_{\min}^{\text{seeds}} + \frac{f(x) - f_{\min}}{f_{\max} - f_{\min}} \left(N_{\max}^{\text{seeds}} - N_{\min}^{\text{seeds}} \right), \tag{16.2}$$

where $f_{\max}$ and $f_{\min}$ are the maximum and minimum fitnesses in the current population, and $f(x)$ is the fitness of plant x.

Seed Dispersal

Exploration of the search space is obtained via a simulated seed dispersal mechanism. The seeds associated with each plant are dispersed by generating a random displacement vector and applying this to the location of their parent plant. The displacement vector has n components corresponding to the n dimensions of the search space and is obtained by generating n normally distributed random numbers, with a mean of 0 and a standard deviation σ_t calculated using

$$\sigma_t = \left(\frac{t_{\max} - t}{t_{\max}} \right)^n (\sigma_{\max} - \sigma_{\min}) + \sigma_{\min}, \tag{16.3}$$

where t is the current algorithm iteration number, $t_{\max}$ is the maximum number of iterations, $\sigma_{\max}$ and $\sigma_{\min}$ are the maximum and minimum allowable values for the standard deviation, n is a nonlinear modulation index, and σ_t is the standard deviation used in the current iteration in calculating the seed displacements.

The effect of this formulation is to encourage random seed dispersal around the location of the parent plant, with decreasing variance over time. This results in greater seed dispersal in earlier iterations of the algorithm, promoting exploration of the search space. Later, the balance is tilted towards exploitation as the value of σ_t is reduced. The incorporation of the nonlinear modulation index in (16.3) also tilts the balance from exploration to exploitation as the algorithm runs.

Depending on the scaling of the search space, the same value of σ_t could be applied when randomly drawing each component of the displacement vector. Alternatively, differing values of σ_{initial} and σ_{final} can be set for each dimension if required.

Competition for Resources

Competition between plants is simulated by placing a population size limit p_{max} on the colony. The plant colony starts with a population of size $p_{initial}$. The population size, p, increases as new plants grow in subsequent generations. Once the population limit p_{max} is reached, parent plants compete with their children for survival. The parent and child plants are ranked by fitness, with only p_{max} plants surviving into the next generation. This mechanism ensures that the best solution found to date cannot be lost between iterations (elitism).

Performance of the Algorithm

The IWO algorithm is a conceptually simple, numerical, non-gradient-based optimisation algorithm. So far, owing to its novelty, there has been limited investigation of its effectiveness, scalability and efficiency. Mehrabian and Lucas [397] reported results from the IWO algorithm competitive with a genetic algorithm (GA) and particle swarm optimisation algorithm (PSO), with settings of 10 to 20 weeds, maximum and minimum numbers of seeds per plant of 2 and 0, respectively, and a nonlinear modulation index value of 3. Competitive results for the IWO algorithm are also reported in [33, 398, 471].

The algorithm requires that several problem-specific parameters be set by the modeller, including the maximum and minimum number of seeds that a plant can produce, the values for σ_{max}, σ_{min}, t_{max}, and the initial and maximum population sizes. However, the determination of good values for these parameters is not necessarily a trivial task, particularly in poorly understood problem environments.

In addition to optimisation applications, the IWO algorithm has been applied to clustering, where each individual seed consists of a string of up to n real vectors of dimension d, corresponding to n cluster centre coordinates (in d-dimensional space $\mathbb{R}^d$) [366]. The IWO for has also been modified for application to multiobjective optimisation problems [342].

16.5.2 Foraging Weed Colony Optimisation

An extension of the IWO algorithm, drawing on models from optimal foraging theory, was proposed in [495], with the modification being termed the *foraging weed colony optimisation* (FWCO) algorithm. The algorithm embeds a different balance between exploration and exploitation during the search phase.

As discussed in Sect. 2.1.2, optimal foraging theory [168, 379] seeks to predict how solitary organisms will behave during foraging. The essence of the theory is the perspective that the foraging process can be viewed as a constrained optimisation problem, where the aim is to maximise an objective function payoff such as net energy gained per unit time, subject to constraints.

In real-world foraging, the resources in a patch are often depleted over time owing to consumption, increased predator density in a patch, or seasonal factors. Similar issues arise with plant foraging, as the greater the number of plants in an area, the greater the competition for soil and light resources. In the case of mobile animals, a critical decision facing foragers is when to leave a patch to search for an alternative. As noted in [545], a relevant decision variable in determining when to leave a patch is the current patch residence time.

Assuming that resources in a patch decrease exponentially as the forager spends time there, the total resources consumed as a function of time (denoted as K_t) can be modelled using the following equation [495]:

$$K_t = K(1 - e^{-ct}), \tag{16.4}$$

where t is the time spent in the patch, c is the decay rate for the resource once foraging starts, and K is the initial endowment of resources in the patch. The FWCO algorithm implements this idea, mimicking the concept of resource depletion, in order to avoid the over-exploration of areas around better solutions that can occur in the canonical IWO algorithm.

In the FWCO algorithm, each weed is initially assigned a fixed amount of *information* (so termed in [495]). Therefore, each weed i has three characteristics, a location, a fitness value and an information value κ_i. The value of κ_i is initially set to one for all weeds. As the weed reproduces, its information value is reduced using an exponential decay rate, with [495] providing an exemplar formulation of this as

$$\kappa_i(t) = \kappa_i(t-1) \cdot e^{-0.002 \cdot \Sigma_t p_i(t)}, \quad \text{when } p_i(t) > 0, \tag{16.5}$$

where $\kappa_i(t)$ is the information value of the ith weed at its tth iteration of existence and $p_i(t)$ is the number of seeds produced by the ith weed at its tth iteration of existence.

When the information value is above a threshold value (for example, 0.2), the usual steps of the canonical IWO algorithm are applied in generating and dispersing new seeds. However, once the information value for a parent weed falls below this threshold, the algorithm is modified in order to promote a greater degree of seed dispersal, and therefore more explorative search. The modified mechanism for generation of new seeds contains a stochastic element and also makes use of the best location found thus far by the algorithm, akin to the concept of global best g^{best} in the PSO algorithm [314, 315]. When $\kappa_i < 0.2$ for a parent weed i, then seed j from that parent is initially positioned in a random location $\text{seed}_{i,j} \in \mathbb{R}^d$ based on the maximum and minimum values for the bounds of the d-dimensional search space: for each dimension $k = 1, \ldots, d$, component k of $\text{seed}_{i,j}$ is drawn from the uniform distribution $U(x_k^{\min}, x_k^{\max})$, where $x_k^{\min}$ and $x_k^{\max}$ are, respectively, the minimum and maximum allowable coordinate in dimension k. Seed j of parent i is then moved stochastically in the direction of the best location found thus far (denoted weed$_{\text{best}}$) using

$$\text{seed}_{i,j} = \text{seed}_{i,j} + r \cdot (\text{weed}_{\text{best}} - \text{seed}_{i,j}), \tag{16.6}$$

where r is a random number drawn from a uniform distribution $U(0,1)$. The information value for all the new seeds is initialised at one.

Two other modifications are made to the canonical IWO algorithm:

i. when selecting weeds for survival into the next iteration of the algorithm, the top 95% are selected based on fitness and 5% are selected randomly from the remaining members of the colony; and

ii. the formulation of the seed dispersal process in the canonical algorithm outlined in (16.3) is altered to (16.7).

The impact of the first modification above is to ensure that every member of the colony or seed has a chance to survive into the next generation, thereby weakening selection pressure. In the second modification, the seed dispersal process is altered to

$$\sigma_t = \left(\frac{t_{max} - t}{t_{max}}\right)^n |\cos(t)|(\sigma_{max} - \sigma_{min}) + \sigma_{min}. \tag{16.7}$$

The addition of the $|\cos(t)|$ term adds a variation in the calculation of σ_t which can help promote the quicker exploitation of a good solution once uncovered.

Although it can be argued that the strict application of patch selection theory to plant reproduction is questionable from a biological perspective, the combination helps encourage a greater degree of exploration of the search space and may help the algorithm escape premature convergence. The concept bears some similarities with the concept of niching in the evolutionary computation literature [214, 383].

16.5.3 Paddy Field Algorithm

The *paddy field algorithm* was first proposed by Premaratne et al. [468]. This algorithm (pseudocode is provided in Algorithm 16.2) draws inspiration from aspects of

---

Algorithm 16.2: Paddy field algorithm [468]

Generate an initial population P of p plants, each located randomly in the search space;
Choose values for t_{max} and n;
Set generation counter $t = 1$;
repeat
　　Calculate fitness y_i of each plant $i = 1, \ldots, p$ and add this fitness vector
　　　$y = (y_1, \ldots, y_p)$ to the matrix containing the locations of all p plants;
　　Sort the population in descending order of fitness (assuming the objective is to
　　　maximise fitness);
　　for $i = 1$ **to** n *(the top n plants)* **do**
　　　　Generate seeds for each selected plant;
　　　　Implement pollination step;
　　　　Disperse pollinated seeds;
　　end
　　Replace old population with new plants;
　　Set $t := t + 1$;
until $t = t_{max}$;
Output the best solution found;

---

the plant reproduction cycle, concentrating on the processes of pollination and seed dispersal.

Given a vector $x = (x_1, x_2, \ldots, x_n) \in \mathbb{R}^n$, we view it as a location in an n-dimensional space. Let $y = f(x)$ be the fitness or quality of that location. Each seed i, therefore, has a corresponding location x_i and a corresponding fitness $y_i = f(x_i)$.

The paddy field algorithm manipulates a population of these seeds in an attempt to find a good solution to the optimisation problem of interest. The algorithm consists of five stages, sowing, selection, seeding, pollination and dispersion [468]. Each of these is described below.

Sowing

An initial population P of p seeds is distributed (sown) at random locations in the search space.

Selection

The seeds are assumed to grow into plants. Each plant i has an associated fitness value y_i determined by the output of the underlying objective function f when evaluated at the plant's location. The plants are ranked by fitness and the best n plants are selected to produce seeds.

Seeding

Each plant produces a number of seeds in proportion to its fitness. The fittest plant produces q_{max} seeds and the other plants produce varying amounts of seeds, with the number of seeds produced by plant i being calculated as

$$s_i = q_{max} \frac{y_i - y_t}{y_{max} - y_t}. \tag{16.8}$$

Here, the term y_{max} is the fitness of the best plant in the current population and y_t is the fitness of the lowest ranked plant selected in the previous step. Although the algorithm describes this step as seeding, it can more correctly be considered as the process of growth of flower structures in order to enable pollination.

Pollination

Only seeds which have been pollinated can become viable: to determine this portion, a simulated pollination process is applied whereby the probability that a seed is pollinated depends on the local density of plants around the seed's parent plant (as a greater density of plants locally increases the chances of pollination). A radius a is defined, and two plants are considered to be neighbours if the distance between them is less than a (that is, each lies within a hypersphere of radius a centred on the other). The pollination factor U_i of plant i (where $0 \leq U_i \leq 1$) is then calculated using

$$U_i = e^{v_i / v_{max} - 1}, \tag{16.9}$$

where v_i is the number of neighbours of plant i and v_{max} is the number of neighbours of the plant with the largest number of neighbours in the population. The effective number of viable seeds produced by a plant i from its initial s_i seeds is therefore

$$s_i^{\text{viable}} = U_i \cdot s_i. \tag{16.10}$$

Dispersion

The s_i^{viable} pollinated seeds are then dispersed from the location of their parent plant i such that the location of the new plant (grown from the dispersed seed) is determined using a normal distribution $N(x_i, \sigma)$, where x_i is the location of the parent plant and σ is a user-selected parameter.

Combination of Steps

The above five steps are iterated until a termination condition is reached. In summary, the fittest plants give rise to the greatest number of seeds, and search is intensified around the better regions of the landscape uncovered thus far. Variants on the paddy field algorithm include one presented in [325].

16.5.4 Flower Pollination Algorithm

The pollination process of flowering plants was used as the inspiration for the design of the *flower pollination algorithm* (FPA) by Yang [641]. In the FPA, the process of pollination is split in stylised form into two scenarios.

The first scenario is loosely modelled on biotic processes and cross-pollination, with this being considered to generate a *global pollination* process in which pollen can potentially be transferred a long distance. The second scenario is based on abiotic processes and self-pollination, with this being considered as *local pollination*. In essence, the algorithm stochastically uses both mechanisms to create a balance of exploration and exploitation during the search process.

In the FPA, it is assumed that each plant has one flower and that each flower produces one pollen gamete, or solution. Therefore, x_i is the solution vector associated with flower or pollen item i in the population. In each iteration of the algorithm, every pollen item is displaced to a new location using either a global or a local pollination process. If the fitness of the new location is better than that of the previous position, the pollen item moves to the new location. The choice of pollination process is stochastic and is governed by a parameter p whose value is chosen by the modeller. All flowers are assumed to be of the same species and it is assumed that pollen from any one flower can fertilise any other flower.

Global Pollination Step

The global pollination step is modelled using

$$x_i(t+1) = x_i(t) + \gamma L(\lambda)(g^* - x_i(t)), \tag{16.11}$$

where $x_i(t)$ is the location of pollen grain i at iteration t, g^* is the location of the best solution found so far across the entire population and γ is a scaling factor. In this operation, each solution is displaced from its current location in the direction of the best-so-far solution, using a step size which is determined by a Lévy flight $L(\lambda)$. The global pollination step, therefore, bears some inspiration from the Lévy flight hypothesis (discussed in Sect. 2.5.2) concerning animal movements, where most animal movement step sizes are relatively small, with occasional long jumps. The Lévy flight step sizes are based on a draw from a Lévy distribution [641]:

$$L \sim \frac{\lambda \Gamma(\lambda) \sin(\pi\lambda/2)}{\pi} \cdot \frac{1}{s^{1+\lambda}}, \quad s \gg s_0 > 0, \tag{16.12}$$

where $\Gamma(\lambda)$ is the standard gamma function and s is a large step size. The value of the step size is determined using the Mantegna algorithm [385], which employs a random draw from two Gaussian distributions and a subsequent transformation, as follows [649]:

$$s = \frac{U}{|V|^{1/\lambda}}, \quad U \sim N(0, \sigma^2), \quad V \sim N(0, 1) \tag{16.13}$$

where U and V are two numbers randomly drawn from Gaussian distributions. The term $U \sim N(0, \sigma^2)$ indicates a sample drawn from a Gaussian distribution with a mean of zero and a variance of σ^2, where the variance is calculated using [649]

$$\sigma^2 = \left[\frac{\Gamma(1+\lambda)}{\lambda\Gamma((1+\lambda)/2)} \cdot \frac{\sin(\pi\lambda/2)}{2^{(\lambda-1)/2}} \right]^{1/\lambda}. \tag{16.14}$$

While this formulation looks complex, it reduces to a number for any given choice of value for λ.

Local Pollination Step

The local pollination step is operationalised using

$$x_i(t+1) = x_i(t) + \epsilon(x_j(t) - x_k(t)), \tag{16.15}$$

where $x_j(t)$ and $x_k(t)$ are the locations of two randomly chosen flowers in the population, and ϵ is a random number drawn uniformly from the interval $[0,1]$. Therefore, in the local pollination step, a solution is moved locally from its current location using a step size which depends stochastically on the dispersion of the population in the search space. Initially, when the population is quite dispersed, this step size can be large. As the population converges later in the algorithm, the step size will tend to reduce.

Parameters and Algorithm

Parameter values of $n = 25$, $\lambda = 1.5$, $\gamma = 0.1$, and $p = 0.8$ have been used in a number of applications of the FPA, including [648, 649]. Pseudocode for the FPA is provided in Algorithm 16.3.

Algorithm 16.3: Flower pollination algorithm [641]

Generate an initial population of n flowers (pollen gametes), each located randomly in
  the search space;
Find the best solution g^* in the initial population;
Select a switch probability $p \in [0,1]$;
Set the maximum number of iterations t_{max};
while $t < t_{max}$ **do**
  for $i = 1$ **to** n *(all n flowers in the population)* **do**
    Select r from a uniform distribution on $[0,1]$;
    if $r < p$ **then**
      Select a d-dimensional step vector L which obeys a Lévy distribution;
      $x_i(t+1) = x_i(t) + \gamma L(\lambda)(g^* - x_i(t))$ (global pollination);
    else
      Select ϵ from a uniform distribution on $[0,1]$;
      Randomly select two solution indices, j and k;
      $x_i(t+1) = x_i(t) + \epsilon(x_j(t) - x_k(t))$ (local pollination);
    end
    Evaluate new solution $x_i(t+1)$;
    Replace old solution with new solution, if new solution is better;
  end
  Find best solution in new population and update g^* if necessary;
  Set $t := t+1$;
end
Output the best solution found;

Discussion

The FPA has the benefit of having relatively few parameters. Initial applications
of the canonical FPA produced reasonable results when tested on a range of basic
benchmark functions but recent, more detailed, analysis of the algorithm has ques-
tioned whether its results are fully competitive with alternative state-of-the-art optimisation algorithms. Readers are referred to [161] for a discussion of this.

It remains an open question as to whether improved variants of the FPA will exhibit stronger performance. Alternative algorithm designs could include using multiple pollen grains per flower or plant, or more complex pollination schemes. A number
of papers have also developed hybrid algorithms, where the canonical version of the
FPA is hybridised with another search algorithm.

The use of a Lévy flight to drive the search process is not unique to the FPA,
with other algorithms, including cuckoo search (Sect. 7.1) and the seed-based plant
propagation algorithm (Sect. 16.5.6), also making use of this approach.

16.5.5 Strawberry Plant Algorithm

Although many plants propagate using seeds, some employ a system of *runners*, or horizontal stems, which grow outwards from the base of the plant (Fig. 16.6). At

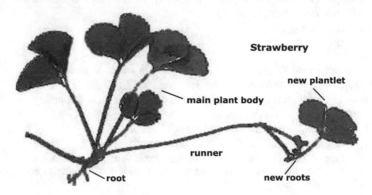

Fig. 16.6. Strawberry plant with runner stems

variable distances from the parent plant, if suitable soil conditions are found, new roots will grow from the runner and in turn produce an offspring clone of the parent plant. An example of this behaviour is provided by modern strawberry plants, which can propagate via seed dispersal and by runners. This has inspired the development of an optimisation algorithm, the *strawberry plant algorithm*, by Salhi and Fraga [502] based on this phenomenon. The algorithm draws from the following principles:

- healthy plants in good resource locations generate more runners;
- plants in good resource locations tend to send short runners in order to exploit local resources;
- plants in poorer resource locations tend to send longer runners to search for better conditions; and
- as the generation of longer runners requires more resource investment, plants generating these will create relatively few of them.

The algorithm therefore seeks to balance exploration with exploitation, with increasing local exploration over time as plants concentrate in the locations with the best conditions for growth. Salhi and Fraga report competitive results from this algorithm when it is applied to a number of real-valued benchmark optimisation problems. Algorithm 16.4 presents an adapted version of the algorithm based on [502].

16.5.6 Seed-Based Plant Propagation Algorithm

The *seed-based plant propagation algorithm* (SBPPA) developed by Sulaiman and Salhi [557] simulates an optimisation process, drawing inspiration from the foraging activities of frugivores. A queueing metaphor is adopted whereby the foragers

---

Algorithm 16.4: Strawberry plant algorithm (adapted from [502])

Generate an initial population of m plants, each located randomly in the search space at
    locations $x_1,\ldots,x_m$;
Choose value for y, a user-defined parameter which defines the intensity of local search
    around each of the fitter plants;
Choose value for t_{max};
Set generation counter $t = 1$;
repeat
    Calculate fitness $f(x_i)$ of each plant and store in vector N, where the ith
        component of N is $N_i = f(x_i)$, $i = 1,\ldots,m$;
    Sort the components $N_1,\ldots,N_m$ of N into descending order (assuming the
        objective is to maximise fitness);
    for $i = 1$ **to** $m/10$ *(top 10% of plants)* **do**
        Generate y/i short runners for each plant;
        if *any of the new locations has higher fitness than that of the parent plant* **then**
            Move the parent plant to the new location with the highest fitness (set
              $x_i := r_i$);
        else
            Discard the new locations: the parent plant stays at its current location;
        end
    end
    for $i = m/10 + 1$ **to** m *(the remaining plants)* **do**
        Generate one long runner for each plant not in the top 10% and select the
        location of the end point r_i for that runner randomly in the search space;
        if *the new location has higher fitness than that of the parent plant* **then**
            Move the parent plant to the new location (i.e. set $x_i := r_i$);
        else
            Discard the new location: the parent plant stays at its current location;
        end
    end
    Set $t := t + 1$;
until $t = t_{max}$;
Output the best solution found;

(agents) are assumed to arrive at a certain rate at a plant, harvest resources from (get
served by) the plant at a certain rate and, in turn, disperse the seeds in the environment. Unlike the other plant propagation algorithms already described, the SBPPA
explicitly incorporates the role of foragers in plant seed dispersal.

Following the description of the algorithm in [557], it is assumed that agents
arrive at a plant following a Poisson process. Let X be the random variable representing the number of arrivals per unit of time; then the probability of k arrivals over
an interval t under this process is

$$P(X = k) = \frac{(\lambda t)^k e^{-\lambda t}}{k!}, \tag{16.16}$$

where λ is the mean arrival rate of foragers per interval of time t. The time taken by agents to eat the fruit before leaving to disperse its seeds (bearing parallel with the service rate in a queuing system) is assumed to follow an exponential distribution as follows:

$$S(t) = \mu e^{-\mu t}, \tag{16.17}$$

where μ is the average number of agents that can feed in a time interval t. Assume that $\lambda < \mu$ (this mimics the case where some fruit is not consumed by foragers and therefore falls to the ground when ripe) and that the system is in a steady state. Let A be the average number of agents at a plant (some feeding and some waiting to feed), and denote the average number waiting to feed as A_q. If we denote the average number feeding as λ/μ, then, from [365]

$$A = A_q + \frac{\lambda}{\mu}. \tag{16.18}$$

Following the argument in [557], if we assume that the plant wishes to maximise the dispersion of its seeds (which the agents have consumed), the object is to maximise A_q, and the resulting constrained optimisation problem can be stated as

$$\text{Maximise} \quad A_q = A - \frac{\lambda}{\mu}, \tag{16.19}$$

$$\text{subject to} \quad g_1(\lambda,\mu) = \lambda < \mu + 1,$$

$$\lambda > 0, \ \mu > 0.$$

If the following parameter values are assigned, $A = 10$ (assumed population size), $\lambda > 0$ and $\mu \leq 100$, the optimal solution to the problem per [557] is $\lambda = 1.1, \mu = 0.1$ and $A_q = 1$.

The model giving rise to the SBPPA also assumes that the steps in the foraging movement of the agents (frugivores) follow a Lévy distribution (Sect. 2.5.2). Under this distribution, the step lengths (jump sizes) of each movement around the landscape by a forager have a heavy-tailed power-law distribution, which can be expressed as

$$L(s) \sim |s|^{-1-\beta}, \tag{16.20}$$

where $L(s)$ is a Lévy distribution with an index $\beta \in (0, 2)$.

Algorithm

In the SBPPA, it is assumed that the foraging agents arrive at the plants according to a Poisson distribution and, following (16.19), that the mean arrival rate λ is 1.1 and the population size N_P is 10. The algorithm moves between exploration and exploitation stochastically during the search process based on a threshold value, such that the probability $\text{Poisson}(\lambda) < 0.05$.

If the value is below this threshold, then the algorithm engages in local search around the current point:

$$x_{i,j}^* = \begin{cases} x_{i,j} + \xi_j(x_{i,j} - x_{l,j}), & \text{if } R \le 0.8, \\ x_{i,j} & \text{otherwise,} \end{cases} \quad j = 1,2,\ldots,n, \; i,l = 1,2,\ldots,N_P, \; i \ne l,$$

$$(16.21)$$

where R is the rate of dispersion of the seeds locally around the parent plant p, $x_{i,j}^*$ and $x_{i,j} \in [a_j, b_j]$ are the jth coordinates of the seed locations x_i^* and x_i respectively, a_j and b_j are the jth lower and upper bounds of the search space, $\xi \in [-1,1]$, and the indices i and l are different.

Alternatively, if Poisson(λ) ≥ 0.05, then exploration is emphasised in the algorithm and the seeds are potentially more widely dispersed, using

$$x_{i,j}^* = \begin{cases} x_{i,j} + L_i(x_{i,j} - \theta_j) & \text{if } R \le 0.8; \; \theta_j \in [a_j, b_j], \\ x_{i,j} & \text{otherwise,} \end{cases} \quad j = 1,2,\ldots,n \; i = 1,2,\ldots,N_P,$$

$$(16.22)$$

where L_i is a step drawn from a Lévy distribution and θ_j is a random coordinate within the search space.

Therefore, (16.21) and (16.22) perturb the current solution either locally or globally. The choice of the threshold value for Poisson(λ) is important, as it influences the balance between local and global search in the algorithm.

A parallel can be drawn with real-world seed dispersal, where some seeds or fruits fall from the plant and colonise the nearby environment, while other seeds are dispersed further away from the parent plant owing to the actions of frugivores.

Initially, the algorithm undertakes a random search process to generate N_P good starting solutions, and these are stored in a memory structure, P_{best}. It is assumed that each plant p produces one fruit and each of these has one seed. The current position of the ith seed in the population P is represented by $x_i = (x_{i,j} : j = 1,2,\ldots,n)$, where $x_{i,j} \in [a_j, b_j]$ is the jth coordinate of solution (seed to be dispersed) x_i, and a_j and b_j are the bounds of the jth dimension of the search space. The initial random population of N_P seeds is generated using

$$x_{i,j} = a_j + (b_j - a_j)\eta_j, \quad j = 1,2,\ldots,n, \tag{16.23}$$

where $\eta_j \in (0,1)$. Pseudocode for the SBPPA is provided in Algorithm 16.5. In this pseudocode, rand represents a random number drawn from a uniform distribution $U(0,1)$, and R is the dispersion (perturbation) rate.

Discussion

The SBPPA differs from the propagation algorithms previously described as it is based on a more comprehensive model of the seed dispersal process. The seed dispersal process is modelled as a mix of local search, simulating fruit dropping from the parent plant onto the ground, and longer-range dispersal of seeds by the foragers which utilise a Lévy flight movement process. As such, the algorithm is the first to attempt to develop an optimisation process based on animal seed dispersal mechanisms.

Algorithm 16.5: Seed-based plant propagation algorithm

Set $N_P = 10$ (population size);
Set n = number of dimensions;
Set $t = 0$ (number of trials when creating the starting population);
Set t_{max} = maximum number of iterations;
for $t = 1$ **to** t_{max} **do**
   if $t \leq N_P$ **then**
      Create a random population of seeds, $P = \{x_i, i = 1, 2, \ldots, N_P\}$ using (16.23)
         and collect the best solution from each trial run in P_{best};
      Evaluate the fitness of population P;
   end
   if $t > N_P$ **then**
      Use updated population P_{best};
   end
   while *stopping criteria not satisfied* **do**
      for $i = 1$ **to** N_P **do**
         if Poisson$(\lambda)_i < 0.05$ *(local or global seed dispersal)* **then**
            for $j = 1$ **to** n **do**
               if rand $\leq R$ **then**
                  Update the current entry using (16.21);
               end
            end
         else
            for $j = 1$ **to** n **do**
               if rand $\leq R$ **then**
                  Update the current entry using (16.22);
               end
            end
         end
      end
      Update current best location;
   end
end
Output the best solution found;

As yet, only a few studies, including [557, 558], have explored the algorithm, indicating competitive results for the SBPPA. The parameter settings adopted in [557] include $N_P = 10$, maximum number of iterations $t_{max} = 20,000n/N_P$, $R = 0.8$ and use of a threshold value of Poisson$(\lambda) = 0.05$. Further work is required to understand the characteristics of the algorithm and to assess its utility across a wider range of test problems.

16.6 Light-Foraging Algorithms

The process of a plant growing above ground in an effort to harvest light has been used as the inspiration to design optimisation algorithms. Below we introduce two of these algorithms, the *plant growth simulation algorithm* [574] and the *plant growth optimisation algorithm* [97].

16.6.1 Plant Growth Simulation Algorithm

Plants exhibit a considerable degree of phenotypic plasticity which can be generally described as the 'response of organisms to environmental conditions or stimuli' [438]. One aspect of this is *developmental plasticity* which is defined here as 'the developmental changes that follow the perception and integration of environmental information' [438]. Developmental plasticity provides an important adaptive capability to both animals and plants in the face of heterogeneous environmental conditions. Given the sessile nature of plants it is of particular importance in facilitating adaptation to the very specific features of their local environment.

The ability of plants to adapt to changing environmental conditions via the direction of shoot, leaf and root growth provides a rich vein of metaphorical inspiration for the design of optimisation algorithms based on plant resource-foraging behaviours. In this subsection we describe an algorithm, the *plant growth simulation algorithm* (PGSA), developed by Tong et al. [574], which is inspired by the light-foraging process.

A key aspect of plant growth is that the initial stem of the plant gives rise over time to branches and leaves as it grows. The location and number of branches and leaves are (at least in part) a function of the resources in the plant's environment.

Plants can display different growth patterns above ground. These can be broadly classified as being either *monopodial* or *sympodial* (Fig. 16.7). Monopodial growth occurs when the plant's growth is led by the development of its stem. While lateral branches and leaves can be expressed off the stem, these are subordinate to the stem, which continues to grow upwards over time. In contrast, sympodial growth occurs when plant growth is led by new leader shoots which branch off the original stem.

The actual growth process in plants is driven by the production of new cells in *meristem tissue* (embryonic undifferentiated cells), which can be found just below the shoot tip (the shoot apical meristem), at the tips of branches (lateral meristems) or just inside the root cap in root tips (apical meristems). As the new cells are produced, the stem, branch or root grows longer.

Metaphorically, the plant growth process can be considered as being the exploration of an environment in a search for a good *architecture* which allows the plant to capture resources effectively. Tong et al. [574] drew inspiration from this idea to develop the PGSA. This initial work inspired a number of follow-on studies and applications of the algorithm, including [233, 563, 575, 609]. The PGSA constructs a virtual plant growth simulation in which a simulated plant grows in the search space and attempts to find the optimal solution (the light source) [575]. The growth of the plant is driven by a pseudophototropic mechanism whereby the degree of growth of

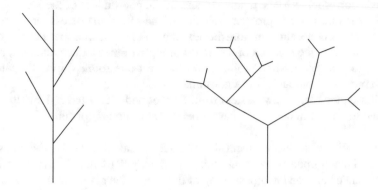

Fig. 16.7. Illustration of monopodial growth with alternate branching (left) and sympodial growth (right)

the main shoot and branches depends on the quality of the solution at various points on each (a proxy for the level of light). In essence, each branch metaphorically undertakes a search in a local area, with the plant's growth pattern being biased towards regions which display higher fitness.

The Algorithm

The Hungarian biologist Aristid Lindenmayer [358], in studying plant growth, developed L-systems to model recursive growth patterns. A key item of L-systems is the notion of rewriting, whereby a complex structure can be generated from a starting *seed* by successively replacing parts of the growing structure using a set of production or rewriting rules.

The PGSA takes inspiration from L-systems in terms of growth and branching-system design. New branches are assumed to grow from nodes on the main stem, or from nodes on previously generated branches. Each new branch is assumed to have a turning angle of 90° (i.e. it grows perpendicularly to its parent branch) and the definition of the length of the new branch (the branching length) is determined by the nature of the optimisation problem at hand. For example, in the initial application of the algorithm to an integer programming problem [574], the branch length was set as 1. The number of new branches grown from a selected node (corresponding to a tip with meristem cells) in each iteration of the algorithm is $2n$, where n is the dimensionality of the search space. Therefore, starting from a seed point (the initial solution), the plant grows and branches at nodes on the growing plant.

The next issue is how to select a node for branching in each iteration of the algorithm. This process is loosely based on elements of a *morphogenetic model of plant development*, in which the concentration of a growth hormone (morphactin) at a node determines whether the cells at that node will start to grow and produce branches. In the design of the optimisation algorithm, the concentration of the growth hormone at a node corresponds to the relative fitness of the coordinates of the location of that node for the optimisation problem of interest. The selection process for choosing the next branching node, and therefore the next region of the search space to be explored, is biased towards the location of the nodes of higher fitness. Each node is a possible solution to the optimisation problem and the best one found during the simulated growth process is output by the algorithm.

Initially, the plant grows a stem from its root node (denoted B_0) and the stem is assumed to have k nodes which have a better environment (or fitness) than the root node.

The quality of the environment at each node i is calculated using the fitness function g, and the morphactin concentration at each node B (denoted C_B) is calculated as the difference between the fitness $g(B_0)$ of the root of the plant and the fitness $g(B_{Mi})$ of that node i (assuming a minimisation problem) divided by the sum of these differences for all nodes $i = 1, 2, \ldots, k$. Only the k nodes with fitness greater than that of the root are considered in the calculation (16.24):

$$C_{Mi} = \frac{g(B_0) - g(B_{Mi})}{\Delta_1}, \qquad i = 1, 2, \ldots, k,$$

$$\text{where } \Delta_1 = \sum_{j=1}^{k} \left(g(B_0) - g(B_{Mj}) \right).$$

(16.24)

The fitness of each node is calculated relative to the root and relative to that of all other nodes. By inspection of (16.24), the sum of the scaled fitnesses must be 1 ($\sum_{i=1}^{k} C_{Mi} = 1$).

To select the node from which the next branch will be grown, the fitnesses of all nodes are laid out in the range $[0, 1]$ and a random draw is made from this interval, with the node corresponding to that interval being selected as the preferential node from which a branch will be grown in the next iteration of the algorithm. Assume that there are q nodes on the new branch that have a better environment than the root node (B_0). The originating node from which the branch grew (node 2 in (16.25)) has its morphactin concentration set to 0 and is ignored for the rest of the algorithm, and the morphactin concentrations for all nodes on the plant are recalculated using (16.25). The term Δ_1 in (16.25) calculates the sum of the fitnesses of all nodes on the plant's stem (omitting node 2, from which the new branch grew) relative to that of the root node, and Δ_2 in (16.25) calculates the sum of the fitnesses of all nodes on the new branch relative to that of the root node:

$$C_{Mi} = \frac{g(B_0) - g(B_{Mi})}{\Delta_1 + \Delta_2}, \qquad i = 1, 2, \ldots, k,$$

$$C_{mj} = \frac{g(B_0) - g(B_{mj})}{\Delta_1 + \Delta_2}, \qquad j = 1, 2, \ldots, q,$$

$$\text{where } \Delta_1 = \sum_{l=1, l\neq 2}^{k} (g(B_0) - g(B_{Ml}))$$

$$\text{and } \Delta_2 = \sum_{l=1}^{q} (g(B_0) - g(B_{ml})).$$

(16.25)

As before, the sum of the scaled fitnesses must be 1 ($\sum_{l=1, l\neq 2}^{k} C_{Ml} + \sum_{l=1}^{q} C_{ml} = 1$). The branching process iterates until a preset terminating number of branching iterations is reached.

Pseudocode for the PGSA is given as Algorithm 16.6.

Algorithm 16.6: Plant growth simulation algorithm (adapted from [574])

Choose values for t_{max} and branch length;
Define a mechanism for node placement on the stem and subsequent branches;
Set generation counter $t = 1$;
Generate an initial root at location B_0;
Define node locations on stem;
Calculate fitness of each node on stem using $g(B_i)$ and store location of highest-fitness node;
Calculate morphactin concentration at each node using (16.24);
repeat
    Select branching node (stochastically but biased by concentration);
    Add $2n$ lateral branches at this node;
    Calculate fitness of new nodes on added branches;
    if *any of these locations have higher fitness than best found to date* **then**
        | Replace best location;
    end
    Update morphactin concentrations at each node on the plant using (16.25);
    Set $t := t + 1$;
until $t = t_{max}$;
Output the best solution found;

In summary, the algorithm traverses a search space by implementing a search process which is loosely modelled on plant growth mechanisms. The plant covers a region of the search space, with the nodes on the plant representing possible solutions to the problem and the value $g(B_i)$ being the objective function value at a node i. The algorithm biases its search process towards the regions of higher fitness (exploitation) whilst maintaining some explorative capability.

While competitive results have been reported in a number of studies using the algorithm, one aspect of the algorithm that has received less attention thus far is its computational efficiency. As each new branch is added, the morphactin concentrations for all nodes must be recalculated as they depend on relative measures (the fitness values need not be recalculated). Implementation of the algorithm also requires the determination of suitable problem-specific branch lengths and the implementation of a mechanism to determine the number and placement of nodes on the main stem and subsequent branches.

Variants on the Plant Growth Simulation Algorithm

There are many ways to operationalise the growth and phototropic mechanisms in the PGSA, and each choice gives rise to an optimisation algorithm with different characteristics. A variable branch growth mechanism is implemented in [664] with the rate of growth of an individual branch in each iteration being determined by its level of photosynthetic activity, which in turn is determined by the light intensity on the branch (proxied by its fitness) and its rate of respiration. The higher the rate of simulated photosynthesis in the branch, the faster its rate of growth. In this version of the algorithm, branches also move in the search space towards increasing light intensity, simulating a *bending to light*. The effect of the two mechanisms is to promote greater exploitation of higher-fitness regions of the search space.

16.6.2 Plant Growth Optimisation Algorithm

A more complex set of growth and branching behaviours is adopted in the *plant growth optimisation* (PGO) *algorithm* developed by Cai et al. in [97]. As in the PGSA, the branching behaviour at growth nodes (corresponding to a location or a solution to the problem of interest) depends on their relative fitness, with sympodial branching occurring if a node is of high fitness relative to other nodes on the plant (i.e. the stem or branch elongates from the node in the current direction of the stem/branch, and child branches are grown from the node in a lateral direction), and monopodial branching occurring if the node is of moderate fitness (i.e. the stem or branch is extended further in its current direction). If a node is of low relative fitness, no growth or branching occurs at the node.

This approach allows for intensive search via branching around the higher-fitness points, growth out of moderate-fitness regions for branches of moderate fitness, and no growth or branching at low-fitness nodes.

The differential branching mechanism is supplemented by a simulated leaf growth process, whereby the region around the end points of new branches is explored using a local search algorithm (simulating leaf growth around the branch end point), with the location of the highest-fitness point in this region (denoted a *leaf point*) being recorded.

The high-level mechanisms of the algorithm are described in [97] as follows.

Morphogen Concentration

As with the PGSA, the determining factor as to whether growth and branching will take place at a potential growing point is the morphogen concentration at that point. For a minimisation problem, the concentration of morphogen at point i (denoted as A_i) can be defined as

$$
A_i = \begin{cases} 1 - \dfrac{f(x_i) - f_{\min}}{\sum_{j=1}^{N}(f(x_j - f_{\min})}, & f(x_i) > f_{\min}, \\ 1, & f(x_i) = f_{\min}, \end{cases} \quad i = 1, 2, \ldots, N, \qquad (16.26)
$$

where x_i is the location of the ith branch point, N is the total number of points, $f(x_i)$ is the objective function value at the ith point and $f_{\min}$ is the minimum objective function value across all N points. From the above formulation it can be observed that all values of morphogen are in the range $[0, 1]$, with the best location (minimum value of objective function) receiving a value of 1, and poorer locations (those with higher objective function values) obtaining a lower morphogen value.

Branching

The branching behaviour for each branch point is controlled by two parameters, α and β, which are randomly selected so that $0 \leq \alpha < \beta \leq 1$. Depending on the morphogen value of a point, its growth behaviour is as follows:

$$\begin{aligned} &\text{Mode 1: sympodial,} && A_i > \beta; \\ &\text{Mode 2: monopodial,} && \alpha < A_i < \beta; \\ &\text{Mode 3: no branching,} && A_i < \alpha. \end{aligned} \qquad (16.27)$$

If the morphogen concentration of a potential branch point is high, it engages in sympodial branching, the stem tip elongates in its current direction and new child shoots are produced in lateral positions along the newly extended branch segment. As the value of β is determined randomly, even a good branch point may not undertake sympodial branching every iteration. If the morphogen concentration is moderate, the branch point engages in monopodial branching: the branch elongates in its current direction, producing a new terminal branch tip, but does not produce new child shoots. Finally, if the morphogen concentration of a potential branch point is low, the branch tip does not grow and only a leaf growth process is performed (see below for a description of this).

In addition to the three branching behaviours, a number of randomly generated new growth locations are generated in each iteration of the algorithm in order to avoid premature convergence of the search process.

Leaf Growth

After the new growth points are generated in the branching process, a leaf growth process is simulated by engaging in a local search in the region around each point. If a better location than the growth point is found, it is designated as a *leaf point*; otherwise (no better location is found), the growth point is termed *mature*.

Selection Mechanism

As new branch (and random) points are generated using the above mechanisms, a selection process is applied to reduce the population of growth points in the next iteration of the algorithm back to N.

Rather than apply a greedy selection process by selecting the points with the lowest value of the objective function (which could lead to premature convergence of the algorithm), the selection process takes account of crowding. If there are multiple points in close proximity, a new point is generated in the centre of this area, and the grouping of points (including the new point) is *thinned out* (in the limit, to a single point) by removal of points with poorer values of the objective function.

The implementation of a crowding operator attempts to maintain diversity in the population.

Maturity Mechanism

If a growth point has grown for a number of iterations and has not found a better location, it is considered *mature* and does not take part in the growth process.

Algorithm

Bringing the above elements together, high-level pseudocode for the PGO algorithm is provided in Algorithm 16.7.

Algorithm 16.7: Plant Growth Optimisation Algorithm

Set $N_G = 0$ (iteration counter) ;
Set $N_C = 0$ (convergence counter);
Set $N_M = 0$ (mature points counter);
Set upper limit for N (number of branch points);
Set values for parameters including N_G^{max}, N_C^{max}, N_M^{max};
Select N_0 branch points at random and perform leaf growth;
while $N_G < N_G^{max}$ **and** $N_C < N_C^{max}$ **and** $N_M < N_M^{max}$ **do**
    Evaluate morphogen concentration at each branch point using (16.26);
    Select values for α and β randomly in the interval $[0, 1]$;
    Produce new points using (16.27) and also select m random locations;
    Undertake leaf growth at all branch points;
    Select the mature branch points and denote the number of these as k, $0 \le k \le N$,
       using maturity mechanism;
    Set $N_M = N_M + k$;
    Produce a new point in the centre of crowded area and implement crowding
       operator to reduce number of points in that region;
    Eliminate poorer points in population to leave N branch points for next iteration;
    Compare current points with mature points and determine the best fitness value
       f_{min};
    Set $N_G = N_G + 1$;
    if $f_{min} > f_{min,old}$ **then**
       Set $f_{min} = f_{min,old}$;
       if $|f_{min} - f_{min,old}| < \epsilon$ **then**
          Set $N_C = N_C + 1$;
       else
          Set $N_C = 0$;
       end
    else
       Set $N_C = N_C + 1$;
    end
end
Output the best solution found;

Discussion

The PGO algorithm presents an alternative implementation of a plant-growth-inspired algorithm for optimisation purposes. Thus far, the algorithm has only been pilot tested on a small set of benchmark optimisation functions in [97] and further work remains to be undertaken in order to assess the effectiveness and efficiency of the approach. While the parameter settings are likely to be problem-specific, the settings used in [97] included $N = 20$ and the number of random points generated in each iteration was $N/2$. Better results were obtained as the number of initial growth points (N_0) was increased.

The description of the algorithm in [97] is concise and a number of aspects of the algorithm would require more detail or user decisions for implementation, including the design of the crowding mechanism, the leaf growth mechanism and the method of generating new growth points after branching.

16.7 Root Foraging Algorithms

In addition to the use of inspiration from shoot and/or branch growth processes to design optimisation algorithms, the below-ground foraging processes of roots have also served as inspiration for the design of foraging algorithms. In the following subsections, we introduce two such algorithms, the *root growth algorithm* [660] and the *root growth optimiser algorithm* [246, 248].

16.7.1 Root Growth Algorithm

The *root growth algorithm* (RGA) was first proposed by Zhang [660]. In this algorithm, which bears some similarities with the PGSA, a root growth process is simulated whereby the root structure grows towards good regions of the search space, bearing a parallel with the process by which plants put more root biomass into soil regions which contain water and required nutrients.

As already discussed in Sect. 16.3, below-ground foraging by plants is a complex process as there are a multitude of factors which affect this, including the distribution of nutrients, soil biology and interactions with other organisms. In the RGA, a combination of concepts drawn both from this literature and from Lindenmayer L-systems [358] are used to design an algorithm which simulates root growth. At a high level in the RGA,

- a *seed* germinates in the *soil* (search environment),
- new root hairs are assumed to grow from the root tips of old root hairs (this behaviour simulates the branching of the root system), and
- the growing root system is composed of numerous root hairs and root tips with similar structure.

As the nutrient concentration in soil is uneven, plant root hairs grow in varying directions but preferentially in the direction of greatest nutrient concentration. The growth process is driven by the concentration of the growth hormone morphactin, with new root hairs being more likely to emerge from root tips with larger morphactin concentrations.

The RGA assumes that the morphactin level in a plant is constant (set to an arbitrary value of 1). Assuming n root tips, for each root tip $i = 1, 2, \ldots, n$, we denote its location by x_i (a D-dimensional vector) and its morphactin concentration by E_i, where

$$E_i = \frac{1/f(x_i)}{\sum_{j=1}^{n} 1/f(x_j)}. \tag{16.28}$$

In (16.28), $f(x_i)$ is the evaluation of the objective function at the location x_i of root tip i, forming an analogue with the spatial distribution of nutrients in the soil. The morphactin concentration of each root tip is determined relative to the total value of the objective function across all root tips, and as new root hairs grow (as the algorithm iterates), this value will change. Note that the *fitness* of root tip i correspond to its morphactin concentration E_i, which drives root tip selection and growth, rather than to the raw objective function value $f(x_i)$ (the values must be positive).

Growth of Root Tips

In the RGA, the root growth process originates from a seed (a single location) which is located randomly in the search space. In each subsequent iteration of the algorithm, a root growth process is simulated, and this in turn will incrementally increase the number of root tips in the population up to a maximum population size. During the growth process, the root tips with higher morphactin concentrations are selected and new root hairs originate from them, simulating a root *branching* process. In order to promote spatial diversity in the search process, the selected root tips should be separated by a minimum distance. Once the number of root tips selected for growth in an individual iteration of the algorithm reaches a modeller-determined parameter, the selection step terminates. Algorithm 16.8 outlines the selection process.

Algorithm 16.8: Root tip selection process

Select root tip with highest fitness value;
repeat
    Select root tip with highest fitness value not already selected;
    if *distance from this root tip to already selected root tips is less than threshold*
    then
       | Discard selected root tip;
    else
       | Add root tip to selected group;
    end
until *required number of root tips (S in total) are selected;*

A new growing point is determined from the old root tips in memory using

$$p_{ij} = \begin{cases} x_{ij} + (2 \cdot \phi_{ij} - 1), & j = k, \\ x_{ij}, & j \neq k, \end{cases} \qquad (16.29)$$

where $j, k \in \{1, 2, \ldots, D\}$ are randomly chosen indices, $p_i, i = 1, 2, \ldots, S$, are S new growing points and ϕ_{ij} is a random number drawn uniformly from $[-1, 1]$. Therefore, the new growth points for the selected root tips are displaced from the location of the old root tip. In the canonical version of the RGA as outlined in [660], each selected root tip generates a number of branches, and four new growing points are generated for each selected root tip.

Root Hair Growth

After the new growth points have been selected, the next step is to determine the length of the resulting root hair and its growing angle from its growth point. The growth length of each hair is given by $\delta_i : i = 1, 2, \ldots, n$, a parameter in the algorithm. After the growth process, a new root tip results at a location which is determined by the previous location of the root tip, the length of the new root hair and the angle that the new root hair grows at. The new location is calculated using

$$x_i = x_i + \delta_i \varphi_i, \qquad (16.30)$$

where the growth angle of the new root hair is given by the parameter $\varphi_i : i = 1, 2, \ldots, n$. To calculate this angle (a vector) for each growth point, a vector of D random numbers $(\varphi_{i1}, \varphi_{i2}, \ldots, \varphi_{iD})$ is drawn and the angle for each new root hair $i = 1, 2, \ldots, n$ is calculated by normalising:

$$\varphi_i = \frac{(\varphi_{i1}, \varphi_{i2}, \ldots, \varphi_{iD})}{\sqrt{\varphi_{i1}^2 + \varphi_{i2}^2 + \ldots + \varphi_{iD}^2}}. \qquad (16.31)$$

Local Search

The RGA embeds a local search process in the vicinity of the new root hair, simulating *trophotropism* (i.e. movement of cells in response to the availability of nutrients), by continuing the growth of the new root hair as long as it continues to lead into better locations. If the fitness of a newly generated root tip is better than that of the old one, the root tip growth process is continued, up to a maximum number of additional growth cycles, as long as the quality of the root tip continues to improve. Each subsequent root tip is generated using

$$x_i(t) = x_i(t) + \delta_i \varphi_i. \qquad (16.32)$$

This is the same formulation as (16.30) with the addition of a time step t to indicate that the improvement process takes place within the same iteration (t) of the overall algorithm. If, at any point in this process, the fitness value of the new root tip disimproves, the root tip stops growing.

Self-Adaptive Growth

Diversity in the search process is promoted by varying the growth length of root hairs (δ_i) as the algorithm iterates. The length of root hairs is determined as in Algorithm 16.9. The parameter τ relative to the value of E_i determines the growth step length. For values of $\tau > 1$, the effect of the formulation is to reduce the step size over time therefore altering the balance of the algorithm towards more local search around already discovered locations.

Algorithm 16.9: Adaptation of root hair growth length

Set δ_i to initial value;
for *each generation* **do**
    for *each point x_i* **do**
        $\delta_i = |E_i|/|E_i + \tau|$;
    end
end

Main Algorithm

Bringing the above elements together, the pseudocode for the RGA is provided in Algorithm 16.10.

Algorithm 16.10: Root growth algorithm

Set iteration counter $t = 0$;
Generate an initial seed at a random location;
repeat
    Evaluate fitness values of all root tips;
    Select S root tips with larger fitness values;
    Use (16.29) to produce a number of new growing points for each selected root tip;
    For each new point, grow new root hairs (which produce new root tips) using
       (16.30) and apply simulated trophotropism mechanism to each new root tip;
    if *number of root tips exceeds predefined value* **then**
        Remove root tips with lowest fitness values;
    end
    Tune the parameter δ_i using (16.9) (adjust root growth length);
    Set $t := t + 1$;
until $t = t_{\max}$;
Output the best solution found;

Parameters

The modeller is required to set the values of a number of parameters in order to implement the RGA. In [660] the maximum number of root tips (maximum population size) was set to 100, the number of root tips selected for growth (branching) S was set to 4, the number of branches created for each selected root tip was set to 4, and the value of τ was in the range $[0.25, 1.5]$. The initial value of δ_i was set to one, as was the minimum distance threshold between root tips.

The value of τ is an important parameter in determining the performance of the RGA as it impacts on the growth length of root hairs. In [661] a variety of settings for τ, varying between 1 and 1000, were examined on a number of benchmark problems with the results indicating that the best setting for the parameter is problem dependent.

In [660, 661] the RGA produced competitive performance relative to a genetic algorithm (GA), particle swarm optimisation (PSO) and differential evolution (DE) on a number of benchmark optimisation functions.

Discussion

The RGA draws metaphorical inspiration from the process of a root network foraging for resources below ground, in order to develop a search algorithm which can be applied for optimisation purposes. The RGA and PGSA bear some similarities in that both rely on a branching mechanism (root branching in the case of the RGA, shoot branching in the case of the PGSA) to explore the search space. Further work is required to determine the utility of each of these algorithms in comparison with one another and relative to longer-established search heuristics.

16.7.2 Root Growth Optimiser Algorithm

Another optimisation algorithm drawing inspiration from the root growth process is the *root growth optimiser* (RGO) *algorithm* (RGO) devised by He et al. [246] (see also [248]). As with the RGA, the objective function is treated as the soil environment, and a root tip located at a specific point in this space represents a solution. The evaluation of the objective function at that location corresponds to an assessment of the soil environment quality by a root tip.

As with real-world root growth, the amount of root biomass that a plant places in a region of soil depends on the resources in that region, with greater biomass being placed in bountiful regions. In the simulated root growth process which makes up the RGO algorithm, it is assumed that there are three primary growth process for each root tip:

   i. each root tip may elongate forward or sideways in the soil;
  ii. each root tip may branch and produce daughter root tips; or
 iii. a root tip may cease to function as a root tip and become an ordinary piece of root mass.

Hence, a root tip may continue to grow, branch and produce new root tips, or cease growth. Root tips are divided into three categories by the algorithm based on the fitness values associated with each root tip: the group of root tips with the best fitness (called *main roots*), the group with the worst fitness (called *ageing roots*) and the rest of the root mass (called *lateral roots*). Roots assigned to the *ageing roots* category stop growing in the next iteration of the algorithm (simulating the case of real plant root fibres in locations where resources have been exhausted), whereas different growth mechanisms are implemented for main and lateral roots.

Growth Strategy for Main Roots

These roots are assumed to follow a monopodial branching strategy in that the root tip grows in the direction of the best concentration of resources and then branches by producing a number of new root tips in the area around its updated position. An inhibition mechanism is also implemented in order to ensure that the population of new root tips does not grow in an uncontrolled fashion. This growth strategy concentrates on exploitation of existing information as to good regions of the search space, as it is applied to the root tips with the highest current fitness.

Regrowing

In the first of the above processes, the current root tip grows in the direction of greatest resource concentration (i.e. the location which produces the highest objective function value). This is proxied by the location of the root tip which produced the best objective function value in the previous iteration of the algorithm, denoted as $x_{\text{loc}}^{\text{best}}$ (or local best). The position update for a root tip currently located at x_i is given by

$$x_i(t) = x_i(t-1) + l \cdot \text{rand} \odot (x_{\text{loc}}^{\text{best}} - x_i(t-1)). \tag{16.33}$$

In this update, $x_i(t-1)$ is the current position of the ith root tip coming forward from the last iteration of the algorithm and $x_i(t)$ is the position it will move to. The parameter l is the local learning constant or step size, which controls the distance the root tip will move in the direction of the $x_{\text{loc}}^{\text{best}}$ root tip. This step is stochastic, governed by a vector, the term rand, whose components are determined by random draws from a uniform $[0, 1]$ distribution on each dimension. The symbol $\odot$ denotes componentwise product of vectors (see Sect. 6.1.3). It can be noted that (16.33) bears passing similarity to the position update equation in PSO. However, unlike the PSO position update, (16.33) does not embed a personal history factor (p^{best}), nor is the position update necessarily driven by the best ever location found by the algorithm (g^{best}).

Branching

This operation generates a number of new root tips around the current root tip. The number of new root tips depends on the quality of the solution produced by the parent root tip, with better (fitter) parent root tips giving rise to a greater number of offspring. The number of new root tips generated in each case is constrained to lie in the range (s_{min}, s_{max}) where s_{min} and s_{max} are modeller-defined parameters. The number of new root tips generated is determined using

$$w_i = \frac{f - f_{min}}{f_{max} - f_{min}} \cdot (s_{max} - s_{min}) + s_{min}. \tag{16.34}$$

The terms f_{max} and f_{min} are the best and worst fitness values in that iteration, and f is the fitness of the parent root tip. If a root tip is the fittest in the population of root tips, it generates s_{max} new root tips. On the other hand, if a root tip has the poorest fitness in the population, it generates only s_{min} new root tips.

The new root tips are generated around the parent root tip, with the location of each being obtained by a random draw from a normal distribution $N(x_i(t), \sigma^2)$, where the standard deviation σ is calculated using

$$\sigma_i = \left(\frac{i_{max} - i}{i_{max}} \right)^n \cdot (\sigma_{init} - \sigma_{fin}) + \sigma_{fin}, \tag{16.35}$$

where i_{max} and i are the maximum iteration number and the current iteration number, respectively, σ_{init} is the initial parameter setting for the standard deviation, and σ_{fin} is the final standard deviation which is set depending on the required level of accuracy of the algorithm. From this formulation, we can note that as the algorithm iterates, the distribution of new root tips will tend to move closer to the parent root tip over time, as the value of σ_i will reduce gradually from σ_{init} to σ_{fin}.

Inhibition Mechanism

This mechanism is designed to control the allocation of new root tips in order to promote diversity in the search process. Under the branching process, the better current root tips give rise to more new root tips, and this may promote overexploitation of a small region of the search space.

The local standard deviation, $\sigma_{local}(f)$, for all the new root tips generated from the same parent is calculated, and some of the new root tips are removed. The number to be removed is determined using

$$w_i^{remove} = \alpha \cdot \left(1 - \frac{\sigma_{local}(f)}{f_{max} - f_{min}} \right) \cdot w_i, \tag{16.36}$$

where α is a user-defined parameter.

The smaller the value of $\sigma_{local}(f)$, the greater the number of root tips that will be removed in the next generation. The new root tips to be removed for parent root tip i, are the w_i^{remove} root tips with the poorest fitness values from all the w_i root tips generated for that parent. This corresponds to a greedy selection process.

Growth Strategy: Lateral Roots

In this mechanism, the root tip produces a new branch laterally, rather than first growing. The parent root tip is replaced by the newly grown one. The new root tip is located at a random position around the original root tip using:

$$x_i(t) = \text{rand} \cdot \beta \cdot x_i(t-1), \tag{16.37}$$

where rand is a random number drawn from $U(0, 1)$. The value of β (a random growing angle) is determined using:

$$\beta = \frac{\lambda_i}{\sqrt{\lambda_i^T \lambda_i}}, \tag{16.38}$$

where λ_i is a random vector with the same dimensionality as the search space. This growth mechanism generates a new root tip in a random position around the old root tip with a random angle. The mechanism promotes a degree of exploration of the search space.

Algorithm

The pseudocode for the RGO algorithm is presented in Algorithm 16.11.

Algorithm 16.11: Root growth optimiser algorithm

Initialise the position of the root tips randomly in the search space;
Calculate the fitness value of each root tip;
repeat
> Divide the root tips into main roots, lateral roots and ageing roots based on their fitness;
> **for** *each main root tip* **do**
>> Regrow using root regrowth operator;
>> Branch using the root branching operator;
>> Evaluate fitnesses of new root tips;
>> Calculate standard deviations and implement inhibition mechanism to reduce number of new root tips;
>
> **end**
> **for** *each lateral root tip* **do**
>> Produce a new root tip replacing the original one;
>
> **end**
> Rank root tips and label roots as main, lateral or wastage;

until *Terminating condition*;
Output the best solution found;

Parameters and Discussion

In [248] the population size (number of root tips) was set to 50, the number of root tips assigned to the main group was set to 30% of the selected (highest-fitness) root tips in each generation, and the values of s_{max} and s_{min} were set to 3.0 and 1.0. The parameters α and l were set to 1.

The RGO algorithm takes loose inspiration from the iterative process of root growth in which greater amounts of root biomass are placed in soil with suitable nutrients and water, in order to design an optimisation heuristic. The algorithm draws a distinction between two types of roots (main roots and lateral roots) and implements a different growth (search) strategy for each. The essence of this is that the better root tips in the current population spawn more new root tips in their vicinity, with poorer-quality root tips generating one or no new root tips. Therefore, the algorithm focuses attention on search of good regions uncovered thus far but maintains some search even in less promising regions in order to reduce the chance of premature search convergence.

So far, only a couple of studies have assessed the performance of this algorithm. Results reported in [246, 248] indicate that the RGO algorithm produced competitive results on a series of benchmark functions when compared with other search heuristics.

A variant on the RGO algorithm which employs a branch-and-leaf growth rather than a root growth metaphor is the *branch–leaf growth algorithm* (BLGA) developed in [247]. In this variant, it is branch apices (tips) rather than root tips which engage in monopodial or sympodial branching, using the same mechanisms as described in the RGO algorithm for main and lateral root growth. After branch tip growth, a leaf growth process at each branch tip is simulated, which produces a local search around each branch tip. The best leaf position becomes a branch tip in the next iteration of the algorithm, with all other leaves being discarded at the end of each iteration of the algorithm. Thus, the search process consists of an iterated process of branch and leaf growth. Readers are referred to [247] for further information on this algorithm.

16.8 Summary

Plants possess a wide array of sensory capabilities which allow them to monitor many characteristics of their environment, and to flexibly adapt their morphology in response to environmental conditions. Despite the fact that our understanding of plant behaviours has increased notably over the last few decades, there are still a significant number of open questions concerning the sensory and information-processing capabilities of plants. In this chapter we have attempted to show that these organisms are heavily interlinked with their environment and are capable of complex foraging behaviours.

Thus far, only a relatively small number of algorithms have been designed which take inspiration from plant foraging behaviours. Most of these algorithms are still experimental, as they have not yet been subject to significant analysis and testing.

However, these algorithms pave the way for future advances in this area and we can expect that further algorithms will emerge as we gain greater insight into the underlying biological mechanisms.

Part VI

Algorithms Derived from Formal Models of Foraging

In the preceding chapters we have introduced a wide span of algorithms which have been inspired by the foraging activities of organisms of varying complexity. In this part of the book, we adopt a different perspective and consider whether ideas from some of the formal models of foraging outlined in Chap. 2 can be used directly as a source of algorithmic design inspiration. Examples of a number of such algorithms are discussed in Chap. 17.

Foraging Models and Algorithms

In this chapter we describe five algorithms. First, we survey the *optimal foraging algorithm* [666] which derives inspiration from *optimal foraging theory* (Sect. 2.1). We then describe the *group search optimiser algorithm* [244] and the *predatory search algorithm* [359, 360] which are inspired by the *producer–scrounger foraging model* (Sect. 2.3.2) and *area-restricted search behaviour* (Sect. 2.5.2) respectively. We also introduce the *predator–prey optimisation algorithm* [528] which is developed from a synthesis of a predator–prey concept (Sect. 2.4) and the particle swarm optimisation (PSO) algorithm. Finally, we outline the *animal migration optimisation algorithm* [354] which is loosely inspired by an animal migration metaphor. The study of animal migration falls within the movement ecology literature (Sect. 2.5).

17.1 Optimal Foraging Algorithm

The *optimal foraging algorithm* (OFA) was introduced by Zhu and Zhang in [666], and we follow this in our discussion of it below. The authors describe the OFA as 'a new stochastic search algorithm, ... used to solve global optimisation problems following animal foraging behavior'. Conceptually, the work translates three questions from the OFT domain to the algorithm domain:

i. 'Where would an animal look for food?' becomes 'Where would the algorithm search for the optimal solution?'
ii. 'When would an animal transfer from a foraging patch?' becomes 'When would the algorithm leave the local optimal solution?'
iii. 'What type of food would an animal choose?' becomes 'How can we verify if the obtained solution is better?'

and the three questions are addressed in this order, below.

For a global minimisation problem with objective function f, we seek $x^* = (x^*_1, \ldots, x^*_d) \in \mathbb{R}^d$ such that $f(x^*) = \min_{x \in R}\{f(x)\}$, where $R = \{x = (x_1, \ldots, x_d) \in \mathbb{R}^d : x_i^L \leq x_i \leq x_i^U, i = 1, 2, \ldots, d\} \subseteq \mathbb{R}^d$. In the OFA, an individual is regarded as a foraging animal whose position $x = (x_1, \ldots, x_d) \in \mathbb{R}^d$ in a patch is a candidate solution. This

© Springer Nature Switzerland AG 2018
A. Brabazon, S. McGarraghy, *Foraging-Inspired Optimisation Algorithms*,
Natural Computing Series, https://doi.org/10.1007/978-3-319-59156-8_17

solution is taken to be the current optimal position of the individual obtained through foraging. Viewed through the lens of optimal foraging theory, the optimisation process can be seen as an animal foraging in various patches to find the optimal patch where the net rate of energy intake can be maximised. Once the optimal patch is found, the animal will search for an optimal position within the patch following the best prey model.

The number of near-optimal solutions in an area is related to the resource abundance, with a larger number meaning a better resource. The function value of the solution is viewed in OFT terms as the energy of the solution, with a lower value (for minimisation) meaning a higher quality of the solution.

17.1.1 Searching for an Optimal Solution

The individual must first decide where to continue the search. According to OFT, animals will forage in the area of greatest food abundance, and will allocate the most foraging time to the patch with most food. The OFA assumes that the current forging position and its neighbourhood comprise a patch with abundant food, so the algorithm will search for an optimal solution near the current position, as nearby solutions could be near-optimal solutions. Let $x(t)$ denote the position of an individual at iteration (time) t. Then its position at iteration $t + 1$ is given by

$$x(t+1) = x(t) - kr_1 \odot \Delta x(t) + kr_2 \odot \Delta x(t), \quad t = 1, 2, \ldots \tag{17.1}$$

where k is a scale factor, r_1 and r_2 are two vectors each of whose components are randomly drawn from a uniform distribution $U(0, 1)$, $\Delta x(t) := x(t) - x(t-1)$ is the vector increment (or delta) from iteration $t-1$ to iteration t, and $\odot$ denotes componentwise product of vectors (see Sect. 6.1.3). Even though r_1 and r_2 only have nonnegative components, the subtraction and addition mean that $x(t+1)$ can be anywhere around $x(t)$ depending on the relative sizes of components, and so an OFA individual can move to forage in an arbitrary position near the current position.

17.1.2 Leaving the Local Optimal Solution to Continue the Search

Often, when foraging, cooperation among animals is needed. When an animal finds prey, it will recruit others, and this behaviour ensures that animals tend towards the patch of greatest prey. The OFA makes use of the concepts of *group* and *recruitment*. A *group* is a set of $N \geq 2$ animals dispersed in different patches of differing quality, giving rise to a set of solutions. The individuals are ranked by objective function value from best $x_{\text{best}}(t)$ (rank 1) to worst $x_{\text{worst}}(t)$ (rank N). Following OFT, the OFA works on the principle that individuals of worse objective function value tend to move towards individuals of better objective function value (from lower quality patches to higher quality patches). We say that an individual with better value *recruits* an individual with worse value. After this recruitment process, the best individual is moved in the direction of the worst, in order to enhance exploration and to avoid being trapped at a local optimum. These ideas are codified as

$$\Delta x_j(t) = x_b(t) - x_j(t), j = 2,\ldots,N, \quad b < j, b \text{ random}, t = 1,2,\ldots \qquad (17.2)$$

$$\Delta x_{\text{best}}(t) = x_{\text{worst}}(t) - x_{\text{best}}(t), \qquad (17.3)$$

where $\Delta x_j(t)$ denotes the vector increment of x_j owing to x_j being recruited (moving towards) a random individual x_b of better objective function value (so $b < j$ because of the best to worst ranking). Combining (17.1) with (17.2–17.3) gives

$$x_j(t+1) = x_j(t) - kr_1 \odot (x_b(t) - x_j(t)) + kr_2 \odot (x_b(t) - x_j(t)), \qquad (17.4)$$

$$j = 2,\ldots,N, b < j, t \geq 1,$$

$$x_{\text{best}}(t+1) = x_{\text{best}}(t) - kr_1 \odot (x_{\text{worst}}(t) - x_{\text{best}}(t)) + kr_2 \odot (x_{\text{worst}}(t) - x_{\text{best}}(t)), \ t \geq 1.$$
$$(17.5)$$

17.1.3 Verifying if the Obtained Solution Is Better

To answer the question 'Is the new position better or worse and can this new position be used in the subsequent search?', [666] adopt the model of prey choice developed by Krebs et al. [332], which assumes that only two types of prey are available: profitable (type 1) and unprofitable (type 2). In this model, the unprofitable prey (type 2) will be ignored by the animal provided that the following is satisfied:

$$\frac{\lambda_1 E_1}{1 + \lambda_1 h_1} > \frac{E_2}{h_2}, \qquad (17.6)$$

where E_1 is the net energy gain of profitable prey, E_2 is the net energy gain of unprofitable prey, h_1 is the handling time (total time spent searching for, capturing, killing and eating the prey item) for profitable prey, h_2 is the handling time for unprofitable prey, and λ_1 is the encounter rate of the profitable prey.

The Krebs et al. model is translated into OFA terms as follows. The encounter rate of profitable prey is taken as a random number $\lambda_j \in [0,1]$. The position x_j is viewed as the prey item and the corresponding objective function value $f_j := f(x_j)$ is viewed as the energy of the prey item. The group's foraging time (number of iterations) t is viewed as the handling time. Zhu and Zhang [666] assume that a better position (which is interpreted as finding a profitable prey item) can be found by increasing the foraging time. Thus, when going from iteration t to iteration $t+1$:

- the position $x_j(t+1)$ obtained at iteration $t+1$ is viewed as the profitable prey and the corresponding objective function value $f_j(t+1) := f(x_j(t+1))$ is viewed as the energy E_1 of the profitable prey;
- the position $x_j(t)$ obtained at iteration t is viewed as the unprofitable prey and the corresponding objective function value $f_j(t)$ is viewed as the energy E_2 of the unprofitable prey.

These translations modify (17.6) into OFA terms as follows:

$$\frac{\lambda_j(t+1)f_j(t+1)}{1 + (t+1)\lambda_j(t+1)} > \frac{f_j(t)}{t} \qquad (17.7)$$

for maximisation problems, and (as maximising f is equivalent to minimising $-f$)

$$\frac{\lambda_j(t+1)f_j(t+1)}{1+(t+1)\lambda_j(t+1)} < \frac{f_j(t)}{t} \qquad (17.8)$$

for minimisation problems.

If (17.8) is satisfied, the position obtained after $t+1$ iterations can be used by the individual for the next search. Otherwise, the position obtained after t iterations is kept for the next search. These ideas are operationalised as Algorithm 17.1.

The OFA was tested on twenty global optimisation benchmark functions of varying difficulty [666]. The performance of the OFA was compared to that of a (real-coded) Genetic Algorithm [261, 212, 213, 255], Differential Evolution [550, 551], Particle Swarm Optimisation [314, 315], the Bees Algorithm (Sect. 10.2.1) [457, 458, 456], the Bacterial Foraging Optimisation Algorithm (Sect. 14.3) [450, 451] and the Shuffled Frog-leaping Algorithm [172]. The experimental results indicated that the quality of solution performance of the OFA was superior to the other algorithms, in terms of ability to converge to an optimal or near-optimal solution, and the statistical performance of the OFA was second best among all algorithms examined, in terms of the Kruskal–Wallis analysis of variance (ANOVA) test.

17.2 Group Search Optimiser

The *group search optimiser algorithm* (GSOA) was introduced by He et al. [244] and further developed and tested in [245]. The algorithm is inspired by the producer–scrounger foraging model (Sect. 2.3.2), which examines the outcomes that occur when only a portion of the population actively seek new resources (the *producers*) and the remainder of the population (the *scroungers*) seek to parasitise these resources once found, without having borne the costs of finding them. The algorithm also embeds a basic sensory perception mechanism.

17.2.1 Algorithm

The population is assumed to consist of three types of members, *producers*, *scroungers* and *rangers*. In the canonical algorithm there is a single producer whose location is denoted as x_p. The other members of the group either join the resource found by the producer (i.e. are scroungers) or search randomly for resources (i.e. are rangers). The role accorded to group members is not fixed and individuals can vary their behaviour between producing, scrounging and ranging as the algorithm progresses.

In each iteration of the GSOA, the producer scans locations around its current position, simulating a visual sensory mechanism, in order to assess whether they offer better resources. The locations scanned correspond to straight ahead (a scan at a zero degree angle in the current direction faced by the producer) and a randomly sampled point to the left or right of this.

If any of these locations yields a better fitness than the current position, the producer will move there. Otherwise, the producer remains in its current position but

---

Algorithm 17.1: Optimal foraging algorithm [666]

Choose maximum number of iterations t_{max};
Let $t = 1$;
Generate a random group $P = \{x_j(1) : j = 1,\ldots,N\}$ of N individuals, uniformly
  distributed throughout the search space;
for $j = 1$ **to** N *(each individual)* **do**
   | Compute objective function value $f_j(1) := f(x_j(1))$;
end
Sort the group P in ascending order by objective function value to get the sequence
  $(x_j(1) : j = 1,\ldots,N)$ with $f_1(1) \leq f_2(1) \leq \cdots f_N(1)$;
Let $x_{best}(1) = x_1(1)$ and $f_{best}(1) = f_1(1)$;
Let $x_{worst}(1) = x_N(1)$ and $f_{worst}(1) = f_N(1)$;
while $t < t_{max}$ **do**
    for $j = 2$ **to** N *(each individual except best)* **do**
       Choose a random integer $b \in \{1, 2, \ldots, j-1\}$;
       Update $x_j(t)$ to $x_j(t+1)$ using (17.4);
       if $x_j(t+1)$ *is outside the bounds of the search space* **then**
         | Amend $x_j(t+1)$ to bring it back within the search space;
       end
       Compute objective function value $f_j(t+1) := f(x_j(t+1))$;
    end
    Update $x_{best}(t)$ to $x_{best}(t+1)$ using (17.5);
    if $x_{best}(t+1)$ *is outside the bounds of the search space* **then**
       | Amend $x_{best}(t+1)$ to bring it back within the search space;
    end
    Compute objective function value $f_1(t+1) := f(x_{best}(t+1))$;
    for $j = 1$ **to** N *(each individual)* **do**
       Choose a random number $\lambda_j(t+1) \in [0, 1]$;
       if *Equation* (17.8) *is satisfied* **then**
         | Leave $x_j(t+1)$ and $f_j(t+1)$ unchanged;
       else
         | Let $x_j(t+1) := x_j(t)$ and let $f_j(t+1) := f_j(t)$;
       end
    end
    Sort the group P in ascending order by objective function value to get the
      sequence $(x_j(t+1) : j = 1,\ldots,N)$ with $= f_1(t+1) \leq \cdots f_N(t+1)$;
    if $f_1(t+1) < f_{best}$ **then**
       | Let $x_{best}(t+1) := x_1(t+1)$ and let $f_{best}(t+1) := f_1(t+1)$;
    end
    Set $t := t+1$;
end
Output the best solution x_{best} found, together with its objective function value f_{best};

reorientates its head to point in a new direction and scans again. If no better location is found after a iterations of the algorithm, the producer turns its head back to the initial zero angle.

In real-world examples of scrounging, animals may engage in *area-copying behaviour* (i.e. move to the area around a producer and search for resources there), *following behaviour* (i.e. follow another animal around without engaging in personal search) or *snatching behaviour* (i.e. take a resource directly from a producer) [244]. In the GSOA, scrounging behaviour is simulated using an area-copying strategy, and scroungers move stochastically towards the location of the producer (x_p).

In each iteration of the algorithm, the group member in the location with the highest fitness is designated as the producer. The algorithm consists of a blend of local search around the current best point by both the current producer and the scroungers, and exploration of new randomly chosen locations by the rangers. The algorithm embeds elitism, as the current best location is maintained until a new, better location is uncovered.

Pseudocode for the algorithm is provided in Algorithm 17.2. A number of steps are required to operationalise the algorithm, and these are discussed next.

Operationalisation of Algorithm

In the algorithm the population of searching agents is termed a *group*, with each individual in the group being termed a *member*. Each individual has a current location in the n-dimensional search space, $x_i(t) \in \mathbb{R}^n$ (where t is the iteration number), a head angle $\varphi_i(t) = (\varphi_{i1}(t), \ldots, \varphi_{i(n-1)}(t)) \in \mathbb{R}^{n-1}$ and a search direction $D_i(t) = (d_{i1}(t), \ldots, d_{in}(t)) \in \mathbb{R}^n$, which can be calculated from $\varphi_i(t)$ using a polar-to-Cartesian coordinate transformation:

$$d_{i1}(t) = \prod_{q=1}^{n-1} \cos(\varphi_{iq}(t)),$$

$$d_{ij}(t) = \sin(\varphi_{i(j-1)}(t)) \cdot \prod_{q=j}^{n-1} \cos(\varphi_{iq}(t)) \qquad j = 2, \ldots, n-1,$$

$$d_{in}(t) = \sin(\varphi_{i(n-1)}(t)). \tag{17.9}$$

In the subsequent text, we will occasionally have need to clarify which head angle $\varphi(t)$ that a particular $D_i(t)$ is a function of, and will make this explicit by using notation of the form $D_i(\varphi(t))$, or $D_i(\varphi(t) + \alpha)$ with α being another angle.

Scanning

The visual mechanism is simulated as follows. The field of vision of the producer is characterised by a maximum pursuit angle of $\theta_{\max} \in \mathbb{R}^{n-1}$ and a maximum pursuit distance of $l_{\max} \in \mathbb{R}$, as illustrated in Fig. 17.1.

In iteration t of the algorithm, the producer x_p randomly scans three locations as follows. A straight ahead (zero degree) location is scanned using

---

Algorithm 17.2: Canonical group search optimiser algorithm

Set values for $\theta_{max}, l_{max}, \alpha_{max}$ and a;
Set $t = 1$;
Generate randomly an initial group $x_i(1), \varphi_i(1) : i = 1,\ldots,n$ (both locations and head angles);
Evaluate fitness $f(x_i(1))$ of each member of the group;
repeat
    repeat
        Determine the producer $x_p(t)$ in the group;
        Create new points using the producer via random sampling of points in its scanning field (17.10–17.12);
        Evaluate the fitness of each generated point;
        if *the best point is better than the current position* **then**
            Producer moves to the new best point;
        else
            Producer stays in its current position and turns its head to a new randomly generated angle using (17.10);
            if *the producer cannot find a better location after a iterations* **then**
                It turns its head back to an angle of zero;
            end
        end
        Select a number of group members as scroungers;
        Each scrounger moves as per (17.15) to exploit resources found by the producer;
        Select rest of group members as rangers and disperse them;
        Dispersed members walk randomly as per (17.18);
    until *each member i in the group is visited*;
    Set $t := t + 1$ (iteration counter) ;
until *terminating condition*;
Output the best solution found;

$$x_z = x_p(t) + r_1 l_{max} D_p(\varphi(t)), \tag{17.10}$$

a point in the right hand side hypercube is scanned using

$$x_r = x_p(t) + r_1 l_{max} D_p(\varphi(t) + r_2 \theta_{max}/2), \tag{17.11}$$

and a point in the left hand side hypercube is scanned using

$$x_l = x_p(t) + r_1 l_{max} D_p(\varphi(t) - r_2 \theta_{max}/2), \tag{17.12}$$

where $r_1 \in \mathbb{R}$ is a random number drawn from a normal distribution with a mean of 0 and a standard deviation of 1, and $r_2 \in \mathbb{R}^{n-1}$ is a random vector with all elements drawn uniformly from $[0, 1]$.

If any of these three points has higher fitness than the producer's current location, the producer relocates to the new position. Otherwise, the producer remains in its current location and turns its head to a new angle as follows:

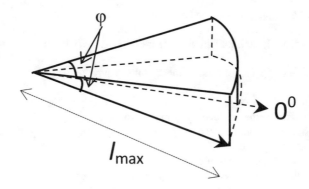

Fig. 17.1. Scanning field in three dimensions for a producer (straight ahead is shown as $0°$, φ is the maximum pursuit angle θ_{max} and l_{max} is the maximum pursuit distance)

$$\varphi(t+1) = \varphi(t) + r_2\alpha_{max},\qquad(17.13)$$

where α_{max} is the maximum turning angle. If after a iterations (a constant parameter) the producer cannot find a better location, it turns its head back to zero degrees:

$$\varphi(t+a) = \varphi(t).\qquad(17.14)$$

Scrounging

The scrounging process is simulated as a random walk from the position of the ith scrounger at the tth iteration (with the location being denoted by $x_i(t)$) towards the location of the producer as follows:

$$x_i(t+1) = x_i(t) + r_3 \cdot (x_p(t) - x_i(t)),\qquad(17.15)$$

where $r_3 \in \mathbb{R}^n$ is a random vector with all elements drawn uniformly from $[0,1]$.

Ranging

If the ith group member is selected to be a ranger during the tth iteration, a random head angle φ_i is generated for that individual:

$$\varphi_i(t+1) = \varphi_i(t) + r_2\alpha_{max},\qquad(17.16)$$

where α_{max} is the maximum turning angle. Next, a random distance is selected using

$$l_i = ar_1 l_{max}\qquad(17.17)$$

and the ranging member moves to the new location:

$$x_i(t+1) = x_i(t) + l_i D_i(\varphi(t+1)).\qquad(17.18)$$

17.2.2 Parameter Setting

While the best choice of parameters will be problem specific, guidance in [244] suggests that 80% of the population should be assigned to behave as scroungers (one will be a producer and circa 20% will therefore be rangers), and suggests a population size of 48 members, an initial head angle φ_i for each individual i of $\pi/4$ and a value for a of round($\sqrt{n+1}$). The maximum pursuit angle θ_{max} is set to π/a^2. The maximum turning angle α is set to $\pi/(2a^2)$ and the maximum pursuit distance l_{max} is calculated from:

$$l_{max} = \|U_i - L_i\| = \sqrt{\sum_{i=1}^{n}(U_i - L_i)^2}, \tag{17.19}$$

where L_i and U_i are the upper and lower bounds for the ith dimension.

Variants of the GSOA have been applied to a number of problems, including constrained optimisation [523]. It has also been applied to create a hybridised invasive weed optimisation algorithm (Sect. 16.5.1) for multimodal optimisation [496].

17.3 Predatory Search Algorithm

A common behaviour in many foraging organisms is that of *area-restricted search behaviour*, whereby their search process changes on detection of a prey item or on detection of evidence of a prey item being in the vicinity (Fig. 17.2) [307]. At this point, foraging search is restricted to a local area. When engaging in area-restricted search, the predator typically slows down and changes its direction of movement more frequently, in order to remain within the local area. One consequence of individuals engaging in this behaviour is that populations of such predators will tend to flow towards regions of high prey density even without explicit communication between them as to prey location [307]. If, after a period of time, no other prey is found in the restricted area, the predator will give up on that area and move elsewhere to search for prey.

Area-restricted search has been documented in several organisms including birds, insects, lizards [359] and could be expected to be a useful strategy when prey items are clustered in a habitat. The strategy will tend to ensure that the time predators spend in an area is in proportion to its prey resources and therefore, the strategy is adaptive to diverse prey distributions [534].

Based on the concept of area-restricted search, a series of papers by Alexandre Linhares in 1998 and 1999 [359, 360, 361] developed the *predatory search algorithm* and applied it to a number of NP-hard, multimodal combinatorial optimisation problems including the travelling salesman problem and the gate matrix layout problem (a VLSI (very large-scale integrated circuit) physical layout problem).

The essence of the algorithm is that once a new best-so-far solution is uncovered, search is intensified in a small neighbourhood around that solution, mimicking an area-restricted search behaviour, with the size of this neighbourhood being gradually

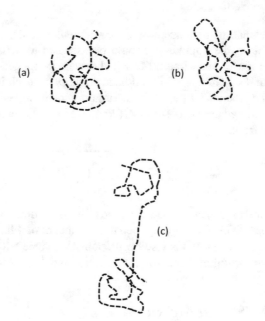

Fig. 17.2. Area-restricted search, with (a) and (b) representing foraging animals finding prey in a localised region, and (c) representing a forager that migrated from one local region to another after the former was exhausted of prey

increased if no better solutions are discovered. After a period of time, if no better solution is found, the restricted-area search is abandoned and the search process moves elsewhere in the search space.

17.3.1 Algorithm

The background to the algorithm is presented as follows in [359, 360, 361]. A combinatorial optimisation problem can be generally described as a pair (Ω, f), where Ω is the set of solutions and the function $f : \Omega \rightarrow \mathbb{R}$ maps each solution to a corresponding cost. The aim of the optimisation process (assuming a minimisation problem) is to find a solution $s^* \in \Omega$ such that $f(s^*) \leq f(s)$ for all $s \in \Omega$. For each solution s, we define a neighbourhood $N(s) \subset \Omega$ around s; then a transformation from s to a solution in $N(s)$ can be termed a *move*. Given two solutions $s_0, s_k \in \Omega$, we say that s_k *can be reached from* (or *is reachable from*) s_0 if there is a suitable sequence or *path* $s_0, s_1, \ldots, s_k$ of intermediate moves such that $s_{i+1} \in N(s_i)$ for each $i = 0, \ldots, k-1$; that is, each of these moves produces a solution with an associated cost.

For the example of the travelling salesman problem (TSP), a solution is a tour of all cities, with the solution cost being the sum of all arc costs along the tour, and a natural definition of a move is the reversal (or transposition) of a pair of cities in a solution.

In the context of a general combinatorial optimisation problem (Ω, f), with s_k reachable from s_0 via the sequence $s_0, s_1, \ldots, s_k$, suppose that for some $R \in \mathbb{R}$ we have $f(s_i) \leq R$ for all $0 \leq i < k$. Then we say that there is a path from solution s_0 to s_k that *respects the restriction R*. We denote this by $s_k \in A(s_0, R)$. This gives rise to a function $A : \Omega \times \mathbb{R} \longrightarrow 2^{\Omega} : (s, R) \longmapsto S$, that is, A maps a pair (s, R) of solution s and restriction R to a subset S of Ω. Here, the notation 2^{Ω} means the set of all subsets of Ω, also called the *power set* of Ω. The meaning of the notation $A(s, R)$ is the set of all solutions reachable from s for which the path from s respects the restriction R.

Thus, given a solution $s \in \Omega$ and a real number $R \in \mathbb{R}$, the function $A : \Omega \times \mathbb{R} \longrightarrow 2^{\Omega}$ defines a restricted search area around s, namely, the set of solutions $A(s, R)$. If a low value is chosen for R, then a small neighbourhood of potential moves is defined. In contrast, if R is set to a very large value, then $A(s, R) = \Omega$.

Predatory search considers at each step the transformation $s := s' \in A(b, R)$, where s is the current solution, b is the best solution found by the algorithm to that point and R is the restriction.

In order to implement a gradually increasing search area around the best solution found, predatory search does not use a single value for R. Instead, it uses an ordered list of $n + 1$ restriction levels, $(Res[L] : L = 0, 1, \ldots, n)$, where n is the number of cities and L is called the level of the restriction $Res[L]$. The restriction values in this list are monotonically increasing, such that $Res[0] = f(b)$ and $Res[i] \leq Res[i + 1]$ for all $0 \leq i \leq n - 1$. The values of $Res[1], \ldots, Res[n]$ are n samples taken from $N(b)$, where b is the best solution found by the algorithm. The restriction level is correlated with the size of the search area.

Pseudocode for the predatory search algorithm is outlined in Algorithm 17.3, after [359], where it is described as follows. In the main loop of the algorithm, the neighbourhood $N(s)$ of the best solution s is sampled (here an arbitrary 5% sampling level is assumed) to produce $N'(s)$. If the best solution in this subsample has a lower cost than the restriction L, the best solution is updated and the process iterates again.

After each move, a counter (denoted by k in Algorithm 17.3) is incremented that tracks the number of iterations spent in an area. When this counter reaches a predefined threshold (defined as $3n$ in Algorithm 17.3), the restriction level L is incremented in order to increase the size of the neighbourhood being searched. If a threshold for this incrementing process is reached (when $L = \lfloor n/20 \rfloor$), the algorithm gives up on area-restricted search and sets a high value for L (i.e. $L = n - \lfloor n/20 \rfloor$). The search process then proceeds into a new region away from the previously restricted area.

Whenever a new improved solution is found, it is stored in b; the list of restriction levels is recomputed from the neighbourhood of b; and L is set to zero so that the algorithm will seek improvements around b. The value of L is increased gradually thereafter, in turn increasing the size of the search space, if no better solutions are found. Eventually, if no better solution is found, the value of L is significantly increased, in order to move the search process away from the current restricted region.

The list of restriction levels needs to be recomputed when a new best overall solution is found, as restricted search is implemented around this improved solution. In a TSP implementation (for n cities), the restriction levels $1, \ldots, n$ are computed from a sample of n solutions on the neighbourhood $N(b)$ (where b is the new improved solution). Res[0] carries the new value of $f(b)$. Res[1] carries the smallest jump possible to escape b and, as the restriction level L grows, the search area $A(b, \text{Res}[L])$ is increased.

Algorithm 17.3: Predatory search algorithm [359]

TSP application for n cities: choose initial solution s, let $b := s$ and let $k := 0$;
while $L < n$ **do**
> Construct proposal set $N'(s)$ by sampling 5% of $N(s)$;
> Choose proposal solution to be $x \in N'(s)$ such that $f(x) \le f(y)$ for all $y \in N'(s)$;
> **if** $x \in A(b, \text{Res}(L))$ **then**
> > Let solution $s := x$;
> > **if** $z(s) < z(b)$ **then**
> > > Recompute list of restriction levels $(\text{Res}[L] : L = 0, 1, \ldots, n)$;
> > > $L := 0$;
> > > $b := s$;
> > **end**
> **else**
> > $k := k + 1$;
> > **if** $k > 3n$ **then**
> > > $k := 0$;
> > > $L := L + 1$;
> > > **if** $L = \lfloor n/20 \rfloor$ **then**
> > > > $L := n - \lfloor n/20 \rfloor$;
> > > **end**
> > **end**
> **end**
end
Output the best solution found;

17.4 Predator–Prey Optimisation Algorithm

The *predator–prey optimisation* (PPO) *algorithm* was developed by Silva et al. [528]. As discussed in Sect. 2.4, foraging takes place in an ecology, with most animals being simultaneously both predators of organisms lower in the food chain and prey of animals higher up the food chain.

The PPO algorithm adds a predator–prey concept to the canonical PSO algorithm, so that the search process in canonical PSO is supplemented by simulated predators chasing prey around the search space. As in canonical PSO, all the agents in the PPO algorithm are termed *particles* and the population of particles is split into two mutually exclusive groups. A subset of the particles are considered as *predators* with the remaining particles being classed as *prey*. The predators are attracted to the best individuals in the swarm of prey particles, whereas the prey particles are repelled by predator particles, thereby generating movement in the swarm (Fig. 17.3).

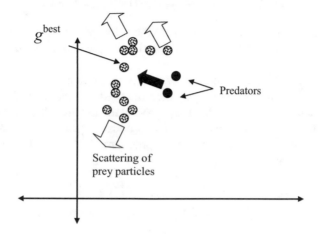

Fig. 17.3. Two predator particles chasing the prey particle located at g^{best}

The biological motivation for the PPO algorithm is that prey tend to gather around locations with good resources, such as places with plentiful food or water. Prey, which are located at resource-rich locations, therefore have little motivation to seek out alternative resource locations. However, if the flock of prey is attacked and scattered by predators, they will be forced to flee and seek out alternative predator-free locations. These new locations may turn out to offer even richer resources than the original location. In terms of optimisation, good resource locations can be considered as local optima.

Initialisation

Each particle i has a number of properties, including a current location, denoted by x_i, a current velocity v_i and a personal best position p_i^{best} in the search space (i.e. a memory of the best location that particle has ever uncovered during its movement around the search space). Each particle also knows the location of the global best solution found by all the particles up to iteration t (this is termed $g^{best}(t)$).

At the start of the algorithm, the p^{best} for each particle is set to the initial location of that particle and g^{best} is set to the location of the best of the p^{best}s. During each iteration of the algorithm, the velocity and location of each particle are updated according to the rules below.

Implementation of Movement

The predator–prey metaphor can be implemented in a variety of ways. The approach taken in [528] is to have a single predator, which is attracted towards the current g^{best} location. The velocity and position update equations for the predator particles are:

$$v_{\text{predator}}(t+1) = \alpha(g^{\text{best}}(t) - x_{\text{predator}}(t)), \tag{17.20}$$

$$x_{\text{predator}}(t+1) = x_{\text{predator}}(t) + v_{\text{predator}}(t+1), \tag{17.21}$$

where $v_{\text{predator}}(t+1)$ and $x_{\text{predator}}(t+1)$ are the velocity and location, respectively, of the predator. The parameter α controls how fast the predator moves towards the g^{best} location.

The influence of the predator on each prey particle depends on how close the predator is to the prey particle. The closer the predator, the more the prey particle reacts by changing its velocity in order to avoid the predator. To capture this effect, a repulsion term $D(d)$ is added to the velocity update equation for the prey particles. Therefore, for each dimension j of each prey particle i, the velocity and position update equations are

$$v_{ij}(t+1) = Wv_{ij}(t) + c_1 r_{1j}(t)(p_{ij}^{\text{best}}(t) - x_{ij}(t))$$

$$+ c_2 r_{2j}(t)(g_j^{\text{best}}(t) - x_{ij}(t)) + c_3 r_{3j} D(d)(t) \tag{17.22}$$

$$x_{ij}(t+1) = x_{ij}(t) + v_{ij}(t+1) \tag{17.23}$$

where the repulsion is calculated using an exponentially decreasing function, defined as $D(d) = ae^{-bd}$. The parameter d is the Euclidean distance between the predator and the prey, a controls the maximum effect the predator can have on the prey's velocity along any dimension, and b is a scaling factor. The repulsion term produces a more violent reaction by the prey if the predator is very close. For example, if the predator and prey are in the same location, $d = 0$, and the repulsion effect is a (since $e^0 = 1$). As the distance d tends to ∞, the repulsive effect tends to 0 (since $e^{-z} \to 0$ as $z \to \infty$). The terms r_1, r_2 and r_3 in (17.22) are drawn (separately for each dimension in particle i's velocity vector) from a uniform distribution in the range $[0,1]$.

Equation (17.22) is used to update each element of a prey's position vector based on a fear threshold, P_f. For each dimension, if a random draw from $U(0,1) < P_f$, then (17.22) is used to update $x_{ij}(t)$. Otherwise, the standard PSO velocity update (17.24) for each dimension j without the repulsion component is used:

$$v_{ij}(t+1) = v_{ij}(t) + c_1 r_1 (p_{ij}^{\text{best}}(t) - x_{ij}(t)) + c_2 r_2 (g_j^{\text{best}}(t) - x_{ij}(t)). \tag{17.24}$$

In the early iterations of the PPO algorithm, most particles will not be close to the predator; therefore, the predator–prey term in the velocity update vector will have limited effect. As the swarm starts to converge towards the best-so-far g^{best}, the repulsion term helps ensure continued diversity in the swarm. Later in the search process, the influence of the predator should be decreased by reducing the fear threshold P_f or by reducing the value of a in order to permit finer search around g^{best}.

A variant on the above predator–prey approach was proposed in [259], in which there are multiple predators which behave as normal PSO particles in that they are both drawn to the g^{best} location and also influenced by their own p^{best} location. Predator particles employ the standard PSO velocity update equation. In contrast, the velocity update equation for the prey contains a repulsion term, whereby a prey particle responds to its nearest predator by moving away from it. Both predators and prey use the same g^{best} information, and both particles can update the position and value of g^{best}. Predators therefore tend to search around g^{best}, with the prey particles engaging in more diverse exploration of the search space.

Antipredator Behaviours

In the PPO algorithm, the response of prey particles to the presence of simulated predator particles is to flee (i.e. avoid the predators). Flight from a predator is only one of a suite of antipredator behaviours which an animal can exhibit. Other such behaviours include hiding (becoming difficult to detect), attempting to fight the predator or the production of predator-deterrent signals [617]. Predator-deterrent signals may be used to indicate to the predator that attempts at prey capture will be costly. Examples of signals include behaviours which informs the predator that it has been spotted by the prey item, that the prey item is in a heightened state of vigilance, or that the potential prey item is in good physical condition and therefore able to produce a robust defence or flee. To date, apart from the PPO algorithm, few attempts have been made to integrate ideas from the antipredator literature into the design of foraging-inspired algorithms.

17.5 Animal Migration Optimisation Algorithm

As identified in Sect. 1.4, there are many influences on an organism's choice of foraging strategy, including the spatial and temporal distribution of food resources in the environment. When environmental conditions and food availability are seasonal, organisms may engage in migratory behaviour, periodically moving to new locations in order to maintain access to food. Migratory behaviour, a subfield of movement ecology (Sect. 2.5), has been the subject of considerable study, and aspects of this literature could plausibly contribute to the design of novel search algorithms.

One exemplar of this is provided by the *animal migration optimisation algorithm* (AMOA) introduced by Li et al. [354], which is loosely inspired by an animal migration metaphor. The canonical version of the algorithm embeds two main components, the first simulating each individual's movement and the second simulating changes in the population of migratory animals over time. During the movement process, each individual is influenced by a local or *peer* group following three basic rules:

   i. move in the same direction as your neighbours;
  ii. remain close to your neighbours; and
 iii. avoid collisions with your neighbours.

The algorithm assumes that the population size remains constant during the algorithm and, hence, an existing member of the population must depart before a new animal can join the migrating group.

17.5.1 Algorithm

In this section, we outline the AMOA following the description provided in [354]. The population size is set at N_{pop} and each member i of the population ($i = 1,\ldots,N_{\text{pop}}$) has a D-dimensional real-valued position vector $x_i = (x_i^1,\ldots,x_i^D) \in \mathbb{R}^D$. During initialisation, each member of the population is randomly positioned within the minimum and maximum bounds set for each dimension, which are given by $x_{\text{min}}^j \leq x_i^j \leq x_{\text{max}}^j, j = 1,\ldots,D$, for each population member i. Therefore, the jth component of the vector representing the ith individual can be obtained using

$$x_i^j = x_{\text{min}}^j + \text{rand}_{i,j} \cdot (x_{\text{max}}^j - x_{\text{min}}^j), \quad i = 1,\ldots,N_{\text{pop}}, j = 1,\ldots,D \qquad (17.25)$$

where each $\text{rand}_{i,j}$ is drawn from a uniform distribution $U(0,1)$ on the interval $[0,1]$.

Movement Process

In each iteration t of the algorithm an individual moves to a new location, influenced by the location of its peer group. The peer (or neighbouring) group for each individual is based on a ring topology, with [354] using a neighbourhood size of five. Each individual in the population is assigned an index number during initialisation and this remains static as the algorithm iterates. Assuming an individual is assigned index i, then its neighbours will be the individuals with indices $i-2$, $i-1$, $i+1$ and $i+2$. Therefore each individual exists in a local group of size five (including itself). The neighbourhoods of successive individuals (for example, i and $i+1$) will overlap substantially. As the topology forms a ring, the neighbourhood around individual 1 will consist of the individuals with the indices $N_{\text{pop}}-1$, N_{pop}, 2 and 3.

When the position of the ith individual is being updated, one of its neighbours is selected randomly, and the position update is calculated using

$$x_i(t+1) = x_i(t) + \delta \cdot (x_{\text{neighbour}}(t) - x_i(t)), \qquad (17.26)$$

where $x_i(t+1)$ is the new position of the ith individual, $x_{\text{neighbour}}(t)$ is the current position of the randomly selected neighbour and δ is a parameter which determines the influence of this neighbour on the position update. This parameter can be operationalised in various ways, with [354] adopting a random draw from a Gaussian distribution.

Changes in Population

During the running of the algorithm, newly created individuals can replace existing members of the population. This process is governed by a parameter $0 \leq a \leq 1$, where better members of the population are less likely to be selected for replacement. A simple means of determining the value of a for each member of the population is to set its value to 1 for the best member of the population, decreasing linearly to $1/N_{pop}$ for the worst member of the population.

Algorithm 17.4 outlines the replacement process. For each member i of the pop-

---

Algorithm 17.4: Replacement process for AMOA

for $i = 1$ **to** N_{pop} **do**
    for $j = 1$ **to** D **do**
        if rand $> a$ **then**
            $x_i(t+1) = x_{r_1}(t) + \text{rand} \cdot (x_{best}(t) - x_i(t)) + \text{rand} \cdot (x_{r_2}(t) - x_i(t))$;
        end
    end
end

---

ulation, the jth component of its location vector is changed if a random draw (rand) from a $U(0, 1)$ distribution is greater than a for that individual, otherwise the jth component of its location vector remains unchanged. The new component for dimension j is influenced by the corresponding component from a randomly selected member of the population $x_{r_1}(t)$, and stochastically by the corresponding component of the individual being altered $x_i(t)$, relative to that of a second randomly chosen member of the population $x_{r_2}(t)$ and the best-located (fittest) member of the population $x_{best}(t)$, where r_1 and r_2 are randomly selected indices (where $r_1 \neq r_2 \neq i$).

From the above formulation, we see that if the value of a for the ith individual is 1 (i.e. it is the best individual in the current population), then that individual will pass unchanged into the next iteration of the algorithm. Conversely, if the individual corresponds to the worst individual in the population, it is likely that most components of the vector containing its position will be altered.

After the new individual $x_i(t+1)$ is fully created (i.e. all components of its location are specified), its fitness is compared with that of the existing individual $x_i(t)$, and the new individual replaces the existing individual if it has better fitness. Otherwise, the existing individual is not replaced.

Bringing the above two steps together, the canonical version of the animal migration optimisation algorithm is outlined in Algorithm 17.5.

Discussion

The AMOA is a fairly simple algorithm which has demonstrated competitive results on a range of test functions. Variants on the canonical algorithm have been developed in [99, 376, 377]. As with some other algorithms in this text, the AMOA is of recent vintage and further work is required in order to gain deeper insight into its performance and computational efficiency.

Algorithm 17.5: Animal migration optimisation algorithm

Set value of N_{pop} and set $t = 0$;

Initialise the position x_i of each individual $i = 1, \ldots, N_{\text{pop}}$ to random locations in the search space;

Evaluate fitness of each individual $i = 1, \ldots, N_{\text{pop}}$ and determine location x_{best} of best fitness in population;

repeat

    for $i = 1$ **to** N_{pop} **do**

        for $j = 1$ **to** D **do**

            $x_i(t+1) = x_i(t) + \delta \cdot (x_{\text{neighbour}}(t) - x_i(t))$;

        end

    end

    for $i = 1$ **to** N_{pop} **do**

        Evaluate fitness of offspring at location $x_i(t+1)$;

        if $x_i(t+1)$ *is fitter than* $x_i(t)$ **then**

            Replace $x_i(t)$ with $x_i(t+1)$;

        end

    end

    Update location of best fitness in population x_{best} if necessary;

    Calculate a for each individual in the population;

    for $i = 1$ **to** N_{pop} **do**

        for $j = 1$ **to** D **do**

            Randomly select indices r_1, r_2, (r_1, r_2 and i all different) ;

            if rand $> a$ **then**

                $x_i(t+1) = x_{r_1}(t) + \text{rand} \cdot (x_{\text{best}}(t) - x_i(t)) + \text{rand} \cdot (x_{r_2}(t) - x_i(t))$;

            end

        end

    end

    for $i = 1$ **to** N_{pop} **do**

        Evaluate fitness of offspring at location $x_i(t+1)$;

        if $x_i(t+1)$ *is fitter than* $x_i(t)$ **then**

            Replace $x_i(t)$ with $x_i(t+1)$;

        end

    end

    Update location x_{best} of best fitness in population if necessary;

    Set $t := t + 1$;

until *terminating condition*;

Output the best solution found;

While the search process implemented in the AMOA is loosely inspired by a migration metaphor, an alternative narrative concerning the algorithm which is provided in [354] is that it can be considered as a swarm intelligence algorithm (Sect. 5.4). There are some similarities in the neighbourhood topology structure of the AMOA to local best versions of PSO, although there are also distinct differences between the operation of the two algorithms. Further work is required in order to determine the significance of the similarities and differences between the AMOA and existing swarm algorithms. It can also be noted that the movement rules for individuals in the AMOA bear some similarity with those in Reynolds' boids simulation [484] (Sect. 6.1.4). However, the focus of the boids simulation is on generating insights into flocking behaviours rather than being designed for optimisation purposes.

More generally, we could pose the question as to whether it is appropriate to apply a migratory metaphor to the design of optimisation algorithms. In search algorithm applications, we generally have limited knowledge concerning the likely regions in which good solutions may exist and, typically, the algorithms emphasise initial exploration of the search space, focusing on greater exploitation of good regions as the algorithm progresses. In contrast, in migratory settings, animals may have a priori experiential or imprinted knowledge of the region or exact location to which they will migrate. Hence, migratory movements often embed a high degree of directionality rather than corresponding to a general search for areas with food availability. Nonetheless, while acknowledging the limits of applicability of a migratory metaphor to a search process, it is evident that considerable scope exists to develop optimisation algorithms which draw from the many theoretical models and empirical studies of animal movement contained in the wider movement ecology literature.

17.6 Summary

In Chap. 2, a subset of models from the theoretical foraging literature were introduced. This chapter describes a number of algorithms which draw on this literature, based on the producer–scrounger model and on the idea of area-restricted search, and an algorithm based on a predator–prey concept. The final algorithm introduced draws from the movement ecology literature. Obvious scope exists to design alternative algorithms drawing from these and other formal models of foraging.

Part VII

Evolving a Foraging Strategy

In the chapters thus far, we have studied an extensive range of optimisation and search algorithms derived from foraging behaviours of diverse classes of organisms, including mammals, birds, fish, invertebrates such as insects, and nonneuronal organisms such as bacteria, moulds and plants; along with foraging algorithms based on formal models of foraging. Each foraging behaviour derives from a mix of influences, including developmental lifetime learning and genetics.

Another much-studied family of biologically-inspired algorithms is the family of *evolutionary algorithms*, which simulate an evolutionary process on a computer, assigning fitness values to individuals and so affecting which individuals may transmit genetic information to the next generation; such algorithms, when applied to a problem we wish to solve, *evolve* a potential solution. Chapter 18 introduces the idea of using an evolutionary algorithm to create a foraging strategy or foraging-inspired algorithm.

Evolving Foraging Algorithms

In previous chapters, a wide range of algorithms inspired by various foraging behaviours have been described. Each of the underlying behaviours has arisen through
a mix of influences, such as lifetime learning and genetics, which determine the
physical and cognitive capabilities of organisms.

Apart from foraging-inspired algorithms, several other families of biologically-
inspired algorithms exist, with *evolutionary algorithms* being one of the best known
of these families. Evolutionary algorithms simulate an evolutionary process on a
computer in order to evolve a potential solution to a problem of interest.

Starting from an initial randomly generated population of potential solutions,
the higher-quality (or *fitter*) of these are preferentially selected for reproductive purposes. New individuals are generated from these parents using a combination of
information reuse (simulating DNA crossover in biological reproduction) and the
generation of novelty (simulating mutation events in biological reproduction).

The mechanisms of fitness-driven selection and crossover promote the retention
of information from the better solutions in the current population (i.e. exploitation
of existing information) and the mutation operator allows for the discovery of novel
solution elements (i.e. generating exploration). Iterating this process on a computer
over multiple simulated generations can produce ever better quality solutions to the
problem at hand. Algorithm 18.1 outlines a general evolutionary metaheuristic.

The solutions generated by the evolutionary algorithm could be a vector of real
numbers, as is the case in real-valued optimisation problems or, more generally,
could be a structure such as a mathematical model or a list of instructions such as
a foraging strategy. Typically, the evolution of solutions for optimisation purposes
where the nature of the structure to be optimised is known in advance (such as a
fixed-length vector of real numbers where these parameterise some model) is tack-
led using a genetic algorithm (GA) [212, 261]. For more complex settings where the
structure and content of the solution is not known a priori, open ended generative
methods such as genetic programming (GP) [330] or grammatical evolution (GE)
[143, 442, 500] are more suitable.

In this chapter we combine a foraging metaphor with an evolutionary metaphor
to evolve foraging strategies. More generally, this highlights the ability to use an evo-

© Springer Nature Switzerland AG 2018
A. Brabazon, S. McGarraghy, *Foraging-Inspired Optimisation Algorithms*,
Natural Computing Series, https://doi.org/10.1007/978-3-319-59156-8_18

Algorithm 18.1: Evolutionary algorithm

Initialise the population of candidate solutions (perhaps randomly);
repeat
  Select better individuals (parents) for breeding from the current population;
  Generate new individuals (offspring) from these parents using simulated crossover
    and mutation operators;
  Replace some or all of the current population with the newly generated individuals;
until *terminating condition*;
Output best individual (solution) found.

---

lutionary algorithm to create powerful new search algorithms, drawn from foraging concepts, crafted for a specific problem environment. We begin by overviewing a study which used GP to evolve a foraging strategy for an individual organism and then proceed to discuss how evolutionary methodologies can be applied to gain insight into key features of real-world foraging scenarios. Finally, we introduce some aspects from the field of robotics. A canonical problem in robotics is the design of collective behaviours to coordinate the search activities of a swarm of robots, a task which bears close parallel with foraging and therefore with many of the topics discussed in this book.

18.1 Using Genetic Programming to Evolve a Foraging Strategy

Koza [331] applied GP to the problem of discovering a food foraging strategy for the Caribbean *Anolis* lizard. These lizards are sit-and-pursue predators which feed on insects. The lizards sit on tree trunks and pursue insects that they see on the ground.

The Model

In [331], a mathematical model of the optimal foraging strategy for the lizard, which determines whether a prey item should be pursued once seen, is initially derived. The assumptions underlying the model are that a lizard has a 180° viewing angle from its perch (located at the coordinates $(0,0)$ in the simulated search space), that all prey chases start from this location, that a potential prey insect does not move from the location where it is first sighted by the lizard, and that the energy values of all prey items caught are identical. To simplify the analysis, it is also assumed that the energy consumption per unit of time during the waiting phase is equivalent to the energy expended per unit of time during the chase phase.

Under these assumptions the derived optimal strategy is a function of three items, the probability that a prey item appears in a square metre of the visual range per second (termed the *abundance* and denoted by a), the speed of the lizard when moving (v metres per second) and the location of the insect within the visual range of the lizard (its x and y coordinates). The optimal strategy is therefore derived as [331]:

$$\text{sign}\left(\sqrt[3]{\frac{3v}{\pi a}} - \sqrt{x^2 + y^2} \right), \tag{18.1}$$

where the insect appears at position (x, y) in the lizard's viewing area and the function sign is the sign function. This function returns a value of $+1$ if its argument is positive (indicating that the prey item should be pursued) and -1 if its argument is negative (indicating that the prey item should be ignored).

Therefore, the lizard should chase a prey item that appears within a semicircle of radius r^* (calculated as $\sqrt[3]{3v/(\pi a)}$) centred on the lizard's location at $(0,0)$. Intuitively, if insects are scarce, the lizard should chase any insect once seen; if insects are abundant, the lizard should only chase nearby insects.

On selecting specific values for a and v, a foraging process can be simulated for a defined period of time t. The results from applying (18.1) in such a simulation provide a benchmark which can be used to assess how well alternative (nonoptimal) foraging strategies perform.

Evolving a Strategy

GP was used to evolve a population of foraging strategies for the above setting. When applying GP to a particular problem, the modeller must first define the basic building blocks which GP can use in constructing and evolving its potential solutions. These building blocks consist of terminals (i.e. any item which requires no additional information to evaluate it, for example, a real number) and functions. Functions require additional information or arguments in order to be evaluated. For example, the mathematical operator + requires two arguments.

A common representation in GP is a syntax tree, each individual in the population consisting of a distinct syntax tree. Hence, each such tree is a potential solution whose worth can be estimated by applying the solution represented in the tree to the problem of interest. Figure 18.1 illustrates a syntax tree for a simple mathematical function which evaluates to a real number. Readers requiring further detail on the GP methodology are referred to [77].

In [331], the terminal set for GP includes the position of a prey item and the values of a and v. The function set includes the basic mathematical operators of +, −, *, protected divide, the two-argument exponentiation function (this function raises the absolute value of its first argument to the power of its second argument) and an **if** branching condition.

For each simulation, values of a and v were set and the fitness of a GP individual (i.e. a foraging strategy which describes a rule governing the movement of the lizard during a simulation) was determined by the number of insects that it captured during the simulation. As the results of single experiments varied, since insects appeared stochastically during the simulation, fitness was measured over 72 simulations, in which the values of a and v were varied systematically to produce 36 distinct combinations, with two experiments being undertaken for each parameter-setting combination. The resulting best-of-run evolved strategy produced a total score over all 72 simulations of 1652 (insects captured) as against the 1671 achieved by the optimal

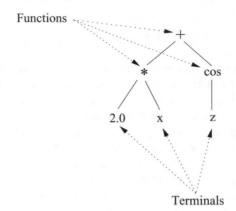

Fig. 18.1. Terminals correspond to the leaves of the syntax tree. Functions can have differing arities and can take either other functions or terminals as inputs. Here, the tree can be rewritten as $2.0x + \cos(z)$ where x and z are variables, with the entire tree outputting a real number for particular choices of x and z

strategy, indicating that GP was able to evolve a foraging strategy which produced results close to those of the optimal strategy.

The study also examined the capability of GP to evolve a foraging strategy for a more complex setting where the prey insect might not be present at its initially observed location by the time the lizard reached that location, adding a further stochastic element to the determination of the foraging strategy. The results indicated that GP was able to evolve a quality foraging strategy in this setting also.

In summary, this work illustrates how an evolutionary methodology can be used to create a high-quality foraging strategy for a case where there is a single organism, foraging in isolation.

18.2 Understanding Real-World Foraging

Another stream of related research is the application of evolutionary methodologies to help us better understand real-world foraging processes. These approaches seek to evolve the decision rules (strategy) for a foraging task in a simulated environment, and allow the undertaking of sensitivity analysis in this closed simulated world in order to gain insight into the importance of various features of the foraging process. Iterating the process of design of the simulated environment, identification of environmental and other physical constraints, and the evolution of the foraging strategies themselves can help biologists identify how key environmental variables and constraints impact on the development of specific foraging strategies found in the real world.

One example of this work is a study [28] in which the foraging strategies of a population of agents are allowed to evolve. Unlike the previous study described in this chapter, no explicit fitness function is defined. Agents are assigned an initial level of resources, which can be increased if they forage successfully and which otherwise decrease over time owing to the costs of metabolism or the undertaking of a reproductive process. The reproductive process is an action which is available to an agent within this simulation, rather than being imposed exogenously by means of an explicit evolutionary algorithm.

If the level of an agent's resources falls below zero, it dies and is removed from the simulation. Resources are made available at a constant rate during the simulation, so an increase in population will result in more competition for available resources. The agents are also permitted to deposit pheromone (like ants) which subsequently evaporates at a constant rate. The agents can take actions such as move, turn left, turn right, eat, reproduce and deposit pheromone. They are also capable of perceiving environmental information such as pheromone level and have visibility of resources within a determined distance of their current location.

Initially, the decision rules used by agents are randomly assigned. When an agent reproduces (one parent giving rise to one child), a mutation operator is applied to the parent's list of decision rules, and this operator can change a rule, a component of a rule or the order of rules. An example of a single line of a longer nested rule from [28] is:

$$\textbf{if } energy < 10 \textbf{ and } reach \ resource \ is \ \textbf{true then } eat \qquad (18.2)$$

and under this rule fragment an organism will eat any resource that is within range of its current position if its current energy level is less than 10 units.

The results of the simulations indicate that the agents evolve coherent foraging strategies over the course of the simulation, leading to increases in the average age of agents (i.e. they succeed in surviving for longer) and in the average energy levels of agents. As reproductive behaviours become common throughout the population, population dynamics is observed, with increases in population leading to pressure on the average resource capture by individual agents. The study illustrates that agents' foraging strategies (i.e. rule lists for decision making) can be evolved in an open-ended simulation.

In similar work [446] strongly typed GP is applied to discover foraging behaviours that are capable of efficiently finding food resources and then returning to a nest, in a population of simulated ants. In the simulation, ants are capable of depositing and sensing two distinct types of pheromone in the environment, one type when travelling to food (which is located at a fixed place in the environment) and one type when returning to the nest.

Each ant decision rule consists of two GP trees, which, when evolved, control the amount and type of pheromone an ant will deposit in each iteration, and also determine where the ant will move in the next time step. The results of the simulations indicate that the ants learn to deposit pheromone and to use the information from the outward and inward pheromone trails to navigate the environment more quickly.

In this case, both the movement and pheromone deposit behaviours of ants were evolved. In contrast, an earlier study [505] used evolutionary approaches to learn the rules by which simulated ants could usefully exploit the information in deposited pheromones, where the deposit process itself was a hard-coded (i.e. nonevolved) behaviour.

18.3 Robotics

There is a natural connection between many of the topics discussed in this book and the field of robotics. Robotics is a significant area of research and practical interest, with robots varying in complexity from those which implement a fixed program (e.g. assembly robots in a factory) to autonomous robots which can adapt their behaviour in response to environmental context and past learning. Robot implementations can also extend beyond individual robots to encompass multiagent (robot) systems where a group of robots seek, through interaction, to jointly solve a task. Research in robotics also ranges from *physically embodied agents* (i.e. robots implemented in hardware that physically interact with a real-world physical environment) to research concerning virtual robots that exist solely within a environment simulated on a computer.

Robots typically consist of three key subsystems (Fig. 18.2). A robot collects in-

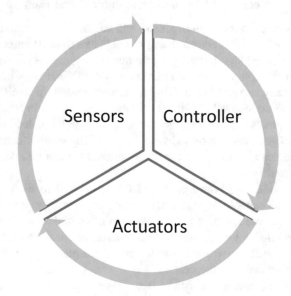

Fig. 18.2. Three key robot subsystems

formation about its environment by means of various sensors (for example, cameras or infrared proximity sensors), processes this information in its controller (for example, using an artificial neural network, where the network inputs are taken from its sensors and the network outputs drive the robot's actuators) and takes consequent action via its actuators, which may power wheels to move the robot or which may move its arms to accomplish a task such as picking up an object (see Fig. 18.3). The form

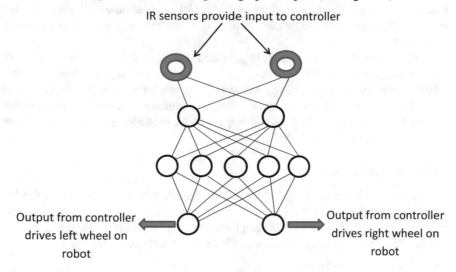

Fig. 18.3. Sensors provide environmental data inputs into the neural network controller, the output of which results in a robot behaviour, here the powering of the robot's wheels

of controller can range from a basic feedforward network to much more complex structures which embed recurrent links in order to implement memory [584].

The parallels between these subsystems and those in biological organisms are clear. Unsurprisingly, significant inspiration has been taken from biology in the design of robotic systems.

Multirobot Systems

A particular problem in designing multirobot systems is that the interactions between robots can give rise to complex, emergent, system-level behaviours which are difficult to anticipate. This problem increases rapidly as the size of the system grows [447]. One of two approaches to designing multirobot systems which output a desired behaviour is usually adopted [79]:

i. behaviour based design; and
ii. automatic design.

In the first case, the design process is bottom-up, whereby individual robot behaviours are implemented and the resulting populational behaviour is observed. Individual behaviours are then altered on a trial and error basis in order to generate the desired behaviour at populational level.

In contrast, automatic design methods typically use approaches such as reinforcement learning or evolutionary learning. The latter approach falls within the subfield of *evolutionary robotics*, which applies the principles of selection, variation and heredity to the design of robots [437, 584]. Evolutionary robotics can seek to evolve the controller and/or the morphologies for real or simulated autonomous robots [108, 252].

An illustration of this approach is provided in Fig. 18.4. The relevant aspects of each robot's characteristics which are to be evolved are encoded in a binary or real-valued string (in Fig. 18.4 a binary string is illustrated). Each string is decoded into a robot whose fitness in the real or simulated environment of interest is determined. The fitnesses corresponding to each string drives the evolutionary process in order to uncover an optimal design for the robot.

18.3.1 Swarm Robotics

Swarm robotics is a bottom-up approach to collective robotics that takes inspiration from the self-organised behaviours of some social animals with the aim of designing robust collective behaviours which can coordinate the activities of large numbers of robots [79]. The main characteristics of a swarm robotic system are [79]:

  i. robots are autonomous;
 ii. robots are situated in the environment and can act to modify it;
iii. robots' sensing and communication capabilities are local;
 iv. robots do not have access to centralised control or to global knowledge; and
  v. robots cooperate to tackle a given task.

Hence, the behaviour of the collection of robots is emergent, arising from simple rules and local interactions. Figure 18.5 outlines some of the collective behaviours (behaviours of the swarm as a whole) considered in the swarm robotics literature.

A number of general problem settings have become common test beds for the development of multirobot behaviours, including [447]:

  i. predator–prey pursuit (a prey item is pursued around an environment by a group of robots);
 ii. foraging;
iii. box pushing (a group of robots are required to collaborate to push one or more box objects to a particular location);
 iv. robot soccer (two teams of robots compete in a soccer match);
  v. cooperative navigation around an environment (robots are required to organise and coordinate their movements in order to navigate to a desired location); and
 vi. cooperative target observation (where a target is tracked as it moves around an environment).

18.3.2 Foraging Behaviours in Swarm Robotics

A standard benchmark task in the swarm robotics literature is a search for resources in a simulated environment. Typically, these resources are created stochastically during the simulation at a random location in a source region in the environment and

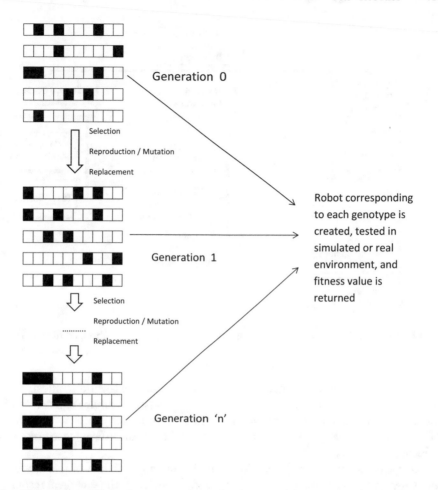

Fig. 18.4. A population of genotypes is randomly created in generation 0. The fitness of each genotype is then assessed and an evolutionary process implemented in order to uncover genotypes which correspond to good robot characteristics/behaviours in the environment of interest

the object of the robots is to find the resources and move them to a central, or home, location, simulating a central place foraging behaviour. The foraging task may consist of a single target object (simple foraging) or may be a multiforaging task, where there are several resource objects and the swarm needs to decide how to allocate its members for the foraging task. The object is to uncover robot behaviours that lead to an effective and efficient collective foraging process [98].

Foraging and related tasks are important research topics in robotics, as they facilitate study of a range of subproblems such as collective exploration of an environment and collective decision making as to how best to allocate resources to find,

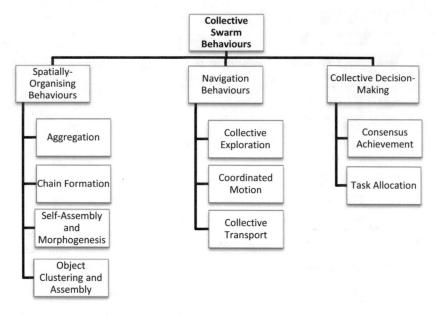

Fig. 18.5. Taxonomy of collective behaviours in swarm robotics (abridged from [79])

subdue and transport a prey item [79]. There are many real-world applications of robot foraging including search and rescue, toxic waste clean-up, explosives disposal, surveying a new environment and collection of terrain samples. The design of software to enable groupings of autonomous robots to undertake these tasks is of evident importance.

One approach is to draw inspiration from the foraging activities of real-world organisms, such as ants or honeybees, to create robot swarms with emergent resource-finding capabilities [461, 554]. The emergent capabilities facilitate a flexible behavioural response by a robot swarm to dynamic environments.

As with real-world social foraging, communication mechanisms between individual robots can be direct (via light, sound or radio signal) or indirect. For example, ant-inspired robot foraging could mark a terrain by depositing a chemical which is detectable using suitable sensors on other robots to create ant-like swarm foraging behaviours on the part of the robots. Alternatively, *virtual pheromone* could be used whereby chemical signals are simulated using light projected on the floor of the environment and where robots decide on their action based on the colour or intensity of the projected light. Pheromone levels are recorded and updated using an external computer, which in turn drives a camera which projects a suitable colour image onto the search space (detectable via a light sensor on top of each robot) [556].

Another recruitment mechanism, bearing parallel with recruitment by honeybees at a hive, could be implemented via light signals in order to allow successful robot foragers to communicate the location of resources to other robots at a central point. A review of some of relevant literature is provided in [461].

Foraging mechanisms from nature, such as those described in earlier chapters, can provide new ideas for the development of robust and adaptive robot control mechanisms. Robotic environments, in turn, can provide a valuable simulation tool for the testing of foraging theories from biology. We can expect to see a continuing close synergy between research in the two domains [343]

18.4 Summary

In this chapter we have introduced a number of studies which have applied evolutionary methodologies to evolve foraging strategies. Depending on the approach, evolutionary methodologies in conjunction with a simulated environment can be used to parameterise a predefined foraging strategy, or can be applied to uncover a complete foraging strategy from basic building blocks. The evolved strategies can range from those of a single forager to more complex, group-level strategies which allow for social communication between group members. We have also introduced some links between the literature on foraging-inspired algorithms and that on robotics.

Part VIII

Foraging Algorithms: The Future

Although there have been many successes in applying optimisation algorithms derived from a foraging metaphor, there are limits—as with all metaphors—to its appropriateness, and these limits depend on the context or setting. In Chap. 19 we clarify some of the differences between foraging and optimisation settings, so as to more clearly delineate limits to the foraging metaphor. Finally, we make some concluding comments and describe open research opportunities.

19

Looking to the Future

The metaphor of foraging as search provides a bountiful source of inspiration for the design of computational algorithms for optimisation and other applications. An extensive research literature has emerged over the past twenty years based on this. During the same period, there has been a significant research effort into theoretical and empirical aspects of the foraging behaviours of many species, as well as in the areas of sensory ecology and neuroscience. These advances, in turn, have generated additional ideas and insights which have fed into the design of foraging-inspired algorithms.

Despite the undoubted success of many applications of optimisation algorithms which have been derived from a foraging metaphor, as with all metaphors there are limits to its appropriateness. In this chapter we outline some of the differences between foraging and optimisation settings in order to more clearly delineate some limits to the metaphor. This, in turn, leads to a discussion of potential avenues for future research and, finally, to some concluding remarks.

19.1 Is Foraging Equivalent to Optimisation?

While an ability to efficiently extract required resources from the environment will enhance survival and successful reproduction, it does not follow that each individual element of a foraging strategy taken in isolation must itself be optimal. As noted in Sect. 1.3, foraging can be defined as a 'a repeated sequence of actions: search, encounter, decide' [545], with the sequence of net energy outcomes of these activities being what matters to the foraging organism. These outcomes are jointly determined by several factors only one of which is a search process. Foraging success is not *just* due to efficient search of an environment.

From previous chapters it is evident that foraging frequently takes place in environments where resources are distributed in patches and that the level of resources available in a particular location can alter quickly based on consumption or degradation. Real-world environments are generally replete with potential predators and

© Springer Nature Switzerland AG 2018
A. Brabazon, S. McGarraghy, *Foraging-Inspired Optimisation Algorithms*,
Natural Computing Series, https://doi.org/10.1007/978-3-319-59156-8_19

other organisms that are competitors for the desired resources. These factors produce a dynamic environment.

Foraging strategies produce characteristic payoff distributions (Sect. 1.5) and have been adapted over evolutionary timescales to fit specific ecological niches. Notwithstanding the requirement for foraging strategies to be robust to changing conditions, there will be limits to this. For example, simple sensory-gradient-climbing foraging strategies, such as bacterial chemotaxis, will work well for finding a strong source in a nonturbulent environment [308]. They will work less well in an environment characterised by turbulent flows, where the sensory signal is discontinuous and patchy.

A foraging strategy which works well in one ecological niche cannot necessarily be expected to perform well in another. The same is true for any foraging-inspired algorithm. There are a wide variety of optimisation scenarios (Sect. 5.1), and no search algorithm will perform well in all of them. As pointed out by Wolpert and Macready [627], there is no 'free lunch' available.

19.1.1 How Faithful Are the Algorithms?

Looking at foraging-inspired algorithms, it is evident that they are very simplified representations of real-world foraging behaviours. In most algorithms the actions of a single population of agents are simulated during the search process, so important influences on foraging activities, including predation risk (fear) and competition for resources, are omitted. A host of other factors described in Sect. 1.4 which influence the choice of foraging strategy are also generally ignored in algorithmic design. Other important omissions include concepts in sensory ecology such as the integration of information from multiple sensory modes, perception, selective attention, the importance of internal state and the use of multiple foraging strategies by all but the simplest organisms (Sects. 3.1 and 3.2).

We must caution against claiming that foraging algorithms are *reverse engineered* from real-world foraging processes. At best, they are loosely inspired by aspects of these foraging activities.

Of course, the fact that an algorithm omits important elements of a foraging process does not in itself invalidate its application to an optimisation task. An algorithm is ultimately a tool for solving (or uncovering a sufficiently good solution to) a problem, and its performance on that task determines whether it is considered useful. It need not necessarily be an accurate representation of the foraging process which has inspired it, in order to be practically useful.

19.1.2 Reinventing the Wheel

In common with other families of metaheuristic algorithms, the literature of foraging-inspired algorithms has exhibited a tendency to focus on new algorithms which are based on interesting foraging narratives. In some cases, unduly strong claims are made for the effectiveness of a 'new' algorithm, based on limited testing against a series of benchmark datasets. As noted by Kenneth Sörensen [539] and other commentators, a more critical perspective should be applied when assessing the worth of new metaheuristics. Further work is also required to consolidate and rigorously test existing foraging algorithms to determine which methods work best for specific problem domains.

New does not necessarily mean better and, indeed, an apparently new algorithm may not actually be so. Algorithms which on the surface have very differing descriptive narratives may in fact have very similar underlying mechanisms on deeper analysis. A useful approach to improving our understanding of existing and proposed algorithms is the development of taxonomies to highlight the similarities and differences between various algorithms. It is notable that few contributions to the literature attempt this, and it is often left to the reader to tease out what the real novelty of a proposed algorithm actually is.

In this book we have illustrated a number of exemplar taxonomies (Sect. 5.2), including the categorisation of algorithms by tree-of-life classification, by primary (simulated) sensory modality and by the mechanisms they use for memory, learning and social communication of information. As illustrated in Sect. 5.3, most foraging algorithms can be encapsulated in a high-level framework, with differing operationalisations of elements of this framework giving rise to alternative algorithms.

19.2 Research Opportunities

In spite of the above cautions, the study of foraging-inspired algorithms presents significant open research opportunities. In earlier chapters we identified some of these in connection with specific foraging algorithms. Here we adopt a higher-level perspective and identify opportunities where there is notable scope to draw on foraging-related literatures for the purposes of algorithmic design.

19.2.1 The Tree of Life

When we consider that every organism has a suite of foraging behaviours, it becomes clear that the range of potential sources of inspiration for the design of optimisation algorithms is truly huge. Even a cursory examination of the various branches of the tree of life indicates that there are wide swathes of organisms whose foraging (and indeed other) behaviours have not yet served as inspiration for computational algorithms. Allied to this issue, there is a long-standing taxonomic bias in animal behaviour research with certain species, particularly vertebrates such as birds and mammals, being much more heavily researched than (for example) arthropods, leading to significant disparities in our understanding of foraging behaviours across biological taxa [493]. There is evident scope for the development of new algorithms based on current and future research into foraging behaviours, particularly when the relevant organisms employ information-processing mechanisms which are distinctly different from those already explored in existing algorithms.

Taking one example, as discussed in Chaps. 14–16, the majority of life, including microbes and plants, is nonneuronal. We are just beginning to tease out the sensory, learning and memory mechanisms for these organisms. Advances in our understanding of these will doubtless open up further potential opportunities for algorithmic design.

19.2.2 Movement and Navigation

In Sect. 2.5, the field of movement ecology which is concerned with empirical and theoretical study into the movement of animals and also of other organisms, was introduced. The ability to figure out where we are, where we have been and where we need to go is key to our survival [415]. In spite of the significant related research effort in robotics concerning goal-directed navigation and movement, surprisingly little inspiration has been drawn from the field of movement ecology to inform the design of foraging-inspired algorithms. This literature provides a fertile medium for exploration in this context.

19.2.3 Balancing Current and Prior Experience

In both foraging and search algorithm design, an important question is the degree of emphasis that should be placed on current information and memory, respectively, in determining what to do next. The relevance of past learning is determined by the stability of good locations in the environment. In cases where an environment is highly dynamic, the value of past knowledge as to good locations is reduced. Foraging takes place in environments with varying dynamics and, consequently, organisms have evolved widely differing learning and memory mechanisms in order to survive. Study of these mechanisms can provide insights for the design of algorithms for application in environments with similar dynamic characteristics.

Taking one example, inspector honeybees (a subset of foraging honeybees) maintain a memory of previously profitable food locations, which they periodically revisit in order to determine whether foraging should be resumed (Sect. 10.3.3). Such a memory mechanism is likely to be adaptive in the scenario where food locations can regenerate after having been previously exhausted. This corresponds to an optimisation setting where good locations *cycle* over time.

19.2.4 Social Learning

As discussed in Sect. 4.2.2, understanding how to take advantage of social information, while managing the risks associated with its use, has become a focus for research on social learning strategies. Such a strategy needs to specify three key components, namely, when an individual should copy, from whom they should learn and what information should be copied. Many learning strategies can be specified drawing from taxonomies in the social learning literature, and scope exists to design search algorithms based on these. Differing specifications of this strategy will lead to algorithms with varying search characteristics. Taking the case of from whom an agent should learn, choices could include the best-ever location found by any member of the population (akin to g^{best} in particle swarm optimisation), implicitly assuming that all members of the population are in sensory range of each other. Alternatively, an agent may be limited to learning from a subset of the population such as its neighbouring agents.

As noted in Sect. 4.2.3, the implementation of social learning concepts in the metaheuristic literature is replete with methodological issues. The term *social learning* is rarely defined in papers concerning foraging-inspired optimisation algorithms, and often means little more than mechanisms by which the search efforts of individual agents are influenced by information uncovered by other agents in the population. This lack of definitional clarity is a particular problem, given the diversity of definitions of the term social learning in the various literatures concerned with learning.

Another issue which arises is the common misapplication of the term social learning to diversity generating operators which operate on solution representations in the metaheuristic literature. Obviously, solution representations cannot be meaningfully said to socially learn from one another, as such learning can only take place at a phenotypic level.

An opportunity exists to enhance the rigor of the application of the social learning metaphor in the foraging-inspired optimisation literature (and, more generally, in the metaheuristic literature). A related open research opportunity is to investigate the potential for the application of a social learning metaphor at a phenotypic level in this domain.

19.2.5 Combining Private and Public Information

A basic question that emerges in both nature and algorithmic design is how an individual can best combine private information that they have uncovered in their own foraging trials with public information that they have obtained from conspecifics or other organisms. A common theme running through many of the foraging-inspired algorithms in this book is a belief that information obtained from social interactions is a valuable input to foraging decisions which can help decrease search time, and decrease the risks in finding new food resources.

However, public information is not invariably useful. It may be noisy, out of date or even deliberately deceptive. Taking the case of honeybee recruitment dances, it has been noted that a dance for the same resource can vary in both distance and directional information (i.e. it is a noisy signal) and recruits may require a number of scouting trips before they successfully locate the advertised resource [232]. The literature concerning producer–scrounger models and Rogers' paradox (Sects. 2.3.2 and 4.2.1) points out further limits on social learning.

As our understanding of real-world foraging strategies has expanded, we increasingly realise that even simple organisms can acquire and use private information based on feedback from their prior foraging expeditions. In addition, assumptions that highly social organisms such as eusocial insects rely only on public information such as pheromone trails when foraging are now known to be simplistic. The interplay between the use of private and public information is likely to be quite subtle in these scenarios. Taking the case of honeybees again, empirical evidence indicates that the vast majority of foragers in a colony at any point in time actually ignore dance information about alternative food sources and continue to frequent their current foraging location. Interest in alternative locations only increases when current foraging locations become depleted [232].

How best to combine private and public information is of significant interest to biologists studying foraging activity. Research developments in this area have evident potential for application to the design of optimisation algorithms, particularly for application in dynamic environments and/or environments in which it is difficult to determine the worth of a potential solution with accuracy.

19.2.6 Group Decision Making

In some scenarios, foraging may require a consensus decision by a group as foraging is undertaken in a communal fashion. In this case, the balancing of private and public information occurs at group level, and questions of interest concern how the group as a whole reaches a decision as to what to do next.

Individuals may have differing private information and may be able to access differing public information from other group members in their local environment. One recent paper which addresses this issue is that of Karpas et al. [308] in which each artificial agent aims to maximise the information from its own sensory perception and minimise the overlap in information it has with group members (i.e. the information diversity among group members is increased). The resulting model of group foraging is termed the *socialtaxis algorithm*.

As noted in Sect. 8.1.2, distributed sensing and quorum decision processes in animals and insects (Sects. 9.11 and 10.4) can combine to generate swarm intelligence which can substitute for complex cognition at the level of individual agents. Such phenomena have also been identified in simple organisms such as bacteria (Chap. 14). There is tantalising evidence that even viruses can engage in quorum-based decision making [170] (Sect. 14.6.3).

The study of group decision making in animals and simpler organisms, including quorum-based decision making, is a vibrant area of research with evident possibilities for incorporation of developments from this research into the design of populational optimisation algorithms.

Another aspect of group decision making is that of forager personality. Foraging behaviour can be plastic to the social context, with an individual's actions altering depending on whether it forages alone or in a group (Sect. 4.4.3). This implies that agents in a search algorithm could potentially embed differing personalities depending on whether they are acting in isolation or as part of a group, with their search strategy varying as the algorithm iterates.

19.2.7 Leadership

While the significance of leaders for group decision making in species with clearly defined dominance hierarchies is relatively well studied, examination of the role of leadership when such hierarchies do not exist has attracted much less attention. Even when dominance hierarchies do not exist, leaders can still emerge in a group because they have information which is not available to others [614]. Hence, animals with knowledge of food locations may act as leaders, being followed by group members to those locations. Alternatively, an animal may have a stronger motivation, perhaps due to an internal state such as hunger or to a personality tendency towards boldness, to engage in foraging behaviours. In turn, this activity can impact on the behaviour of other animals, altering their propensity to engage in following behaviours (i.e. facilitating their recruitment by a leader).

In all populational search algorithms, a key question is who in the population transmits information (leads) and who listens (follows). To date, the possible linkages between these algorithmic design issues and the literature concerning group leadership in nonhuman animals remain little explored.

19.2.8 Context Dependence

Another important research topic concerns the flexible use of context-dependent foraging strategies by organisms. We have increasing evidence that organisms do not usually employ a single, fixed foraging strategy but rather select from a repertoire of strategies depending on feedback from earlier foraging activity, their internal state and the environment they face [486]. Even employing a small number of relatively simple strategies in a flexible manner can result in quite sophisticated and effective foraging behaviours (Sect. 4.2.2). Scope exists to design algorithms which can likewise switch between search strategies based on feedback.

19.2.9 Stochastic Mechanisms in Foraging Algorithms

Most search heuristics, even if not foraging-inspired, contain a stochastic element (Sect. 5.2.6). Generally, the inclusion of stochastic mechanisms is essential in order to ensure good search performance. These mechanisms can be designed to promote exploitation and/or exploration. An example of the former is embedded in the canonical bees algorithm (Sect. 10.2.1) which intensifies search around already discovered good locations by randomly selecting trial locations within a defined hypersphere around them. In the case of exploration, the canonical bacteria foraging optimisation algorithm (BFOA) (Sect. 14.3) provides an example where the locations of stochastically selected foragers are occasionally reinitialised to new random locations, thereby increasing the degree of exploration in the search process. This mechanism helps to reduce the risk of the search process stagnating prematurely.

Another approach to inclusion of a stochastic mechanism is to allow the weights assigned to different elements of a foraging strategy, such as the degree of reliance on private versus public information, to be randomly selected as the algorithm runs.

This ensures that simulated foragers effectively employ more than one (fixed) foraging strategy. The mechanism bears some parallel with the behaviour of real-world foragers who typically use multiple foraging strategies, although it does not explicitly embed context dependence.

Curiously, despite the importance of stochastic mechanisms for algorithmic performance, virtually no research has been undertaken concerning how best to design them for specific algorithmic settings. Too much exploration tends towards random search, as little attention is paid to prior knowledge gained during the search process. Too much exploitation risks premature convergence of the algorithm to a local optimum. Depending on the choice of the other mechanisms in an algorithm, stochastic mechanisms need to be designed accordingly, in order to create a suitable exploration–exploitation balance. In most cases, it would appear that the design of stochastic mechanisms is ad hoc, with little, if any, theoretical underpinning. Greater understanding of how best to design these mechanisms for specific settings is required.

19.2.10 Empirical Comparison of Algorithms

As commented on in Sect. 19.1.2, the number of algorithms claiming inspiration from foraging processes has increased dramatically over time, with research tending to focus on new narratives rather than on rigorous testing of extant algorithms. Further research is required to consolidate and rigorously test foraging-inspired algorithms (and, more generally, algorithms from any source of natural inspiration) to determine which methods work best for specific problem domains. It is also notable that many algorithms have surface similarities to each other and further rigorous review is required to better determine which algorithms are actually little more than variants of earlier algorithms.

19.2.11 Modelling of Foraging Processes

Apart from the utility of foraging-inspired algorithms for optimisation and other computational applications, such algorithms can also serve as models of foraging processes themselves. Simulation studies using simplified models of foraging can potentially produce insight into real-world foraging processes. Simulation provides an opportunity to undertake closed-world experiments, allowing investigation of issues such as the impact of different cognitive mechanisms or the impact of different mechanisms of social transmission of information on foraging outcomes. In turn, this could support the development of new foraging theories or help to delineate the limits of existing theories.

19.3 Concluding Remarks

The study of foraging algorithms is a fascinating, active and highly multidisciplinary field of research. Aspects of foraging speak directly to important problems surrounding search, reward seeking and information processing, in a multitude of fields, as well as providing inspiration for the design of optimisation algorithms. We can expect to see a rich, continuing flow between advances in our understanding of foraging behaviours and of their underlying mechanisms, and the development of computational algorithms for various applications.

References

1. Aanen D, Eggleton P, Rouland-Lefevre C, Guldberg-Froslev T, Rosendahl S, Boomsma J (2002) The evolution of fungus-growing termites and their mutualistic fungal symbionts. Proceedings of the National Academy of Sciences 99(23):14887–14892
2. Adamatzky A (2014) Route 20, Autobahn 7, and Slime Mold: Approximating the Longest Roads in the USA and Germany with Slime Mold on 3-D Terrains. IEEE Transactions on Cybernetics 44(1):126–136
3. Adams M, Jacobs L (2007) Cognition for Foraging. In: Stephens D, Brown J, Ydenberg R (eds) *Foraging Behavior and Ecology*, pp 105–138, Chicago: University of Chicago Press
4. Aiello L, Wheeler P (1995) The expensive-tissue hypothesis – the brain and the digestive-system in human and primate evolution. Current Anthropology 36(2):199–221
5. Akay B, Karaboga D (2012) A modified artificial bee colony algorithm for real-parameter optimisation. Information Sciences 192:120–142
6. Alim K, Andrew N, Pringle N, Brenner M (2017) Mechanism of signal propagation in Physarum polycephalum, Proceedings of the National Academy of Science, 114(20):5136–5141
7. Allee W (1931) Co-operation among animals. American Journal of Sociology 37(3):386–393
8. Allee W, Bowen E (1932) Studies in animal aggregations: mass protection against colloidal silver among goldfishes. Journal of Experimental Zoology 61(2):185–207
9. Alvarez L, Alvarez W, Asaro F, Michel H (1980) Extraterrestrial cause for the Cretaceous-Tertiary extinction. Science 208(4448):1095–1108
10. Anderson J (1991) Foraging behavior of the American white pelican (*Pelecanus erythrorhyncos*) in western Nevada. Colonial Waterbirds 14(2):166–172
11. Andrews C, Viviani J, Egan E, Bedford T, Brilot B, Nettle D, Bateson M (2015) Early life adversity increases foraging and information gathering in European starlings, *Sturnus vulgaris*. Animal Behaviour 109:123–132
12. Appel H, Cocroft R (2014) Plants respond to leaf vibrations caused by insect herbivore chewing. Oecologia 175(4):1257–1266
13. Applewhite P (1975) Learning in bacteria, fungi and plants. In: Corning W, Dyal J, Williams A (eds) *Invertebrate learning Vol 3: Cephalopods and echinoderms*, pp 179–186, New York: Plenum Press

© Springer Nature Switzerland AG 2018
A. Brabazon, S. McGarraghy, *Foraging-Inspired Optimisation Algorithms*,
Natural Computing Series, https://doi.org/10.1007/978-3-319-59156-8

14. Arbilly M, Motro U, Feldman M, Lotem A (2010) Co-evolution of learning complexity and social foraging strategies. Journal of Theoretical Biology 267(4):573–581
15. Ardiel E, Rankin C (2010) An elegant mind: Learning and memory in *Caenorhabditis elegans*. Learning and Memory 17(4):191–201
16. Arita H, Fenton B (1997) Flight and echolocation in the ecology and evolution of bats. Tree 12(2):53–58
17. Arnaout J P (2014) Worm Optimisation: A novel optimisation algorithm inspired by C. elegans. In: Proceedings of the 2014 IEEE International Conference on Industrial Engineering and Operations Management, pp 2499–2505, New Jersey: IEEE Press
18. Arnaud-Haond S, Duarte CM, Diaz-Almela E, Marbà N, Sintes T, Serrão EA (2012) Implications of Extreme Lifespan in Clonal Organisms: Millenary Clones in Meadows of the Threatened Seagrass Posidonia oceanica. PLoS ONE 7(2):e30454
19. Asenova E, Fu E, Nicolau (Jr) D, Lin H-Y Nicolau D (2016) Space Searching Algorithms Used by Fungi. In: Proceedings of the 9th EAI International Conference on Bio-inspired Information and Communications Technologies (BICT 2015) (formerly BIONETICS) pp 375–380, New York: ACM Press
20. Asenova E, Lin H-Y, Fu E, Nicolau (Jr) D, Nicolau D (2016) Optimal Fungal Space Searching Algorithms. IEEE Transactions on NanoBioscience 15(7):613–618
21. Asensio N, Korstjens A, Aureli F (2009) Fissioning minimizes ranging costs in spider monkeys: a multiple-level approach. Behavioral Ecology and Sociobiology 63(5):649–659
22. Atkinson D, Sibly R (1997) Why are organisms usually bigger in colder environments? Making sense of a life history puzzle. Trends in Ecology and Evolution 12(6):235–239
23. Aupetit S, Monmarché N, Slimane M, Liardet P (2005) An exponential representation in API algorithm for hidden Markov models training. In: Proceedings of the 7th International Conference, Evolution Artificielle (EA 2005), LNCS 3871, pp 61–72, Berlin: Springer
24. Avarguès-Weber A, Lachlan R, Chittka L (2018) Bumblebee social learning can lead to suboptimal foraging choices Animal Behaviour 135:209–214
25. Avilés L, Guevara J (2017) Sociality in Spiders. In: Rubenstein D, Abbot P (eds) Comparative Social Evolution, pp 188–223, Cambridge: Cambridge University Press
26. Balcombe J (2016) What a fish knows: The inner lives of our underwater cousins. New York: Scientific American / Farrar, Staus and Giroux
27. Bansal J C, Sharma H, Jadon S S, Clerc M (2014) Spider Monkey Optimisation algorithm for numerical optimisation. Memetic Computing 6(1):31–47
28. Baptista T, Costa E (2011) The evolution of foraging in an open-ended simulation environment. In: Proceedings of the 15th Portuguese Conference on Artificial Intelligence (EPIA 2011), LNAI 7026, pp 125–137, Berlin: Springer
29. Bar-On Y, Phillips R, Milo R (2018) The biomass distribution on Earth. Proceedings of the National Academy of Sciences 115(25):6506–6511
30. Barclay R (1982) Interindividual use of echolocation calls: Eavesdropping by bats. Behavioral Ecology and Sociobiology 10(4):271–275
31. Barnard C, Sibly R (1981) Producers and scroungers: A general model and its application to captive flocks of house sparrows. Animal Behaviour 29(2):543–550
32. Barron A, Hebets E, Cleland T, Fitzpatrick C, Hauber M, Stevens J (2015) Embracing multiple definitions of learning. Trends in Neurosciences 38(7):405–407
33. Basak A, Pal S, Das S, Abraham A, Snasel V (2010) A Modified Invasive Weed Optimisation Algorithm for Time-Modulated Linear Antenna Array Synthesis. In: Proceedings of the IEEE World Congress on Computational Intelligence (WCCI 2010), pp 372–379, New Jersey: IEEE Press

34. Bastos Filho C, de Lima Neto F, Lins A, Nascimento A, Lima M (2008) A Novel Search Algorithm Based on Fish School Behavior. In: Proceedings of the IEEE International Conference on Systems, Man and Cybernetics (SMC 2008), pp 2646–2651, New Jersey: IEEE Press

35. Bastos Filho C, de Lima Neto F, Sousa M, Pontes M, Madeiro S (2009) On the Influence of the Swimming Operators in the Fish School Search Algorithm. In: Proceedings of the IEEE International Conference on Systems, Man and Cybernetics (SMC 2009), pp 5012–5017, New Jersey: IEEE Press

36. Bates G H (1950) Track making by man and domestic animals. Journal of Animal Ecology 19(1):21–28.

37. Baykasoglu A, Ozbakir L, Tapkan P (2007) Artificial Bee Colony Algorithm and Its Application to Generalized Assignment Problem. In: Chan F and Tiwari M K (eds) Swarm Intelligence, Focus on Ant and Particle Swarm Optimisation, pp 113–144, Rijeka, Croatia: Intech

38. Beauchamp G, Giraldeau L (1996) Group Foraging Revisited: Information Sharing or Producer-Scrounger Game? The American Naturalist 148(4):738–743

39. Bebber D, Hynes J, Darrah P, Boddy L, Fricker M (2007) Biological solutions to transport network design. Proceedings of the Royal Society B 274(1623):2307–2315

40. Bechara J (2015) Bioluminescence: A Fungal Nightlight with an Internal Timer. Current Biology 25(7):R283–R285

41. Beckers R, Deneubourg J L, Goss S (1993) Modulation of trail laying in the ant Lasius niger (Hymenoptera: Formicidae) and its role in the collective selection of a food source. Journal of Insect Behavior 6(6):751–759

42. Beebe W (1921) Edge of the Jungle. New York: Henry Holt and Co

43. Beekman M, Fathke R, Seeley T (2006) How does an informed minority of scouts guide a honeybee swarm as it flies to its new home? Animal Behaviour 71(1):161–171

44. Beer C, Hendtlass T, Montgomery J (2012) Improving Exploration in Ant Colony Optimisation with Antennation, In: Proceedings of the IEEE World Congress on Computational Intelligence 2012 (WCCI 2012), pp 2926–2933, New Jersey: IEEE Press

45. Begall S, Cervený J, Neef J, Vojtech O, Burda H (2008) Magnetic alignment in grazing and resting cattle and deer. Proceedings of the National Academy of Sciences 105(36):13451–13455

46. Behie S, Bidochka M (2014) Nutrient transfer in plant-fungal symbioses. Trends in Plant Science 19(11):734–740

47. Bell W (1990) Searching behavior patterns in insects. Annual Review of Entomology 35:447–467

48. Bell W (1991) Searching Behaviour: The Behavioural Ecology of Finding Resources. London: Chapman and Hall

49. Benoit-Bird K, Au W (2009) Cooperative prey herding by the pelagic dolphin Stenella longirostris. Journal of the Acoustical Society of America 125(1):125–137

50. Bentley-Condit V, Smith E (2010) Animal tool-use: Current definitions and an updated comprehensive catalog. Behaviour 147(2):185–221

51. Berg H, Brown D (1972) Chemotaxis in Escherichia coli analysed by three-dimensional tracking, Nature 239(5374):500–504

52. Berdahl A, Torney C, Ioannou C, Faria J, Couzin I (2013) Emergent Sensing of Complex Environments by Mobile Animal Groups. Science 339(6119):574–576

53. Biedrzycki M, Bais H (2010) Kin recognition in plants: a mysterious behaviour unsolved. Journal of Experimental Botany 61(15):4123–4128

54. Biegler R (2000) Possible uses of path integration in animal navigation. Animal Learning & Behavior 28(3):257–277

55. Biesmeijer J, Seeley T (2005) The use of waggle dance information by honeybees throughout their foraging career. Behavioral Ecology and Sociobiology 59(1):297-306
56. Bilchev G, Parmee I (1995) The ant colony metaphor for searching continuous design spaces. In: Proceedings of the AISB Workshop on Evolutionary Computing, LNCS 993, pp 25–39, Berlin: Springer
57. Bilde T, Lubin Y (2011) Group living in spiders: cooperative breeding and coloniality. In: Herberstein M (ed) Spider Behaviour: Flexibility and Versatility, pp 275–306, Cambridge: Cambridge University Press
58. Blackwell M (2011) The fungi: 1, 2, 3, …5.1 million species? American Journal of Botany 98(3):426-438
59. Blackwell T M, Branke J (2004) Multi-swarm optimisation in dynamic environments. In: Proceedings of EvoSTOC Workshop 2004, LNCS 3005, pp 489–500, Springer: Berlin
60. Blackwell T M, Branke J (2006) Multi-swarms, exclusion and anti-convergence in dynamic environments. IEEE Transactions on Evolutionary Computation 10(4):459–472
61. Boal J (2006) Social recognition: A top down view of cephalopod behaviour. Life and Environment 56(2):69–79
62. Boddy L (1999) Saprotrophic Cord-Forming Fungi: Meeting the Challenge of Heterogeneous Environments. Mycologia 91(1):13–32
63. Boddy L, Hynes J, Bebber D, Fricker M (2009) Saprotrophic cord systems: dispersal mechanisms in space and time. Mycoscience 50(1):9–19
64. Boeckle M, Clayton N (2017) A raven's memories are for the future. Science 357(6347):126–127
65. Boender A, Roubos E, van der Velde G (2011) Together or alone? Foraging strategies in *Caenorhabditis elegans*. Biological Reviews 86(4):853–862
66. Boisseau R, Vogel D, Dussutour A (2016) Habituation in non-neural organisms: evidence from slime moulds. Proceedings of the Royal Society B 283(1829):20160446
67. Bonabeau E, Dorigo M, Theraulaz G (1999) Swarm Intelligence: From Natural to Artificial Systems. Oxford: Oxford University Press
68. Bonfante P, Genre A (2015) Arbuscular mycorrhizal dialogues: do you speak 'plantish' or 'fungish'? Trends in Plant Science 20(3):150–154
69. Bora T, Coelho L, Lebensztajn L (2012) Bat-inspired Optimisation Approach for the Brushless DC Wheel Motor Problem. IEEE Transactions on Magnetics 48(2):947–950
70. Bos N, Sundström L, Fuchs S, Freitak D (2015) Ants medicate to fight disease. Evolution 69(11):2979–2984
71. Boswella G, Davidson F (2013) Modelling hyphal networks. Fungal Biology Reviews 26(1):30–38
72. Box H (1984) Primate Behaviour and Social Ecology. London: Chapman and Hall
73. Boyd C, Grünbaum D, Hunt G, Punt A, Weimerskirch H, Bertrand S (2016) Effectiveness of social information used by seabirds searching for unpredictable and ephemeral prey. Behavioral Ecology 27(4):1223–1234
74. Brabazon A, Agapitos A, O'Neill M (2015) Promoting Better Generalisation in Multilayer Perceptrons Using a Simulated Synaptic Downscaling Mechanism. In: Washington S (ed) New Developments in Evolutionary Computation Research, New York: Nova Science Publishers
75. Brabazon A, Cui W, O'Neill M (2015) Information Propagation in a Social Network: The Case of The Fish Algorithm, In: Krol D, Fay D, Gabrys B (eds) Propagation Phenomena in Real World Networks, pp 27-51, Berlin: Springer
76. Brabazon, A, Cui W, O'Neill M (2016) The Raven Roosting Optimisation Algorithm. Soft Computing 20(2):525–545

77. Brabazon A, O'Neill M, McGarraghy S (2015) Natural Computing Algorithms. Berlin: Springer

78. Bradbury J, Vehrencamp S (2011) Principles of Animal Communication (2nd ed). Sunderland, MA, USA: Sinauer Associates

79. Brambilla M, Ferrante E, Birattari M, Dorigo M (2013) Swarm robotics: a review from the swarm engineering perspective. Swarm Intelligence 7(1):1–41

80. Breitbart M, Rohwer F (2005) Here a virus, there a virus, everywhere the same virus? Trends in Microbiology 13(6):278–284

81. Bremermann H (1974) Chemotaxis and optimisation. Journal of the Franklin Institute 297(5):397–404

82. Britton A, Jones G (1999) Echolocation behaviour and prey-capture success in foraging bats: laboratory and field experiments on Myotis daubentonii. Journal of Experimental Biology 202(13):1793–1801

83. Brockmann H, Barnard C (1979) Kleptoparasitism in birds. Animal Behaviour 27(2):487–514

84. Brothers J R, Lohmann K J (2015) Evidence for geomagnetic imprinting and magnetic navigation in the natal homing of sea turtles. Current Biology 25(3):392–396

85. Brothers L (1990) The social brain: A project for integrating primate behavior and neurophysiology in a new domain. Concepts in Neuroscience 1:27–51

86. Brown J (1964) The evolution of diversity in avian territorial systems. Wilson Bulletin 76(2):160–169

87. Brown D, Berg H (1974) Temporal Stimulation of Chemotaxis in Escherichia coli. Proceedings of the National Academy of Sciences 71(4):1388–1392

88. Brown J, Kotler B (2007) Foraging and the Ecology of Fear. In: Stephens D, Brown J, Ydenberg R (eds) Foraging Behavior and Ecology, pp 437–480, Chicago: University of Chicago Press

89. Brown P (2016) Rocky Mountain Tree Ring Research: oldlist (2016), http://www.rmtrr.org/oldlist.htm

90. Brusatte S, Luo Z-X (2016) Ascent of the mammals. Scientific American 314(6):28–35

91. Buehlmann C, Graham P, Hansson B, Knaden M (2015) Desert ants use olfactory scenes for navigation. Animal Behaviour 106:99–105

92. Buhl J, Hicks K, Miller E, Persey S, Alinvi O, Sumpter D (2009) Shape and efficiency of wood ant foraging networks. Behavioral Ecology and Sociobiology 63(3):451–460

93. Bullnheimer B, Hartl R, Strauss C (1999) A new rank-based version of the Ant-System: A computational study. Central European Journal for Operations Research and Economics 7(1):25–38

94. Burgess J W (1979) Web-signal processing for tolerance and group predation in the social spider Mallos gregalis. Animal Behaviour 27(1):157–164

95. Burt W (1943) Territoriality and home range concepts as applied to mammals. Journal of Mammalogy 24(3):346–352

96. Bushdid C, Magnasco M, Vosshall L, Keller A (2014) Humans Can Discriminate More than 1 Trillion Olfactory Stimuli. Science 343(6177):1370–1372

97. Cai W, Yang W, Chen X (2008) A Global Optimisation Algorithm Based on Plant Growth Theory: Plant Growth Optimisation. In: Proceedings of the 2008 International Conference on Intelligent Computation Technology and Automation (ICICTA 2008), pp 1194–1199, New Jersey: IEEE Press

98. Campo A, Dorigo M (2007) Efficient Multi-Foraging in Swarm Robotics. In: Proceedings of the 9th European Conference on Advances in Artificial Life (ECAL 2007), LNCS 4648, pp 696–705, Berlin: Springer

99. Cao Y, Li X, Wang J (2013) Opposition-Based Animal Migration Optimisation. Mathematical Problems in Engineering, Vol. 2013, Article ID 308250, DOI:10.1155/2013/308250

100. Caraco T, Martindale S, Whittam T (1980) An empirical demonstration of risk-sensitive foraging preferences. Animal Behaviour 28(3):820–830

101. Caro T, Hauser M (1992) Is there teaching in non-human animals? Quarterly Review of Biology 67(2):151–174

102. Cavalcanti-Junior G, Bastos-Filho C, Lima-Neto F (2012) Volitive Clan PSO — An Approach for Dynamic Optimisation Combining Particle Swarm Optimisation and Fish School Search. In: Parpinelli R (ed) Theory and New Applications of Swarm Intelligence, pp 69–86, Rijeka, Croatia: Intech

103. Červený J, Begall S, Koubek P, Nováková P, Burda H (2011) Directional preference may enhance hunting accuracy in foraging foxes. Biology Letters 7(3):355–357

104. Chamovitz D (2012) What a plant smells. Scientific American 306(5):48–51

105. Chapman C (1990) Association patterns of spider monkeys: the influence of ecology and sex on social organisation. Behavioral Ecology and Sociobiology 26(6):409–414

106. Chapman C, Lefebvre L (1990) Manipulating foraging group size: spider monkey food calls at fruiting trees. Animal Behaviour 39(5):891–896

107. Charnov E (1976) Optimal foraging, the marginal value theorem. Theoretical Population Biology 9(2):129–136

108. Chaumont N, Adami C (2016) Evolution of sustained foraging in three-dimensional environments with physics. Genetic Programming and Evolvable Machines 17(4):359–390

109. Chen H, Niu B, Ma L, Su W, Zhu Y (2014) Bacterial colony foraging optimisation. Neurocomputing 137:268–284

110. Chen H, Yunlong Y, Hu K, Ma L (2014) Bacterial colony foraging algorithm: Combining chemotaxis, cell-to-cell communication, and self-adaptive strategy. Information Sciences 273:73–100

111. Chernetsov N (2017) Compass systems. Journal of Comparative Physiology A 203(6–7):447–453

112. Clarke D, Whitney H, Sutton G, Robert D (2013) Detection and Learning of Floral Electric Fields by Bumblebees. Science 340(6128):66–69

113. Clayton N, Dally J, Emery N (2007) Social cognition by food-caching corvids. The western scrub-jay as a natural psychologist. Philosophical Transactions of the Royal Society B 362(1480):507–522

114. Cocroft R, Appel H (2013) Comments on Green symphonies. Behavioral Ecology 24(4):800–801

115. Collins S, Whitehead D (2004) The functional roles of passive electroreception in non-electric fishes. Animal Biology 54(1):1–25

116. Couzin I (2017) Modelling Biodiversity and Collective Behaviour. In: Fritsch J, Borchert Y, Hacker J (eds) Crossing Boundaries in Science: Modelling Nature and Society — Can We Control the World? Documentation of the Workshop of the German National Academy of Sciences Leopoldina 30 June — 2 July 2016 in Weimar, Germany, Nova Acta Leopoldina NF 419:97–107, http://www.leopoldina.org/uploads/tx_leopublication/NAL_Nr419_Gesamt_Internet.pdf#page=99

117. Crist E (2004) Can an insect speak? The case of the honeybee dance language. Social Studies of Science (Sage Publications) 34(1):7–43

118. Cronin T, Johnsen S, Marshall N J, Warrant E J (2014) Visual Ecology. Princeton, New Jersey: Princeton University Press

119. Crowley P, Linton M (1999) Antlion Foraging: Tracking Prey Across Space and Time. Ecology 80(7):2271–2282

120. Cuevas E, Cienfuegos M, Zaldívar D, Pérez-Cisneros M (2013) A swarm algorithm inspired by the behavior of the social spider. Expert Systems with Applications 40(16):6374–6384

121. Cui W, Brabazon A, Agapitos A (2015) Extending the Bat Foraging Metaphor for Optimisation Algorithm Design. International Journal of Metaheuristics 4(1):1–26

122. Cvikel N, Berg K E, Levin E, Hurme E, Borissov I, Boonman A, Amichai E, Yove Y (2015) Bats Aggregate to Improve Prey Search but Might Be Impaired when Their Density Becomes Too High. Current Biology 25(2):206–211

123. Czaczkes T, Franz S, Witte V, Heinze J (2015) Perception of collective path use affects path selection in ants. Animal Behaviour 99:15–24

124. Czaczkes T, Grüter C, Ratnieks F (2013) Negative feedback in ants: crowding results in less trail pheromone deposition. Journal of the Royal Society Interface 10(81):20121009

125. Czaczkes T, Heinze J (2015) Ants adjust their pheromone deposition to a changing environment and their probability of making errors. Proceedings of the Royal Society B 282(1810):20150679

126. Czech-Damal N, Liebschner A, Miersch L, Klauer G, Hanke F, Marshall C, Dehnhardt G, Hanke W (2012) Electroreception in the Guiana dolphin (*Sotalia guianensis*). Proceedings of the Royal Society B 279(1729):663–668

127. Dahlquist F, Elwell R, Lovely P (1976) Studies of bacterial chemotaxis in defined concentration gradients. A model for chemotaxis toward L-serine. Journal of Supramolecular Structure 4(3):329–342

128. Dall S (2002) Can information sharing explain recruitment to food from communal roosts? Behavioral Ecology 13(1):42–51

129. Dasgupta S, Das S, Abraham A, Biswas A (2009) Adaptive Computational Chemotaxis in Bacterial Foraging Optimisation: An Analysis. IEEE Transactions on Evolutionary Computation 13(4):919–941

130. Davidson A (2017) Phages make a group decision. Nature 541(7638):466–467

131. Davies N (2011) Cuckoo adaptations: trickery and tuning. Journal of Zoology 284(1):1–14

132. Davies N, Krebs J, West S (2012) An Introduction to Behavioural Ecology (4th ed). Chichester, UK: Wiley-Blackwell

133. Dawkins R (1996) Climbing Mount Improbable. New York: W.W. Norton

134. Deaner R, Schaik C, Johnson V (2006) Do some taxa have better domain-general cognition than others? A meta-analysis of nonhuman primate studies. Evolutionary Psychology 4(1):149–196

135. de Bono, M, Bargmann C (1998) Natural Variation in a Neuropeptide Y Receptor Homolog Modifies Social Behavior and Food Response in *C. elegans*. Cell 94(5):679–689

136. DeCasien A, Williams S, Higham J (2017) Primate brain size is predicted by diet but not sociality. Nature Ecology & Evolution 1:0112 doi:10.1038/s41559-017-0112

137. Dechmann D, Kranstauber B, Gibbs D, Wikelski M (2010) Group Hunting — A Reason for Sociality in Molossid Bats. PLoS One 5(2):e9012

138. De Houwer J, Barnes-Holmes D, Moors A (2013) What is learning? On the nature and merits of a functional definition of learning. Psychonomic Bulletin and Review 20(4):631–642

139. Delsuc F (2003) Army Ants Trapped by Their Evolutionary History. PLoS Biology 1(2):e37

140. De Marco R, Gurevitz J, Menzel R (2008) Variability in the encoding of spatial information by dancing bees. Journal of Experimental Biology 211(10):1635–1644

141. De Waal F (2001) The ape and the sushi master: cultural reflections of a primatologist. New York, NY: Basic Books

142. De Waal F (2017) Are we smart enough to know how smart animals are? London, UK: Granta Publications

143. Dempsey I, O'Neill M, Brabazon A (2009) Foundations in Grammatical Evolution for Dynamic Environments. Berlin: Springer

144. Deneubourg J, Aron S, Goss S, Pasteels J (1990) The Self-Organizing Exploratory Pattern of the Argentine Ant. Journal of Insect Behavior 3(2):159–168

145. de Oliveira D, Parpinelli R, Lopes H (2011) Bioluminescent Swarm Optimisation Algorithm. In: Kita E (ed) Evolutionary Algorithms, Rijeka, Croatia: Intech

146. Des Marais D (2000) When Did Photosynthesis Emerge on Earth? Science 289(5485): 1703–1705

147. Deygout C, Gault A, Duriez O, Sarrazin F, Bessa-Gomes C (2010) Impact of food predictability on social facilitation by foraging scavengers. Behavioral Ecology 21(6):1131–1139

148. Dicke M, Agrawal A, Bruin J (2003) Plants talk, but are they deaf? Trends in Plant Science 8(9):403–405

149. Diwold K, Beekman M, Middendorf M (2010) Bee nest site selection as an optimisation process. In: Proceedings of ALife XII Conference, pp 626–633, Cambridge, MA: MIT Press

150. Diwold K, Himmelbach D, Meier R, Baldauf C, Middendorf M (2011) Bonding as a Swarm: Applying Bee Nest-Site Selection Behavior to Protein Docking. In: Proceedings of the 12th Genetic and Evolutionary Computation Conference (GECCO 2011), pp 93–100, New York: ACM Press

151. Dodd M, Papineau D, Grenne T, Slack J, Rittner M, Pirajno F, O'Neil J, Little C (2017) Evidence for early life in Earth's oldest hydrothermal vent precipitates. Nature 543(7643):60–64

152. Dor R, Rosenstein S, Scharf I (2014) Foraging behaviour of a neglected pit-building predator: the wormlion. Animal Behaviour 93:69–76

153. Dorigo M (1992) Optimisation, Learning and Natural Algorithms. PhD Thesis. Politecnico di Milano

154. Dorigo M, DiCaro G (1999) Ant colony optimisation: a new meta-heuristic. In: Proceedings of the IEEE Congress on Evolutionary Computation (CEC 1999), pp 1470–1477, New Jersey: IEEE Press

155. Dorigo M, Gambardella L (1997a) Ant Colony System: A cooperative Learning Approach to the Travelling Salesman Problem. IEEE Transactions on Evolutionary Computation 1(1):53–66

156. Dorigo M, Gambardella L (1997b) Ant colonies for the travelling salesman problem. BioSystems 43(2):73–81

157. Dorigo M, Maniezzo V, Colorni A (1996) Ant system: optimisation by a colony of cooperating agents. IEEE Transactions on Systems, Man, And Cybernetics - Part B: Cybernetics 26(1):29–41

158. Dorigo M, Stützle T (2002) The Ant Colony Optimisation Metaheuristic: Algorithms, Applications and Advances. In: Glover F, Kochenberger G (eds) Handbook of Metaheuristics, Volume 57 of International Series in Operations Research & Management Science, Chap. 9, pp 251–285, Boston, MA : Kluwer Academic Publishers

159. Dorigo M, Stützle T (2004) Ant Colony Optimisation, Cambridge, Massachusetts: MIT Press

160. Dorigo M, Stützle T (2010) Ant Colony Optimisation: Overview and Recent Advances. In: Gendreau M, Potvin J-Y (eds) Handbook of Metaheuristics, Volume 146 of International Series in Operations Research & Management Science, pp 227–263, Berlin: Springer

161. Draa A (2015) On the performances of the flower pollination algorithm — Qualitative and quantitative analyses. Applied Soft Computing 34:349–371

162. Dudley S, File A (2007) Kin recognition in an annual plant. Biology Letters 3(4):435–438

163. Dukas R (1999) Costs of Memory: Ideas and Predictions. Journal of Theoretical Biology 197(1):41–50

164. Dunbar R (1998) The social brain hypothesis. Evolutionary Anthropology 6(5):178–190

165. Dunbar R, Shultz S (2017) Why are there so many explanations for primate brain evolution? Philosophical Transactions of the Royal Society B 372(1727):20160244

166. Dussutour A, Latty T, Beekman M, Simpson S (2010) Amoeboid organism solves complex nutritional challenges. Proceedings of the National Academy of Sciences 107(10):4607–4611

167. Emery N (2016) Bird Brain: An exploration of avian intelligence. Princeton, New Jersey: Princeton University Press

168. Emlen J (1966) The role of time and energy in food preference. The American Naturalist 100(916):611–617

169. Emlen J (1968) Optimal Choice in Animals. The American Naturalist 102(926):385–390

170. Erez Z, Steinberger-Levy I, Shamir M, Doron S, Stokar-Avihail A, Peleg Y, Melamed S, Leavitt A, Savidor A, Albeck S, Amitai G, Sorek R (2017) Communication between viruses guides lysis-lysogeny decisions. Nature 541(7638):488–493

171. Erickson C (1991) Percussive foraging in the aye-aye, Daubentonia madagascariensis. Animal Behaviour 41(5):793–801

172. Eusuff M M, Lansey K E (2003) Optimisation of Water Distribution Network Design Using the Shuffled Frog Leaping Algorithm. Journal of Water Resources Planning and Management 129(3):210–225

173. Fagan W, Lewis M, Auger-Méthé M, Avgar T, Benhamou S, Breed G, LaDage L, Schlägel U, Tang W-W, Papastamatiou Y, Forester J, Mueller T (2013) Spatial memory and animal movement. Ecology Letters 16(10):1316–1329

174. Falconer R, Bown J, White N, Crawford J (2005) Biomass recycling and the origin of phenotype in fungal mycelia. Proceedings of the Royal Society B 272(1573):1727–1734

175. Falik O, Reides P, Gersani M, Novoplansky A (2003) Self /non-self discrimination in roots. Journal of Ecology 91(4):525–531

176. Feinerman O, Traniello J (2016). Social complexity, diet, and brain evolution: modeling the effects of colony size, worker size, brain size, and foraging behavior on colony fitness in ants. Behavioral Ecology and Sociobiology 70(7):1063–1074

177. Felix M-A, Braendle C (2010) The natural history of *Caenorhabditis elegans*. Current Biology 20(22):R965–R969

178. Fenton B (2013) Questions, ideas and tools: lessons from bat echolocation. Animal Behaviour 85(5):869–879

179. Fertin A, Casas J (2006) Efficiency of antlion trap construction. Journal of Experimental Biology 209(18):3510–3515

180. Fields R (2007) The Shark's Electric Sense. Scientific American 297(2):74–80

181. Finger N, Bastian A, Jacobs D (2017) To seek or speak? Dual function of an acoustic signal limits its versatility in communication. Animal Behaviour 127:135–152

182. Fister I, Fister I (jr), Yang X-S, Brest J (2013) A comprehensive review of firefly algo-rithms. Swarm and Evolutionary Computation 13:34–46

183. Fister I (jr), Yang X-S, Fister D, Fister I (2014) Cuckoo Search: A brief literature review. In: Yang X-S (ed) Cuckoo Search and Firefly Algorithm: Theory and Applications, Studies in Computational Intelligence 516, pp 49–64, Berlin: Springer

184. Flower T, Gribble M, Ridley A (2014) Deception by Flexible Alarm Mimicry in an African Bird. Science 344(6183):513–516

185. Fountas N, Vaxevanidis N (2013) A Modified Virus Evolutionary Genetic Algorithm for Rough Machining Optimisation of Sculptured Surfaces. Annals of the Faculty of Engineering – Hunedoara 11(3):283–288

186. Fountas N, Vaxevanidis N, Stergiou C, Benhadj-Djilali R (2015) Quality research on the performance of a virus-evolutionary genetic algorithm for optimized sculptured surface CNC machining, through standard benchmarks. In: Proceedings of the 9th International Quality Conference (June 2015), Center for Quality, Faculty of Engineering, University of Kragujevac, http://www.cqm.rs/2015/cd1/pdf/papers/focus_1/025.pdf

187. Franklin E, Franks N (2012) Individual and social learning in tandem-running recruit-ment by ants. Animal Behaviour 84(2):361–368

188. Franks N, Gomez N, Goss S, Deneubourg J (1991) The blind leading the blind in army ant raid patterns: Testing a model of self-organisation (Hymenoptera: Formicidae). Jour-nal of Insect Behavior 4(5):583–607

189. Franks N, Richardson T (2006) Teaching in tandem-running ants. Nature 439(7073):153–153

190. Fretwell S, Lucas H (1970) On territorial behavior and other factors influencing habitat distribution in birds. Acta Biotheoretica 19(1):16–36

191. Fujioka E, Aihara I, Sumiya M, Aihara K, Hiryu S (2016) Echolocating bats use future-target information for optimal foraging. Proceedings of the National Academy of Sci-ences 113(17):4848–4852

192. Gagliano M (2012) Green symphonies: A call for studies on acoustic communication in plants. Behavioral Ecology 24(4):789–796

193. Gagliano M, Mancuso S, Robert D (2012) Towards understanding plant bioacoustics. Trends in Plant Science 17(6):323–325

194. Gagliano M, Renton M, Depczynski M, Mancuso S (2014) Experience teaches plants to learn faster and forget slower in environments where it matters. Oecologia 175(1):63–72

195. Gagliano M, Renton M, Duvdevani N, Timmins M, Mancuso S (2012a) Out of Sight but Not out of Mind: Alternative Means of Communication in Plants. PLoS One 7(5):e37382

196. Gagliano M, Renton M, Duvdevani N, Timmins M, Mancuso S (2012b) Acoustic and magnetic communication in plants: Is it possible? Plant Signaling & Behavior 7(10):1346–1348

197. Gagliano M, Vyazovskiy V, Borbély A, Grimonprez M, Depczynski M (2016) Learning by Association in Plants. Scientific Reports 6:38427

198. Galef B (1976) Social transmission of acquired behavior: a discussion of tradition and social learning in vertebrates. Advances in the Study of Behavior 6:77–100

199. Galef B (1988) Imitation in animals: History, definition, and interpretation of data from the psychological laboratory. In: Zentall T and Galef B (eds) Social learning: Psycho-logical and biological perspectives, pp 3–28, Hillsdale, NJ: Erlbaum

200. Galef B, Buckley L (1996) Use of foraging trails by Norway rats. Animal Behaviour 51(4):765–771

201. Galef B, Dudley K, Whiskin E (2008) Social learning of food preferences in dissatisfied and uncertain Norway rats. Animal Behaviour 75(2):631–637

202. Galland P, Pazur A (2005) Magnetoreception in plants. Current Topics in Plant Research 118(6):371–389

203. Gandomi A, Alavi A (2012) Krill herd: A new bio-inspired optimisation algorithm. Communications in Nonlinear Science and Numerical Simulation 17(12):4831–4845

204. Gandomi A, Yang X-S, Alavi A, Talatahari S (2013) Bat algorithm for constrained optimisation tasks. Neural Computing and Applications 22(6):1239–1255

205. Gao F, Hongwei L, Qiang Z, Cui G (2006) Virus-Evolutionary Particle Swarm Optimisation Algorithm. In: Proceedings of the 2006 International Conference on Natural Computation (ICNC 2006), LNCS 4222, pp 156–165, Berlin: Springer

206. Gawryszewski F, Calero-Torralbo M, Gillespie R, Rodríguez-Gironés M, Herberstein M (2017) Correlated evolution between coloration and ambush site in predators with visual prey lures. Evolution 71(8):2010–2021

207. Genest W, Stauffer W, Schultz W (2016) Utility functions predict variance and skewness risk preferences in monkeys. Proceedings of the National Academy of Sciences 113(30):8402–8407

208. Ghai R, Fugère V, Chapman C, Goldberg T, Davies T J (2015) Sickness behaviour associated with non-lethal infections in wild primates. Proceedings of the Royal Society B 282(1814):20151436

209. Gibbons A (2007) Food for Thought: Did the first cooked meals help fuel the dramatic evolutionary expansion of the human brain? Science 316(5831):1558–1560

210. Ginsburg S, Jablonka E (2009) Epigenetic learning in non-neuronal organisms. Journal of Biosciences 34(4):633–645

211. Giraldeau L A, Caraco T (2000) Social Foraging Theory. Princeton, New Jersey: Princeton University Press

212. Goldberg D E (1989) Genetic Algorithms in Search, Optimisation and Machine Learning. Boston: Addison-Wesley

213. Goldberg D E (1991) Real-coded Genetic Algorithms, Virtual Alphabets, and Blocking. Complex Systems 5:139–167

214. Goldberg D E, Richardson J (1987) Genetic algorithms with sharing for multimodal function optimisation. In: Proceedings of the Second International Conference on Genetic Algorithms, pp 41–49, Hillsdale, NJ: L. Erlbaum Associates Inc.

215. Goldbogen J A, Friedlaender A S, Calambokidis J, McKenna M F, Simon M, Nowacek D P (2013). Integrative approaches to the study of baleen whale diving behavior, feeding performance, and foraging ecology. BioScience 63:90–100

216. Goldfield, E (2018) Bioinspired Devices: Emulating Nature's Assembly and Repair Processes. Cambridge, MA: Harvard University Press

217. Gotelli N (1996) Ant Community Structure: Effects of Predatory Ant Lions. Ecology 77(2):630–638

218. Gotelli N (1997) Competition and coexistence of larval ant lions. Ecology 78(6):1761–1773

219. Gould J (2004) Animal Navigation. Current Biology 14(6):R221-R224

220. Gould J (2008) Animal Navigation: The Evolution of Magnetic Orientation. Current Biology 18(11):R482–R484

221. Gould J (2015) Animal Navigation: Birds have Magnetic Maps. Current Biology 25(19):R827–R844

222. Granovskiy B, Latty T, Duncan M, Sumpter D, Beekman M (2012) How dancing honeybees keep track of changes: the role of inspector bees. Behavioral Ecology 23(3):588–596

223. Grassé P-P (1959) La reconstruction du nid et les coordinations inter-individuelles chez Belicositermes natalensis et Cubitermes sp. La théorie de la Stigmergie: Essai d'interprétation du comportement des termites constructeurs. Insectes Sociaux 6:41–80

224. Greene J, Brown M, Dobosiewicz M, Ishida I, Macosko E, Zhang X, Butcher R, Cline D, McGrath P, Bargmann C (2016) Balancing selection shapes density-dependent foraging behaviour. Nature 539(7628):254–258

225. Greggers U, Koch G, Schmidt V, Dürr A, Floriou-Servou A, Piepenbrock D, Göpfert M, Menzel R (2014). Reception and learning of electric fields in bees. Proceedings of the Royal Society B 280(1759):20130528

226. Griffin D (1944) Echolocation by blind men, bats and radar. Science 100(2609):589–590

227. Griffin D (1958) Listening in the Dark. New Haven, CT: Yale University Press

228. Griffiths D (1980) Foraging Costs and Relative Prey Size. American Naturalist 116(5):743–752

229. Grunbaum D, Viscido S, Parrish J (2004) Extracting interactive control algorithms from group dynamics of schooling fish. Coop Control, Lecture Notes in Control and Information Sciences (LNCIS) 309, pp 103–117, Berlin: Springer

230. Grüter C, Ratnieks L (2011) Honeybee foragers increase the use of waggle dance information when private information becomes unrewarding. Animal Behaviour 81(5):949-954

231. Grüter C, Schürch R, Czaczkes T, Taylor K, Durance T, et al. (2012) Negative Feedback Enables Fast and Flexible Collective Decision-Making in Ants. PLoS ONE 7(9): e44501

232. Grüter C, Segers F, Ratnieks F (2013) Social learning strategies in honeybee foragers: do the costs of using private information affect the use of social information? Animal Behaviour 85(6):1443–1449

233. Guney K, Durmus A, Basbug S (2009) A Plant Growth Simulation Algorithm for Pattern Nulling of Linear Antenna Arrays by Amplitude Control. Progress in Electromagnetics Research B 17:69–84

234. Gunji Y, Shirakawa T, Niizato T, Haruna T (2008) Minimal model of a cell connecting amoebic motion and adaptive transport networks. Journal of Theoretical Biology 253(4):659–667

235. Haddock S, Moline M, Case J (2010) Bioluminescence in the sea. Annual Review of Marine Science 2:443–493

236. Halstead L B (1990) Cretaceous-Tertiary (terrestrial). In: Briggs D E G, Crowther P R (eds) Palaeo-biology: A Synthesis, pp 203-207, Oxford, UK: Blackwell Science

237. Hao X, Falconer R, Bradley D, Crawford J (2009) FUNNet - A Novel Biologically-Inspired Routing Algorithm Based on Fungi. In: Proceedings of the Second International Conference on Communication Theory, Reliability, and Quality of Service (CTRQ 2009) pp 97–102, New Jersey: IEEE Press

238. Hanson K, Nicolau D (jr), Filipponi L, Wang L, Lee A, Nicolau D (2006) Fungi Use Efficient Algorithms for the Exploration of Microfluidic Networks. Small 2(10):1212–1220

239. Harari Y N (2014) Sapiens: A Brief History of Humankind. London: Harvill Secker

240. Harel R, Spiegel O, Getz W, Nathan R (2017) Social foraging and individual consistency in following behaviour: testing the information centre hypothesis in free-ranging vultures. Proceedings of the Royal Society B 284(1852):20162654

241. Hata H, Kato M (2006) A novel obligate cultivation mutualism between damselfish and *Polysiphonia* algae. Biology Letters 2(4):593–596

242. Hays G, Bastian T, Doyle T, Fossette S, Gleiss A, Gravenor M, Hobson V, Humphries N, Lilley M, Pade N, Sims D (2012) High activity and Levy searches: jellyfish can search the water column like fish. Proceedings of the Royal Society B 279(1728):465–473

243. He D, Qu L, Guo X (2009) Artificial Fish-school Algorithm for Integer Programming. In: Proceedings of the IEEE International Conference on Information Engineering and Computer Science (ICIECS 2009), pp 1–4, New Jersey: IEEE Press

244. He S, Wu Q, Saunders J (2006) A Novel Group Search Optimizer Inspired by Animal Behavioral Ecology. In: Proceedings of the IEEE World Congress on Computational Intelligence (WCCI 2006), pp 4415–4421, New Jersey: IEEE Press

245. He S, Wu Q, Saunders J (2009) Group Search Optimizer: An Optimisation Algorithm Inspired by Animal Searching Behavior. IEEE Transactions on Evolutionary Computation 13(5):973–990

246. He X, Chen H, Niu B, Wang J (2015) Root Growth Optimizer with Self-Similar Propagation. Mathematical Problems in Engineering, Vol. 2015, Article ID 498626, DOI:10.1155/2015/498626

247. He X, Wang J, Bi Y (2015) A novel branch-leaf growth algorithm for numerical optimisation. In: Proceedings of the 11th International Conference on Intelligent Computing Theories and Methodologies Part II (ICIC 2015), LNCS 9226, pp 742–750, Berlin: Springer

248. He X, Zhang S, Wang J (2015) A Novel Algorithm Inspired by Plant Root Growth with Self-similarity Propagation. In: Proceedings of the 1st International Conference on Industrial Networks and Intelligent Systems (INISCom 2015), pp 157–162, New Jersey: IEEE Press

249. Healy K, McNally L, Ruxton G, Cooper N, Jackson A (2013) Metabolic rate and body size are linked with the perception of temporal information. Animal Behaviour 86(4):685–696

250. Heaton L, Obara B, Grau V, Jones N, Nakagaki T, Boddy L, Fricker M (2012) Analysis of fungal networks. Fungal Biology Reviews 26(1):12–29

251. Hegenauer V, Fürst U, Kaiser B, Smoker M, Zipfel C, Felix G, Stahl M, Albert M (2016) Detection of the plant parasite Cuscuta reflexa by a tomato cell surface receptor. Science 353(6298):478–481

252. Heinerman J, Zonta A, Haasdijk E, Eiben A E (2016) On-line Evolution of Foraging Behaviour in a Population of Real Robots. In: Proceedings of the 19th European Conference on Applications of Evolutionary Computation (EvoApps 2016), LNCS 9598, pp 198–212, Berlin: Springer

253. Hendtlass T (2005) WoSP: a multioptima particle swarm algorithm. In: Proceedings of the IEEE Congress on Evolutionary Computation 2005 (CEC 2005), pp 727-734, New Jersey: IEEE Press

254. Herberstein M, Wignall A (2011) Deceptive signals in spiders. In: Herberstein M (ed) Spider Behaviour: Flexibility and Versatility, pp 190–214, Cambridge: Cambridge University Press

255. Herrera F, Lozano M, Verdegay J L (1998) Tackling Real-Coded Genetic Algorithms: Operators and Tools for Behavioural Analysis. Artificial Intelligence Review 12:265–319

256. Heun M, Schäfer-Pregl R, Klawan D, Castagna R, Accerbi M, Borghi B, Salamini F (1997) Site of Einkorn Wheat Domestication Identified by DNA Fingerprinting. Science 278(5341):1312–1314

257. Heyes C (1994) Social Learning in Animals: Categories and Mechanisms. Biological Reviews 69(2):207–231

258. Hickey D, Noriega L (2008) Insights into Information Processing by the Single Cell Slime Mold *Physarum Polycephalum*. UKACC Control Conference, Manchester, UK, September 2–4, 2008, http://fusion-edu.eu/HAM/Papers_files/p246.pdf

259. Higashitani M, Ishigame A, Yasuda K (2006) Particle Swarm Optimisation Considering the Concept of Predator-Prey Behavior. In: Proceedings of the 2006 IEEE Congress on Evolutionary Computation (CEC 2006), pp 1541–1544, New Jersey: IEEE Press

260. Higginson A, Houston A (2015) The influence of the food-predation trade-off on the foraging behaviour of central-place foragers. Behavioral Ecology and Sociobiology 69(4):551–561

261. Holland J H (1975) Adaptation in Natural and Artificial Systems: An Introductory Analysis with Applications to Biology, Control, and Artificial Intelligence. Ann Arbor, Michigan: University of Michigan Press

262. Hölldobler B, Wilson E (1994) Journey to the Ants: A story of scientific exploration. Cambridge, Massachusetts: Bellknap Press/Harvard University Press

263. Hölldobler B, Wilson E (2009) The Superorganism: The Beauty, Elegance, and Strangeness of Insect Societies. New York: W. W. Norton

264. Hollis K, Cogswell H, Snyder K, Guillette L, Nowbahari E (2011) Specialised Learning in Antlions (Neuroptera: Myrmeleontidae), Pit-Digging Predators, Shortens Vulnerable Larval Stage. PLoS One 6(3):e17958

265. Hollis K, Harrsch F, Nowbahari E (2015) Ants vs. antlions: An insect model for studying the role of learned and hard-wired behavior in coevolution. Learning and Motivation 50:68–82

266. Holmes E (2011) What Does Virus Evolution Tell Us about Virus Origins? Journal of Virology 85(11):5247–5251

267. Hoppitt W, Brown G, Kendal R, Rendell L, Thornton A, Webster M, Laland K (2008) Lessons From Animal Teaching. Trends in Ecology and Evolution 23(9):486–493

268. Hoppitt W, Laland K (2013) Social Learning: An introduction to mechanisms, methods and models. Princeton, New Jersey: Princeton University Press

269. Howe H, Smallwood J (1982) Ecology of Seed Dispersal. Annual Review of Ecology and Systematics 13:201-228

270. Huffman M, Seifu M (1989) Observations on the illness and consumption of a possibly medicinal plant Vernonia amygdalina (Del.), by a wild chimpanzee in the Mahale Mountains National Park, Tanzania. Primates 30(1):51–63

271. Humphries J, Xiong L, Liu J, Prindle A, Yuan F, Arjes H, Tsimring L, Süel G (2017) Species-Independent Attraction to Biofilms through Electrical Signaling. Cell 168(1-2):200–209

272. Hunt G (1996) The manufacture and use of hook tools by New Caledonian crows. Nature 379(6562):249–251

273. Jablonski D, Chaloner W G (1994) Extinctions in the fossil record (and discussion). Philosophical Transactions of the Royal Society of London, Series B. 344(1307):11–17

274. Jack-McCollough R, Nieh J (2015) Honeybees tune excitatory and inhibitory recruitment signalling to resource value and predation risk. Animal Behaviour 110:9–17

275. Jackson C, Hunt E, Sharkh S, Newland P (2011) Static electric fields modify the locomotory behaviour of cockroaches. Journal of Experimental Biology 214(12):2020–2026

276. Jackson D, Ratnieks F (2006) Communication in ants, Current Biology 16(15):R570–574

277. James W (1890) Principles of Psychology. New York: Holt

278. Janecek A and Tan Y (2011) Feeding the Fish — Weight Update Strategies for the Fish School Search Algorithm. In: Proceedings of the Second International Conference on Swarm Intelligence (ICSI 2011), pp 553–562, Berlin: Springer

279. Janson S, Middendorf M, Beekman M (2007) Searching for a new home — scouting behavior of honeybee swarms. Behavioral Ecology 18(2):384–392

280. Janzen, D (1978). Complications in Interpreting the Chemical Defenses of Trees Against Tropical Arboreal Plant-eating Vertebrates, In: Proceedings of the Ecology of Arboreal Folivores: A Symposium Held at the Conservation and Research Center, National Zoological Park, Smithsonian Institution, May 29–31, 1975, pp 73–84, Washington: Smithsonian Institution Press

281. Jarvis E, Güntürkün O, Bruce L, et al. (2005) Avian brains and a new understanding of vertebrate brain evolution. Nature Reviews Neuroscience 6(2):151–159

282. Jaumann S, Scudelari R, Naug D (2013) Energetic Cost of Learning and Memory Can Cause Cognitive Impairment in Honeybees. Biology Letters 9(4):20130149

283. Jensen K, Mayntz D, Toft S, Clissold F, Hunt J, Raubenheimer D, Simpson S (2015) Optimal foraging for specific nutrients in predatory beetles. Proceedings of the Royal Society B 279(1736):2212–2218

284. Jeong M, Shim C, Lee J, et al. (2008) Plant gene responses to frequency-specific sound signals. Molecular Breeding 21(2):217–226

285. Jerozolimski A, Beatriz Ribeiro M, Martins M (2009) Are Tortoises Important Seed Dispersers in Amazonian Forests? Oecologia 161(3):517–528

286. Johnson J H, Wolman A A (1984) The humpback whale, Megaptera novaeangliae. Marine Fisheries Review 46(4):30–37

287. Jones G (2005) Echolocation. Current Biology 15(13):R484–R488

288. Jones G (2008) Sensory Ecology: Echolocation Calls are Used for Communication. Current Biology 18(1):R34–R35

289. Jones G, Siemers B (2011) The communicative potential of bat echolocation pulses. Journal of Comparative Physiology A 197(5):447–457

290. Jones G, Teeling E (2006) The evolution of echolocation in bats. Trends in Ecology and Evolution 21(3):149–156

291. Jones J (2016) Applications of Multi-Agent Slime Mould Computing. International Journal of Parallel, Emergent and Distributed Systems 31(5):420–449

292. Juarez J R C, Wang H-J, Lai Y-C, Liang Y-C (2009) Virus Optimisation Algorithm (VOA): A Novel Metaheuristic for Solving Continuous Optimisation Problems. In: Proceedings of the 2009 Asia Pacific Industrial Engineering & Management Systems Conference (APIEMS 2009), December 14-16, 2009, Kitakyushu, Japan, pp 2166–2174

293. Kabadayi C, Osvath M (2017) Ravens parallel great apes in flexible planning for tool-use and bartering. Science 357(6347):202–204

294. Kagan E, Ben-Gal I (2015) Search and Foraging: Individual Motion and Swarm Dynamics. Boca Raton, FL: CRC Press

295. Kalan A, Mundry R, Boesch C (2015) Wild chimpanzees modify food call structure with respect to tree size for a particular fruit species. Animal Behaviour 101:1–9

296. Kalenscher T, van Wingerden M (2011) Why We Should Use Animals to Study Economic Decision Making – A Perspective. Frontiers in Neuroscience Vol 5, Article 82:1–11 doi:10.3389/fnins.2011.00082

297. Kalmijn A (1971) The electric sense of sharks and rays. Journal of Experimental Biology 55(2):371–383

298. Kandel E, Schwartz J, Jessell T, Siegelbaum S, Hudspeth A (eds) (2013) Principles of Neural Science (5th ed). New York: McGraw-Hill

299. Karban R (2008) Plant Behaviour and Communication. Ecology Letters 11(7):727–739

300. Karban R (2015) Plant Sensing and Communication. Chicago: University of Chicago Press

301. Karban R, Shiojiri K (2009) Self-recognition affects plant communication and defense. Ecology Letters 12(6):502–506
302. Karaboga D (2005) An idea based on honeybee swarm for numerical optimisation. Technical Report TR06, Engineering Faculty, Computer Engineering Department, Erciyes University, http://mf.erciyes.edu.tr/abc/pub/tr06_2005.pdf
303. Karaboga D, Akay B (2009) A survey: algorithms simulating bee swarm intelligence. Artificial Intelligence Review 31(1–4):61–85
304. Karaboga D, Basturk B (2007) A powerful and efficient algorithm for numerical function optimisation: artificial bee colony (ABC) algorithm. Journal of Global Optimisation 39(3):459–471
305. Karaboga D, Gorkemli B, Ozturk C, Karaboga N (2014) A comprehensive survey: artificial bee colony (ABC) algorithm and applications. Artificial Intelligence Review 42(1):21–57
306. Karaboga D, Ozturk C (2011) A novel clustering approach: Artificial Bee Colony (ABC) algorithm. Applied Soft Computing 11(1):652–657
307. Kareiva P, Odell G (1987) Swarms of predators exhibit "preytaxis" if individual predators use area-restricted search. American Naturalist 130(2):233–270
308. Karpas E, Shklarsha A, Schneidmana E (2017) Information socialtaxis and efficient collective behavior emerging in groups of information-seeking agents. Proceedings of National Academy of Science 114(22):5589–5594
309. Katz K, Naug D (2016) Dancers and followers in a honeybee colony differently prioritize individual and colony nutritional needs. Animal Behaviour 119:69–74
310. Kavanau J (1998) Vertebrates that never sleep: Implications for sleep's basic function. Brain Research Bulletin 46(4):269–279
311. Kaveh A, Farhoudi N (2013) A new optimisation method: Dolphin echolocation. Advances in Engineering Software 59:53–70
312. Kaveh A, Hosseini P (2014) A simplified dolphin echolocation optimisation method for optimum design of trusses. International Journal of Optimisation in Civil Engineering 4(3):381–397
313. Kemperman J, Barnes B (1976) Clone size in American aspens. Canadian Journal of Botany 54(22):2603–2607
314. Kennedy J, Eberhart R (1995) Particle swarm optimisation. In: Proceedings of the 1995 IEEE International Conference on Neural Networks, pp 1942–1948, New Jersey: IEEE Press
315. Kennedy J, Eberhart R, Shi Y (2001) Swarm Intelligence. San Mateo, California: Morgan Kaufmann
316. Kerth G, Reckardt K (2003) Information transfer about roosts in female Bechstein's bats: an experimental field study. Proceedings of the Royal Society B 270(1514):511–515
317. Kiefer J (1953) Sequential minimax search for a maximum. Proceedings of the American Mathematical Society 4(3):502–506
318. Kiran M, Findik O (2015) A directed artificial bee colony algorithm. Applied Soft Computing 26:454–462
319. Kirschvink J, Walker M, Diebel C (2001) Magnetite-based magnetoreception. Current Opinion in Neurobiology 11(4):462–467
320. Kirschvink J, Winklhofer M, Walker M (2010) Biophysics of magnetic orientation: strengthening the interface between theory and experimental design. Journal of Royal Society Interface 7(Suppl 2):S179–S191

321. Koen R, Engelbrecht A (2017) Maze exploration using a fungal search algorithm: Part 1 - Algorithm model. In: Proceedings of the 2017 International Conference on Intelligent Systems (ISMSI 2017), Metaheuristics & Swarm Intelligence, pp 40–45, New York: ACM Press

322. Kok P, Mostert P, de Lange F (2017) Prior expectations induce prestimulus sensory templates, Proceedings of National Academy of Science, doi:10.1073/pnas.1705652114

323. Kolarik A J, Cirstea S, Pardhan S, Moore B C J (2014) A summary of research investigating echolocation abilities of blind and sighted humans. Hearing Research 310:60–68

324. Koller D (2011) The Restless Plant. Cambridge, MA: Harvard University Press

325. Kong X, Chen Y-L, Xie W, Wu X (2012) A Novel Paddy Field Algorithm Based on Pattern Search Method. In: Proceedings of the 2012 IEEE International Conference on Information and Automation, pp 686–690, New Jersey: IEEE Press

326. Korobkova E, Emonet T, Vilar J, Shimizu T, Cluzel, P (2004) From molecular noise to behavioural variability in a single bacterium. Nature 428(6982):574–578

327. Kost C, de Oliveira E, Knoch T, Wirth R (2005) Spatio-temporal permanence and plasticity of foraging trails in young and mature leaf-cutting ant colonies (Attaspp.). Journal of Tropical Ecology 21(6):677–688

328. Kountché D A, Monmarché N, Slimane M (2012) The *Pachycondyla Apicalis* Ants Search Strategy for Data Clustering Problems. In: Swarm and Evolutionary Computation, Proceedings of SIDE 2012 and EC 2012 International Symposia, LNCS 7269, pp 3–11, Berlin: Springer

329. Kowalski P, Lukasik S (2014) Experimental Study of Selected Parameters of the Krill Herd Algorithm, In: Proceedings of the Seventh IEEE International Conference Intelligent Systems 2014 (IS 2014), Advances in Intelligent Systems and Computing 322, pp 473–485, Berlin: Springer

330. Koza J R (1992) Genetic Programming. Cambridge, Massachusetts: MIT Press

331. Koza J R, Rice J, Roughgarden J (1992) Evolution of food foraging strategies for the Caribbean Anolis lizard using genetic programming. Adaptive Behavior 1(2):171–199

332. Krebs J R, Erichsen J T, Webber M I, Charnov E L (1977) Optimal prey selection in the great tit (Parus major). Animal Behaviour 25(Part 1):30–38

333. Krishnanand K, Ghose D (2005) Detection of Multiple Source Locations using a Glowworm Metaphor with Applications to Collective Robotics. In: Proceedings of the 2005 IEEE Swarm Intelligence Symposium, pp 84–91, New Jersey: IEEE Press

334. Krishnanand K, Ghose D (2006a) Glowworm swarm based optimisation algorithm for multimodal functions with collective robotics applications. Multiagent and Grid Systems 2(3):209–222

335. Krishnanand K, Ghose D (2006b) Theoretical Foundations for Multiple Rendezvous of Glowworm-inspired Mobile Agents with Variable Local-decision domains. In: Proceedings of the 2006 American Control Conference, pp 3588–3593, New Jersey: IEEE Press

336. Krishnanand K, Ghose D (2009) Glowworm swarm optimisation for simultaneous capture of multiple local optima for multimodal functions. Swarm Intelligence 3(2):87–124

337. Kromdijk J, Glowacka K, Leonelli L, Gabilly S, Iwai M, Niyogi K, Long S (2016) Improving photosynthesis and crop productivity by accelerating recovery from photoprotection. Science 354(6314):857–861

338. Kubota N, Arakawa T, Fukuda T, Shimojima K (1997) Fuzzy manufacturing scheduling by virus-evolutionary genetic algorithm in self-organizing manufacturing system. In: Proceedings of the 1997 IEEE International Conference on Evolutionary Computation, pp 1283–1288, New Jersey: IEEE Press

339. Kubota N, Fukuda T, Arakawa T, Shimojima K (1997) Evolutionary transition on Virus-Evolutionary Genetic Algorithm. In: Proceedings of the 1997 IEEE International Conference on Fuzzy Systems, pp 291–296, New Jersey: IEEE Press

340. Kubota N, Shimojima K, Fukuda T (1996) The role of virus infection in virus-evolutionary genetic algorithm. In: Proceedings of the 1996 IEEE International Conference on Evolutionary Computation, pp 182–187, New Jersey: IEEE Press

341. Kumar S, Sharma V K, Kumari R (2014) Modified Position Update in Spider Monkey Optimisation Algorithm. International Journal of Emerging Technologies in Computational and Applied Sciences 7(2):198–204

342. Kundu D, Suresh K, Ghosh S, Das S, Panigrahi B K, Das S (2011) Multi-objective optimisation with artificial weed colonies. Information Sciences 181(12):2241–2454

343. Labella T, Dorigo M, Deneubourg J-L (2006) Division of labor in a group of robots inspired by ants' foraging behavior. ACM Transactions on Autonomous and Adaptive Systems 1(1):4–25

344. Laland K (2004) Social learning strategies. Learning and Behavior 32(1):4–14

345. Lambinet V, Hayden M, Reigl K, Gomis S, Gries G (2017) Linking magnetite in the abdomen of honeybees to a magnetoreceptive function. Proceedings of the Royal Society B 284(1851):20162873

346. Land M, Nilsson D (2002) Animal Eyes. Oxford: Oxford University Press

347. Latty T, Beekman M (2009) Food quality affects search strategy in the accellular slime mould, *Physarum polycephalum*. Behavioral Ecology 20(6):1160–1167

348. Latty T, Beekman M (2011) Irrational decision-making in an amoeboid organism: transitivity and context-dependent preferences. Proceedings of the Royal Society B 278(1703):307–312

349. Latty T, Ramsch K, Ito K, Nakagaki T, Sumpter DJT, Middendorf M (2011) Structure and formation of ant transportation networks. Journal of the Royal Society Interface 8(62):1298–1306

350. Lewis S, Cratsley C (2008) Flash Signal Evolution, Mate Choice, and Predation in Fireflies. Annual Review of Entomology 53:293–321

351. Li M, , Zhao H, Weng X, Han T (2016) A novel nature-inspired algorithm for optimisation: Virus colony search. Advances in Engineering Software 92:65–88

352. Li W, Wang H, Zou Z J (2005) Function Optimisation Method Based on Bacterial Colony Chemotaxis. Journal of Circuits and Systems (China), 10(1):58–63

353. Li X, Shao Z, Qian J (2002) An optimizing method based on autonomous animats: fish swarm algorithm. Systems Engineering Theory and Practice 22(11):32–38 (in Chinese)

354. Li X, Zhang J, Yin M (2014) Animal migration optimisation: an optimisation algorithm inspired by animal migration behavior. Neural Computing and Applications 24(7):1867–1877

355. Liang Y-C, Juarez J R C (2016) A novel metaheuristic for continuous optimisation problems: Virus optimisation algorithm. Engineering Optimisation 48(1):73–93

356. Liao T, Socha K, Montes de Oca M, Stüzle T, Dorigo M (2014) Ant colony optimisation for mixed-variable optimisation problems. IEEE Transactions on Evolutionary Computation 18(4):503–518

357. Liao T, Stüzle T, Montes de Oca M, Dorigo M (2014) A unified ant colony optimisation algorithm for continuous optimisation. European Journal of Operational Research 234(3):597–609

358. Lindenmayer A (1968) Mathematical Models for Cellular Interaction in Development, Parts I and II. Journal of Theoretical Biology 18(3):280–315

359. Linhares A (1998) State-space search strategies gleaned from animal behavior: a traveling salesman experiment. Biological Cybernetics 78(3):167–173

360. Linhares A (1998) Preying on optima: a predatory search strategy for combinatorial problems. In: Proceedings of the 1998 IEEE International Conference on Systems, Man and Cybernetics, 3:2974–2978, New Jersey: IEEE Press

361. Linhares A (1999) Synthesizing a Predatory Search Strategy for VLSI Layouts. IEEE Transactions on Evolutionary Computation 3(2):147–152

362. Lins A, Bastos-Filho C, Nascimento D, Oliveira Junior M, Lima-Neto F (2012) Analysis of the Performance of the Fish School Search Algorithm Running in Graphic Processing Units. In: Parpinelli R (ed) Theory and New Applications of Swarm Intelligence, pp 17–32, Rijeka, Croatia: Intech

363. Lissmann H (1951) Continuous electrical signals from the tail of a fish *Gymnarchus niloticus* Cuv. Nature 167(4240):201–202

364. List C (2004) Democracy in animal groups: a political science perspective. Trends in Ecology and Evolution 19(4):168–169

365. Little J (1961) A proof for the queuing formula: $L = \lambda W$. Operations Research 9(3):383–387

366. Liu R, Wang X, Li Y (2012) Multi-objective Invasive Weed Optimisation Algorithm for Clustering. In: Proceedings of the 2012 IEEE World Congress on Computational Intelligence (WCCI 2012), pp 1556–1563, New Jersey: IEEE Press

367. Lloyd J (1984) Occurence of Aggressive Mimicry in Fireflies. Florida Entomologist 67(3):368–376

368. Locey K, Lennon J (2016) Scaling laws predict global microbial diversity. Proceedings of the National Academy of Sciences 113(21):5970–5975

369. Lockery S (2009) Neuroscience: A social hub for worms. Nature 458(7242):1124–1125

370. Lonnstedt O, Ferrari M, Chivers D (2014) Lionfish predators use flared fin displays to initiate cooperative hunting. Biology Letters 10(6):20140281

371. Lopes J, Brugger M, Menezes R, Camargo R, Forti L, Fourcassié V (2016) Spatio-Temporal Dynamics of Foraging Networks in the Grass-Cutting Ant Atta bisphaerica Forel, 1908 (Formicidae, Attini). PLoS ONE 11(1):e0146613

372. Lotka A J (1910) Contribution to the Theory of Periodic Reaction. Journal of Physical Chemistry 14(3):271–274

373. Lu C, Li X, Gao L, Liao W, Yi J (2017) An effective multiobjective discrete virus optimisation algorithm for flexible job-shop scheduling problem with controllable processing times. Computers & Industrial Engineering 104:156–174

374. Lu Y, Liu Y, Gao C, Tao L, Zhang Z (2014) A Novel *Physarum*-Based Ant Colony System for Solving the Real-World Travelling Salesman Problem. In: Proceedings of the 5th International Conference on Swarm Intelligence (ICSI 2014), LNCS 8794, pp 173–180, Berlin: Springer

375. Lubbock J (1882) Ants, Bees and Wasps: A Record of Observations on the Habits of the Social Hymenoptera. London: Kegan Paul, Trench & Co

376. Luo Q, Ma M, Zhou Y (2015) A Novel Animal Migration Algorithm for Global Numerical Optimisation. Computer Science and Information Systems 13(1):259–285

377. Ma M, Luo Q, Zhou Y, Chen X, Li L (2015) An Improved Animal Migration Optimisation Algorithm for Clustering Analysis. Discrete Dynamics in Nature and Society, Vol. 2015, Article ID 194792, DOI:10.1155/2015/194792

378. Maák I, Lörinczi G, Le Quinquis P, Módra G, Bovet D, Call J, d'Ettorre P (2017) Tool selection during foraging in two species of funnel ants. Animal Behaviour 123:207–216

379. MacArthur R, Pianka E (1966) On the optimal use of a patchy environment. The American Naturalist 100(916):603–609

380. MacNab R, Koshland D (1972) The Gradient-Sensing Mechanism in Bacterial Chemotaxis. Proceedings of the National Academy of Sciences 69(9):2509–2512

381. Madigan M, Martinko J, Bender K, Buckley D, Stahl D (2014) Brock Biology of Microorganisms (14th ed). Boston: Pearson

382. Madsen P, Aguilar de Soto N, Arranz P, Johnson M (2013) Echolocation in Blainville's beaked whales (*Mesoplodon densirostris*). Journal of Comparative Physiology A 199(5):451–469

383. Mahfoud S (1995) Niching Methods for Genetic Algorithms. PhD Thesis. Department of General Engineering, University of Illinois at Urbana-Champaign, http://www.leg.ufpr.br/~leonardo/artigos/tese_mahfoud.pdf

384. Mallon E, Pratt S, Franks N (2001) Individual and collective decision-making during nest site selection by the ant *Leptothorax albipennis*. Behavioral Ecology and Sociobiology 50(4):352–359

385. Mantegna R (1994) Fast, accurate algorithm for numerical simulation of Lévy stable stochastic processes. Physics Review E 49(5):4677–4683

386. Marichelvam M, Prabaharan T, Yang, X-S (2014) A Discrete Firefly Algorithm for the Multi-Objective Hybrid Flowshop Scheduling Problem. IEEE Transactions on Evolutionary Computation 18(2):301–305

387. Marshall J, Bogacz R, Dornhaus A, Planque R, Kovacs T, Franks N (2009) On optimal decision-making in brains and social insect colonies. Journal of the Royal Society Interface 6(40):1065–1074

388. Martens E, Wadhwa N, Jacobsen N, Lindemann C, Andersen K, Visser A (2015) Size structures sensory hierarchy in ocean life. Proceedings of the Royal Society B 282(1815):20151346

389. Marwan W (2010) Amoeba-Inspired Network Design. Science 327(5964):419–420

390. Marzluff J, Heinrich B (2001) Raven roosts are still information centres. Animal Behaviour 61:F14–F15

391. Marzluff J, Heinrich B, Marzluff C (1996) Raven roosts are mobile information centres. Animal Behaviour 51(1):89–103

392. Masumoto T (1998) Cooperative Prey Capture in the Communal Web Spider, *Philoponella Raffrayi* (Araneae, Uloboridae). The Journal of Arachnology 26(3):392–396

393. Maynard Smith J (1982) Evolution and the Theory of Games. Cambridge: Cambridge University Press

394. Maynard Smith J, Price G (1973) The logic of animal conflict. Nature 246(5427):15–18

395. McDonald N, Rands S, Hill F, Elder C, Ioannou C (2016) Consensus and experience trump leadership, suppressing individual personality during social foraging. Science Advances 2(9):e1600892

396. McGlynn T, Graham R, Wilson J, Emerson J, Jandt J, Hope Jahren A. (2015) Distinct types of foragers in the ant *Ectatomma ruidum*: typical foragers and furtive thieves. Animal Behaviour 109:243–247

397. Mehrabian A R, Lucas C (2006) A novel numerical optimisation algorithm inspired from weed colonisation. Ecological Informatics 1(4):355–366

398. Mehrabian A R, Yousefi-Koma A (2007) Optimal positioning of piezoelectric actuators on a smart fin using bio-inspired algorithms. Aerospace Science and Technology 11(2-3):174–182

399. Mery F, Kawecki T (2005) A Cost of Long-Term Memory in Drosophila. Science 308(5725):1148

400. Meškauskas A, Fricker M, Moore D (2004) Simulating colonial growth of fungi with the Neighbour-Sensing model of hyphal growth. Mycological Research 108(11):1241–1256

401. Miller N, Garnier S, Hartnett A, Couzin I (2013) Both information and social cohesion determine collective decisions in animal groups. Proceedings of the National Academy of Sciences 110(13):5263–5268

402. Miller P, Johnson M, Tyack P (2004) Sperm whale behaviour indicates the use of echolocation click buzzes 'creaks' in prey capture. Proceedings of the Royal Society B 271(1554):2239–2247

403. Mirjalili S (2015) How effective is the Grey Wolf Optimizer in training multilayer perceptrons. Applied Intelligence 43(1):150–161

404. Mirjalili S (2015) Ant Lion Optimizer (ALO) Matlab Code. https://uk.mathworks.com/matlabcentral/fileexchange/49920-ant-lion-optimizer--alo-?s_tid=prof_contriblnk (accessed 29 September 2017)

405. Mirjalili S, Lewis A (2016). The Whale Optimisation Algorithm. Advances in Engineering Software 95:51–67

406. Mirjalili S (2016) The Whale Optimisation Algorithm Matlab Code. https://uk.mathworks.com/matlabcentral/fileexchange/55667-the-whale-optimisation-algorithm?s_tid=prof_contriblnk (accessed 29 September 2017)

407. Mirjalili S, Mirjalili S M, Lewis A (2014). Grey Wolf Optimizer. Advances in Engineering Software 69:46–61

408. Möglich M (1978) Social organisation of nest emigration in *Leptothorax* (Hym., Form.). Insectes Sociaux 25(3):205–225

409. Monismith D, Mayfield B (2008) Slime Mold as a Model for Numerical Optimisation. In: Proceedings of the IEEE Swarm Intelligence Symposium (SIS 2008), pp 1–8, New Jersey: IEEE Press

410. Monmarché N, Venturini G, Slimane M (2000) On how *Pachycondyla apicalis* ants suggest a new search algorithm. Future Generation Computer Systems 16(8):937–946

411. Morand-Ferron J, Sol D, LeFebvre L (2007) Food stealing in birds: brain or brawn? Animal Behaviour 74(6):1725–1734

412. Morawetz L, Spaethe J (2012). Visual attention in a complex search task differs between honeybees and bumblebees. Journal of Experimental Biology 215(14):2515–2523

413. Morris T, Letnic M (2017) Removal of an apex predator initiates a trophic cascade that extends from herbivores to vegetation and the soil nutrient pool. Proceedings of the Royal Society B 284(1854):2017.0111

414. Morteza Jaderyan M, Khotanlou H (2016) Virulence Optimisation Algorithm. Applied Soft Computing 43:596–618

415. Moser M B, Moser E (2016) Where Am I? Where Am I Going? Scientific American 314(1):24–29

416. Mueller U, Schultz T, Currie C, Adams R, Malloch D (2001) The origin of the attine ant–fungus mutualism. The Quarterly Review of Biology 76(2):169–197

417. Müller G, Pilzecker A (1900) Experimentelle Beiträge zur Lehre vom Gedächtniss. Zeitschrift für Psychologie. Ergänzungsband, Vol. 1, JA Barth

418. Müller S, Airaghi S, Marchetto J, Koumoutsakos P (2000) Optimisation algorithms based on a model of bacterial chemotaxis. In: Proceedings of the 6th International Conference on the Simulation of Adaptive Behavior: From Animals to Animats (SAB 2000), pp 375–384, Cambridge, Massachusetts: MIT Press

419. Müller S, Marchetto J, Airaghi S, Koumoutsakos P (2002) Optimisation based on bacterial chemotaxis. IEEE Transactions on Evolutionary Computation 6(1):16–29

420. Muro C, Escobedo R, Coppinger R (2011) Wolf pack (*Canis lupus*) hunting strategies emerge from simple rules in computational simulations. Behavioural Processes 88(3):192–197

421. Muro C, Escobedo R, Coppinger R, et al. (2011) Wolf-pack hunting strategy: An emergent collective behavior described by a classical robotic model. Journal of Veterinary Behavior 6(1):94–94

422. Murphy G, Dudley S (2009) Kin Recognition: Competition and Cooperation in *Impatiens* (Balsaminaceae). American Journal of Botany 96(11):1990–1996

423. Murray R (1959) The response of the ampullae of Lorenzini to combined stimulation by temperature change and weak direct currents. Journal of Physiology 145(1):1–13

424. Murray R (1960) Electrical sensitivity of the ampullae of Lorenzini. Nature 187(4741):957

425. Nakagaki T, Iima M, Ueda T, Nishiura Y, Saigusa T, Tero A, Kobayashi R, Showalter K (2007) Minimum-risk Path Finding by an Adaptive Amoebal Network. Physical Review Letters 99(6):068104

426. Nakagaki T, Kobayashi R, Nishiura Y, Ueda T (2004) Obtaining multiple separate food sources: behavioural intelligence in the *Physarum* plasmodium. Proceedings of the Royal Society London B 271(1554):2305–2310

427. Nakagaki T, Yamada H, Hara M (2004) Smart network solutions in an amoeboid organism. Biophysical Chemistry 107(1):1–5

428. Nakagaki T, Yamada H, Toth A (2000). Maze-solving by an amoeboid organism. Nature 407(6803):470

429. Nakamura R, Pereira L, Costa K, Rodrigues D, Papa J, Yang X-S (2012) BBA: A Binary Bat Algorithm for Feature Selection. In: Proceedings of the XXV SIBGRAPI Conference on Graphics, Patterns and Images, pp 291–297, New Jersey: IEEE Press

430. Nathan R, Katul G, Bohrer G, Kuparinen A, Soons M, Thompson S, Trakhtenbrot A, Horn H (2011) Mechanistic models of seed dispersal by wind. Theoretical Ecology 4(2):113-132

431. Nelson X, Jackson R (2011) Flexibility in the foraging strategies of spiders. In: Herberstein M (ed) Spider Behaviour: Flexibility and Versatility, pp 31–56, Cambridge: Cambridge University Press

432. Neshat M, Sepidnam G, Sargolzaei M, Toosi A N (2012) Artificial fish swarm algorithm: a survey of the state-of-the-art, hybridisation, combinatorial and indicative applications. Artificial Intelligence Review 42(4):965–997

433. Neuweiler G (1990) Auditory adaptations for prey capture in echolocating bats. Physiological Reviews 70(3):615–641

434. Nieh J (2010) A negative feedback signal that is triggered by peril curbs honeybee recruitment. Current Biology 20(4):310–315

435. Nielsen J, Hedeholm R, Heinemeier J, Bushnell P, Christiansen J, Olsen J, Ramsey C, Brill R, Simon M, Steffensen K, Steffensen J (2016) Eye lens radiocarbon reveals centuries of longevity in the Greenland shark (Somniosus microcephalus). Science 353(6300):702-704

436. Niknam T, Azizipanah-Abarghooee R, Zare M, Bahmani-Firouzi B (2013) Reserve Constrained Dynamic Environmental/Economic Dispatch: A New Multiobjective Self-Adaptive Learning Bat Algorithm. IEEE Systems Journal 7(4):763–776

437. Nolfi S, Floreano D (2000) Evolutionary Robotics: The Biology, Intelligence, and Technology of Self-Organizing Machines. Cambridge, Massachusetts: MIT Press

438. Novoplansky A (2002) Developmental plasticity in plants: implications of non-cognitive behavior. Evolutionary Ecology 16(3):177–188

439. Nutman A, Bennett V, Friend C, Van Kranedonk M, Chivas A (2016) Rapid emergence of life shown by discovery of 3,700 million year old microbial structures. Nature 537(7621):535–538

440. Nyffeler M, Olson E, Symondson W (2016) Plant-eating by spiders. Journal of Arachnology 44(1):15–27
441. O'Brien B, Parrent J, Jackson J, Moncalvo J, Vilgalys R (2005) Fungal community analysis by large-scale sequencing of environmental samples. Applied and Environmental Microbiology 71(9):5544-5550
442. O'Neill M, Ryan C (2003) Grammatical Evolution: Evolutionary Automatic Programming in an Arbitrary Language. Boston: Kluwer Academic Publishers
443. O'Shea-Wheller T, Masuda N, Sendova-Franks A, Franks N (2017) Variability in individual assessment behaviour and its implications for collective decision-making. Proceedings of the Royal Society B 284(1848):20162237
444. Oster G, Wilson E (1978) Caste and Ecology in the Social Insects. Princeton, New Jersey: Princeton University Press
445. Pampara G, Engelbrecht A (2011) Binary Artificial Bee Colony Optimisation. In: Proceedings of the 2011 IEEE Swarm Intelligence Symposium (SiS 2011), pp 170–177, New Jersey: IEEE Press
446. Panait L, Luke S (2004) Learning ant foraging behaviors. In: Proceedings of the Ninth International Conference on the Simulation and Synthesis of Living Systems (ALIFE9), pp 575–581, Cambridge, Massachusetts: MIT Press
447. Panait L, Luke S (2005) Cooperative Multi-Agent Learning: The State of the Art. Autonomous Agents and Multi-Agent Systems 11(3):387–434
448. Panyutina A, Kuznetsov A, Volodin I, Abramov A, Soldatova I (2016) A blind climber: The first evidence of ultrasonic echolocation in arboreal mammals. Integrative Zoology 12(2):172–184
449. Parrish J, Viscido S, Grunbaum D (2002) Self-organized Fish Schools: An Examination of Emergent Properties. Biological Bulletin 202(3):296–305
450. Passino K (2000) Distributed Optimisation and Control Using Only a Germ of Intelligence. In: Proceedings of the 2000 IEEE International Symposium on Intelligent Control, pp 5–13, New Jersey: IEEE Press
451. Passino K (2002) Biomimicry of Bacterial Foraging for Distributed Optimisation and Control. IEEE Control Systems Magazine 22(3):52–67
452. Passino K, Seeley T (2006) Modeling and analysis of nest-site selection by honeybee swarms: the speed and accuracy trade-off. Behavioral Ecology and Sociobiology 59(3):427–442
453. Pedersen M, Chipperfield A (2010) Simplifying Particle Swarm Optimisation. Applied Soft Computing 10(2):618–628
454. Perna A, Latty T (2014) Animal transportation networks. Journal of the Royal Society Interface 11(100):20140334
455. Perry C, Baciadonna L, Chittka L (2016). Unexpected rewards induce dopamine-dependent positive emotion–like state changes in bumblebees. Science 353(6307):1529–1531
456. Pham D, Castellani M (2009) The Bees Algorithm: modelling foraging behaviour to solve continuous optimisation problems. Proceedings of Institute of Mechanical Engineering, Part C: Journal of Mechanical Engineering Science 223(12):2919–2938
457. Pham D, Ghanbarzadeh A, Koc E, Otri S, Rahim S, Zaidi M (2005) The Bees Algorithm — Technical Note, Manufacturing Engineering Centre, Cardiff University: Cardiff, UK.
458. Pham D, Ghanbarzadeh A, Koc E, Otri S, Rahim S, Zaidi M (2006) The Bees Algorithm — A novel tool for complex optimisation problems. In: Proceedings of International Production Machines and Systems (IPROMS 2006), pp 454–459, Amsterdam: Elsevier

459. Pianka E (2011) Evolutionary Ecology (7th ed). (ebook) `https://books.google.com.br/books?id=giFL5bonGhQC&lpg=PR3&hl=ptBR&pg=PR3#v=onepage&q&f=false`

460. Pion M, Spangenberg J, Simon A, Bindschedler S, Flury C, Chatelain A, Bshary R, Job D, Junier P (2013) Bacterial farming by the fungus *Morchella crassipes*, Proceedings of the Royal Society B 280(1773):20132242

461. Pitonakova L, Crowder R, Bullock S (2016) Information flow principles for plasticity in foraging robot swarms. Swarm Intelligence 10(1):33–63

462. Pourtakdoust S, Nobahari H (2004) An Extension of Ant Colony System to Continuous Optimisation Problems. In: Proceedings of the 4th International Workshop on Ant Colony Optimisation and Swarm Intelligence, pp 294–301, Berlin: Springer

463. Prat Y, Taub M, Yovel Y (2016) Everyday bat vocalisations contain information about emitter, addressee, context, and behavior. Scientific Reports 6:39419

464. Pratt S (2005) Behavioral mechanisms of collective nest-site choice by the ant *Temnothorax curvispinosus*. Insectes Sociaux 52(4):383–392

465. Pratt S (2010) Nest site choice in social insects. In: Breed M D and Moore J (eds) Encyclopedia of Animal Behavior 2:534–540, Amsterdam: Elsevier

466. Pratt S, Mallon E, Sumpter D, Franks N (2002) Quorum sensing, recruitment, and collective decision-making during colony emigration by the ant *Leptothorax albipennis*. Behavioral Ecology and Sociobiology 52(2):117–127

467. Pratt S, Sumpter D, Mallon E, Franks N (2005) An agent-based model of collective nest choice by the ant *Temnothorax curvispinosus*. Animal Behaviour 70(5):1023–1036

468. Premaratne U, Samarabandu J, Sidhu, T (2009) A New Biologically Inspired Optimisation Algorithm. In: Proceedings of the Fourth International Conference on Industrial and Information Systems (ICIIS 2009), pp 279–284, New Jersey: IEEE Press

469. Prindle A, Liu J, Asally M, Ly S, Garcia-Ojalvo J, Süel G (2015) Ion channels enable electrical communication in bacterial communities. Nature 527(7576):59–63

470. Quesada R, Triana E, Vargas G, Douglass J, Seid M, Niven J, Eberhard W, Wcislo W (2011) Large brains in small spiders: the central nervous system extends into the legs of miniature spiders. Arthropod Structure and Development 40(6):521–529

471. Rad, H S, Lucas C (2007) A recommender system based on invasive weed optimisation algorithm. In: Proceedings of the 2007 IEEE Congress on Evolutionary Computation (CEC 2007), pp 4297–4303, New Jersey: IEEE Press

472. Ramos-Fernández G, Morales J (2014) Unraveling fission-fusion dynamics: how subgroup properties and dyadic interactions influence individual decisions. Behavioral Ecology and Sociobiology 68(8):1225–1235

473. Raynaud X, Nunan N (2014) Spatial Ecology of Bacteria at the Microscale in Soil. PLoS ONE 9(1):e87217

474. Real L, Caraco T (1986) Risk and Foraging in Stochastic Environments. Annual Review of Ecology and Systematics 17:371–390

475. Réale D, Reader S, Sol D, McDougall P, Dingemanse N (2007) Integrating animal temperament within ecology and evolution. Biological Reviews 82(2):291–318

476. Rechenberg I (1965) Cybernetic Solution Path of an Experimental Problem. Royal Aircraft Establishment, Farnborough, Library Translation No. 1122, August

477. Rechenberg I (1973) Evolutionsstrategie: Optimierung technisher Systeme nach Prinzipien der biologischen Evolution, Stuttgart: Frommann-Holzboog Verlag

478. Regular P, Hedd A, Montevecchi W (2013) Must Marine predators always follow scaling laws? Memory guides the foraging decisions of a pursuit-diving seabird. Animal Behaviour 86(3):545–552

479. Reid C, Beekman M (2013) Solving the Towers of Hanoi — how an amoeboid organism efficiently constructs transport networks. The Journal of Experimental Biology 216(9):1546–1551

480. Reid C, Beekman M, Latty T, Dussutor A (2013) Amoeboid organism used extracellular secretions to make smart foraging decisions. Behavioral Ecology 24(4):812–818

481. Reid C, Garnier S, Beekman M, Latty T (2015) Information integration and multiattribute decision making in non-neuronal organisms. Animal Behaviour 100:44–50

482. Reid C, Latty T, Dussutor A, Beekman M (2012) Slime mold uses an externalized spatial 'memory' to navigate in complex environments. Proceedings of the National Academy of Sciences 109(43):17490–17494

483. Reynolds A (2015) Liberating Lévy walk research from the shackles of optimal foraging. Physics of Life Reviews 14:59–83

484. Reynolds C (1986) `http://www.red3d.com/cwr/boids/` includes a link to the original software (accessed 16 November 2015)

485. Reynolds C (1987) Flocks, Herds and Schools: A distributed behavioral model. In: Proceedings of the 14th annual conference on computer graphics and interactive techniques (SIGGRAPH 1987), pp 25–34, New York: ACM

486. Rendell L, Fogarty L, Hoppitt W, Morgan T, Webster M, Laland K (2011) Cognitive culture: theoretical and empirical insights into social learning strategies. Trends in Cognitive Sciences 15(2):68–76

487. Robinson E, Jackson D, Holcombe M, Ratnieks F (2005) Insect communication: 'No entry' signal in ant foraging. Nature 438(7067):442–442

488. Roces F, Tautz J, Hölldobler B (1993) Stridulation in leaf-cutting ants: short-range recruitment through plant-borne vibrations. Naturwissenschaften 80(11):521–524

489. Roces F, Hölldobler B (1995) Vibrational communication between hitchhikers and foragers in leaf-cutting ants (Atta cephalotes). Behavioral Ecology and Sociobiology, 37(5):297–302

490. Rodrigo T (2002) Navigational strategies and models. Psicólogica 23(1):3–32

491. Rogers, A (1988) Does Biology Constrain Culture? American Anthropologist 90(4):819–831

492. Romp G (1997) Game Theory: Introduction and Applications. Oxford: Oxford University Press

493. Rosenthal M, Gertler M, Hamilton A, Prasad S, Maydianne, Andrade M (2017) Taxonomic bias in animal behaviour publications. Animal Behaviour 127:83–89

494. Ross R, Quetin L (1986) How productive are Antarctic krill. BioScience 36(4):264–269

495. Roy G G, Chakroborty P, Zhao S Z, Das S, Suganthan P N (2010) Artificial Foraging Weeds for Global Numerical Optimisation Over Continuous Spaces. In: Proceedings of the 2010 IEEE World Congress on Computational Intelligence (WCCI 2010), pp 1189–1196, New Jersey: IEEE Press

496. Roy S, Islam S M, Das S, Ghosh S (2013) Multimodal optimisation by artificial weed colonies enhanced with local group search optimizers. Applied Soft Computing 13(1):27–46

497. Ruczynski I, Kalko E, Siemers B (2007) The sensory basis of roost finding in a forest bat, *Nyctalus noctula*. Journal of Experimental Biology 210(20):3607–3615

498. Rudy J (2014) The Neurobiology of learning and memory (2nd ed). Sunderland, Massachusetts: Sinauer Associates Inc

499. Rui T, Fong S, Yang X-S, Deb S (2012) Nature-inspired Clustering Algorithms for Web Intelligence Data. In: Proceedings of the 2012 IEEE, WIC, ACM International Conference on Web Intelligence and Intelligent Agent Technology, pp 147–153, New Jersey: IEEE Press

500. Ryan C, Collins J J, O'Neill M (1998) Grammatical evolution: evolving programs for an arbitrary language. In: Proceedings of the First European Workshop on Genetic Programming, LNCS 1391, pp 83–95, Berlin: Springer

501. Rybicki E (1990) The classification of organisms at the edge of life, or problems with virus systematics. South African Journal of Science 86:182-186

502. Salhi A, Fraga E (2011) Nature-inspired Optimisation Approaches and the New Plant Propagation Algorithm. In: Proceedings of the 2011 International Conference on Numerical Analysis and Optimisation (ICeMATH 2011), pp K2–1:K2–8

503. Santorelli C, Schaffner C, Campbell C, Notman H, Pavelka M, Weghorst J, Aureli F (2011) Traditions in Spider Monkeys Are Biased towards the Social Domain. PLoS One 6(2):e16863

504. Saremi S, Mirjalili S M, Mirjalili S (2014). Chaotic krill herd optimisation algorithm. Procedia Technology 12:180–185

505. Sauter J, Matthews R, Van Dyke Parunak H, Brueckner S (2002) Evolving adaptive pheromone path planning mechanisms. In: Proceedings of the First International Joint Conference on Autonomous Agents and Multiagent Systems (AAMAS 2002), pp 434–440, ACM: New York

506. Sayadi M, Ramezanian R, Ghaffari-Nasab N (2010) A discrete firefly meta-heuristic with local search for makespan minimisation in permutation flow shop scheduling problems. International Journal of Industrial Engineering Computations 1(1):1–10

507. Scharf I (2016) The multifaceted effects of starvation on arthropod behaviour. Animal Behaviour 119:37–48

508. Scharf I, Ovadia O (2006) Factors influencing site abandonment and site selection in a sit-and-wait predator: A review of pit-building antlion larvae. Journal of Insect Behavior 19(2):197–218

509. Scheres B, van der Putten W (2017) The plant perceptron connects environment to development. Nature 543(7645):337–345

510. Schneirla T (1944) A unique case of circular milling in ants, considered in relation to trail following and the general problem of orientation. American Museum Novitates 1253:1–26

511. Schönrogge K, Barbero F, Casacci L, Settele J, Thomas J (2016) Acoustic communication within ant societies and its mimicry by mutualistic and socially parasitic myrmecophiles. Animal Behaviour http://dx.doi.org/10.1016/j.anbehav.2016.10.031, published on line 22 November 2016

512. Schwefel H-P (1965) Kybernetische Evolution als Strategie der experimentellen Forschung in der Strömungstechnik. Diploma Thesis. Technical University of Berlin

513. Schwefel H-P (1981) Numerical Optimisation of Computer Models. Chichester, UK: Wiley

514. Scott P (2008) Physiology and Behaviour of Plants. Hoboken, New Jersey: John Wiley and Sons

515. Seeley T (1995) The Wisdom of the Hive: The Social Physiology of Honeybee Colonies. Cambridge, MA: Harvard University Press

516. Seeley T, Mikheyev A, Pagano G (2000) Dancing bees tune both duration and rate of waggle-run production in relation to nectar-source profitability. Journal of Comparative Physiology A 186(9):813–819

517. Seeley T, Morse R, Visscher P (1979) The natural history of the flight of honeybee swarms. Psyche 86(2–3):103–113

518. Seeley T, Visscher P, Passino K (2006) Group Decision Making in Honeybee Swarms. American Scientist 94(3):220–229

519. Selten R, Shmida A (1991) Pollinator foraging and flower competition in a game equilibrium model. In: Selten R (ed) Game Theory in Behavioral Sciences, pp 195–256, Berlin: Springer

520. Senthilnath J, Omkar S, Mani V (2011) Clustering using firefly algorithm: performance study. Swarm and Evolutionary Computation 1(3):164–171

521. Serfass T (1995) Cooperative foraging by North American river otters, *Lutra canadensis*. Canadian Field Naturalist 109(4):458–459

522. Sharma A, Sharma A, Panigrahi B, Kiran D, Kumar R (2016) Ageist Spider Monkey Optimisation algorithm. Swarm and Evolutionary Computation 28:58–77

523. Shen H, Zhu Y, Niu B, Wu Q (2009) An improved group search optimizer for mechanical design optimisation problems. Progress in Natural Science 19(1):91–97

524. Sherry D, Mitchell J (2007) Neuroethology of Foraging. In: Stephens D, Brown J, Ydenberg R (eds) Foraging Behavior and Ecology, pp 61–102, Chicago: University of Chicago Press

525. Shieh K, Wilson W, Winslow M, McBride D, Hopkins C (1996) Short-range orientation in electric fish: an experimental study of passive electrolocation. Journal of Experimental Biology 199(11):2383–2393

526. Shtonda B, Avery L (2006) Dietary choice behavior in *Caenorhabditis elegans*. Journal of Experimental Biology 209(1):89–102

527. Silliman B, Newell S (2003) Fungal farming in a snail. Proceedings of the National Academy of Sciences 100(26):15643–15648

528. Silva A, Neves A, Costa E (2002) An empirical comparision of particle swarm and predator prey optimisation. In: Proceedings of Artificial Intelligence and Cognitive Science (AICS 2002), LNAI 2464, pp 103–110, Berlin: Springer

529. Simard S, Beiler K, Bingham M, Deslippe J, Philip L, Teste F (2012) Mycorrhizal networks: Mechanisms, ecology and modelling. Fungal Biology Reviews 26(1):39–60

530. Simard S, Perry D, Jones M, Myrold D, Durall D, Molina R (1997) Net transfer of carbon between ectomycorrhizal tree species in the field. Nature 388(6642):579–582

531. Singh G P, Singh A (2014) Comparative study of krill herd, firefly and cuckoo search algorithms for unimodal and multimodal optimisation. International Journal of Intelligent Systems and Applications 2(3):26–37

532. Slobodchikoff C, Perla B, Verdolin J (2009) Prairie Dogs: Communication and Community in an Animal Society. Cambridge, Massachusetts: Harvard University Press

533. Smith H (2000) Phytochromes and light signal perception by plants — an emerging synthesis. Nature 407(6804):585–591

534. Smith J (1974) The food searching behavior of two European thrushes. II: The adaptiveness of the search patterns. Behavior 49(1-2):1–61

535. Smotherman M, Knörnschild M, Smarsh G, Bohn K (2016) The origins and diversity of bat songs. Journal of Comparative Physiology A 202(8):535–554

536. Socha K (2004) ACO for continuous and mixed-variable optimisation, In: Proceedings of International Workshop on Ant Colony Optimisation and Swarm Intelligence (ANTS 2004), LNCS 3172, pp 25–36, Berlin: Springer

537. Socha K, Dorigo M (2008) Ant colony optimisation for continuous domains. European Journal of Operational Research 185(3):1155–1173

538. Song Y Y, Zeng R S, Xu J F, Li J, Shen X, Yihdego W G (2010) Interplant Communication of Tomato Plants through Underground Common Mycorrhizal Networks. PLoS ONE 5(10): e13324

539. Sörensen K (2015) Metaheuristics - the metaphor exposed. International Transactions in Operational Research 22(1):3–18

540. Squire L (1987) Memory and the Brain. Oxford: Oxford University Press
541. Stahler D, Heinrich B, Smith D (2002) Common ravens, *Corvus corax*, preferentially associate with grey wolves, *Canis lupus*, as a foraging strategy in winter. Animal Behaviour 64(2):283–290
542. Stander P (1992) Cooperative hunting in lions: the role of the individual. Behavioral Ecology and Sociobiology 29(6):445–454
543. Stefanic P, Kraigher B, Lyons N A, Kolter R, Mandic-Mulec I (2015) Kin discrimination between sympatric *Bacillus subtilis* isolates. Proceedings of the National Academy of Sciences 112(45):14042–14047
544. Stephens D (2007) Models of information use. In: Stephens D, Brown J, Ydenberg R (eds) Foraging Behavior and Ecology, pp 105–138, Chicago: University of Chicago Press
545. Stephens D, Krebs J (1986) Foraging Theory. Princeton, New Jersey: Princeton University Press
546. Stephens M (2013) Sensory Ecology, Behaviour and Evolution. Oxford: Oxford University Press
547. Stephenson S (2010) The Kingdom of Fungi: The biology of mushrooms, molds, and Lichens. London: Timber Press
548. Sterling P, Laughlin S (2015) Principles of Neural Design. Cambridge, MA: MIT Press
549. Stocker S (1999) Models for tuna school formation. Mathematical Biosciences 156(1-2):167–190
550. Storn R, Price K (1995) Differential evolution — a simple and efficient adaptive scheme for global optimisation over continuous spaces. Technical Report TR-95–012: International Computer Science Institute, Berkeley
551. Storn R, Price K (1997) Differential evolution — a simple and efficient heuristic for global optimisation over continuous spaces. Journal of Global Optimisation 11(4):341–359
552. Stützle T, Hoos H (2000) *MAX-MIN* Ant System. Future Generation Computer Systems 16(8):889–914
553. Su H, Zhao Y, Zhou J, Feng H, Jiang D, Zhang K, Yang J (2017) Trapping devices of nematode-trapping fungi: formation, evolution, and genomic perspectives. Biological Reviews Cambridge Philosophical Society 92(1):357–368
554. Suarez J, Murphy R (2011) A survey of animal foraging for directed, persistent search by rescue robotics. In: Proceedings of the 2011 IEEE International Symposium on Safety, Security and Rescue Robotics, pp 314–320, New Jersey: IEEE Press
555. Suddendorf T, Addis D, Corballis M (2009) Mental time travel and the shaping of the human mind. Philosophical Transactions of the Royal Society B 364(1521):1317–1324
556. Sugawara K, Kazama T, Watanabe T (2004) Foraging Behavior of Interacting Robots with Virtual Pheromone. In: Proceedings of the 2004 IEEE / RSJ International Conference on Intelligent Robots and Systems, pp 3074–3079, New Jersey: IEEE Press
557. Sulaiman M, Salhi A (2015) A Seed-Based Plant Propagation Algorithm: The feeding station model. Scientific World Journal, Vol. 2015, Article ID 904364
558. Sulaiman M, Salhi A (2016) A Hybridisation of Runner-based and Seed-based Plant Propagation Algorithms. Nature-Inspired Computation in Engineering, Studies in Computational Intelligence 637, pp 195-215, Berlin: Springer
559. Sumpter, D (2010) Collective Animal Behavior. Princeton, New Jersey: Princeton University Press
560. Sumpter D, Brannstrom A (2008) Synergy in social communication. In: D'Ettorre P, Hughes D P (eds) Social Communication, pp 191–208, Oxford: Oxford University Press

561. Sumpter D, Krause J, James R, Couzin I, Ward A (2008). Consensus Decision Making by Fish. Current Biology 18(22):1773-1777

562. Szyszka P, Gerkin R, Galizia G, Smith B (2014) High-speed odor transduction and pulse tracking by insect olfactory receptor neurons. Proceedings of the National Academy of Sciences 111(47):16925–16930

563. Tavakolian R, Charkari N (2011) Novel Hybrid Clustering Optimisation Algorithms Based on Plant Growth Simulation Algorithm. Journal of Advanced Computer Science and Technology Research 1(2):84–95

564. ten Cate C (2012) Acoustic communication in plants: do the woods really sing? Behavioral Ecology 24(4):799–800

565. Tereshko V, Lee T (2002) How Information-Mapping Patterns Determine Foraging Behavior of a Honeybee Colony. Open Systems and Information Dynamics 9(2):181–193

566. Tero A, Kobayashi R, Nakagaki T (2007) A mathematical model for adaptive transport network in path finding by true slime mold. Journal of Theoretical Biology 244(4):553–564

567. Tero A, Takagi S, Saigusa T, Ito K, Bebber D, Fricker M, Yumiki K, Kobayashi R, Nakagaki T (2010) Rules for Biologically Inspired Adaptive Network Design. Science 327(5964):439–442

568. Thornton A, Raihani N (2008) The evolution of teaching. Animal Behaviour 75(6):1823–1836

569. Thorpe W (1963) Learning and Instinct in Animals. London: Methuen

570. Tietjen W (1982) Influence of activity patterns on social organisation of *Mallos gregalis* (Araneae, Dictynidae). Journal of Arachnology 10(1):75–84

571. Tlalka M, Bebber D P, Darrah P R, Watkinson S C, Fricker M D. (2008) Quantifying dynamic resource allocation illuminates foraging strategy in Phanerochaete velutina. Fungal Genetics and Biology 45(7):1111–1121

572. Todd N (2015) Do Humans Possess a Second Sense of Hearing? American Scientist 103(5):348–355

573. Tolman E (1948) Cognitive maps in rats and men. Psychological Review 55(4):189–208

574. Tong L, Wang C, Wang W, Su W (2005) A global optimisation bionics algorithm for solving integer programming—plant growth simulation algorithm. Systems Engineering — Theory and Practice 25(1):76–85

575. Tong L, Zhong-Tuo W (2008) Application of Plant Growth Simulation Algorithm on Solving Facility Location Problem. Systems Engineering — Theory & Practice 28(12):107–115

576. Toufailia H, Couvillon M, Ratnieks F, Gruter C (2013) Honeybee waggle dance communication: signal meaning and signal noise affect dance follower behaviour. Behavioral Ecology and Sociobiology 67(4):549–556

577. Trewavas A (2014) Plant Behaviour and Intelligence. Oxford: Oxford University Press

578. Tsuda S, Zauner K-P, Gunji Y-P (2006) Robot Control: From Silicon Circuitry to Cells. In: Biologically Inspired Approaches to Advanced Information Technology, Proceedings of the Second International Workshop (BioADIT 2006), pp 20–32, Berlin: Springer

579. Tsuda S, Zauner K-P, Gunji Y-P (2007) Robot control with biological cells. BioSystems 87(2-3):215–223

580. Uexküll J. von (1909). Umwelt und Innenwelt der Tiere. Berlin: Verlag von Julius Springer

581. Valian E, Mohanna S, Tavakoli S (2011) Improved Cuckoo Search Algorithm for Global Optimisation. International Journal of Communications and Information Technology 1(1):31–44

582. van den Bergh F, Engelbrecht A P (2004) A Cooperative approach to particle swarm optimisation. IEEE Transactions on Evolutionary Computation 8(3):225–239

583. van der Heijden M, Dombrowski N, Schlaeppi K (2017) Continuum of root-fungal symbioses for plant nutrition. Proceedings of the National Academy of Sciences 114(44):11574–11576

584. Vargas P, Di Paolo E, Harvey I, Husbands P (2014) The horizons of evolutionary robotics. Cambridge, MA: MIT Press

585. Varshney L, Chen B, Paniagua E, Hall D, Chklovskii D (2011) Structural Properties of the *Caenorhabditis elegans* Neuronal Network. PLoS Computational Biology 7(2):e1001066

586. Vega G C, Wiens J (2012) Why are there so few fish in the sea? Proceedings of the Royal Society B 279(1737):2323–2329

587. Vergassola M, Villermaux E, Shaiman B (2007) 'Infotaxis' as a strategy for searching without gradients. Nature 445(7126):406–409

588. Viana F, Kotinda G, Rade D, Steffan V (2006) Can Ants Design Mechanical Engineering Systems. In: Proceedings of the 2006 IEEE Congress on Evolutionary Computation (CEC 2006), pp 3173–3179, New Jersey: IEEE Press

589. Vickery W, Giraldeau L D, Templeton J, Kramer D, Chapman C (1991) Producers, scroungers, and group foraging. The American Naturalist 137(6):847–863

590. Viswanathan G, Buldyrev S, Havlin S, da Luz M, Raposo E, Stanley H E (1999) Optimizing the success of random searches. Nature 401(6756):911–914

591. Viswanathan G, da Luz M, Raposo E, Stanley H E (2011) The Physics of Foraging: An Introduction to Random Searches and Biological Encounters. Cambridge: Cambridge University Press

592. Viswanathan G, Raposo E, da Luz M (2008) Lévy flights and superdiffusion in the context of biological encounters and random searches. Physics of Life Reviews 5(3):133–150

593. Vitt L, Caldwell J (2014) Herpetology: An Introductory Biology of Amphibians and Reptiles. London: Academic Press

594. Vittori K, Talbot G, Gautrais J, Fourcassie V, Araujo A, Theraulaz G (2006) Path efficiency of ant foraging trails in an artificial network. Journal of Theoretical Biology 239(4):507–515

595. Vogel D, Dussutour A (2016) Direct transfer of learned behaviour via cell fusion in non-neural organisms. Proceedings of the Royal Society B 283(1845):20162382

596. Voigt-Heucke S, Taborsky M, Dechmann D (2010) A dual function of echolocation: bats use echolocation calls to identify familiar and unfamiliar individuals. Animal Behaviour 80(1):59–67

597. Volterra V (1926) Variazioni e fluttuazioni del numero d'individui in specie animali conviventi. Mem. Acad. Lincei Roma 2:31–113

598. von der Emde G (1999) Active Electrolocation of Objects in Weakly Electric Fish. Journal of Experimental Biology 202(10):1205–1215

599. von der Emde G, Amey M, Engelmann J, Fetz S, Folde C, Hollmann M, Metzen M, Pusch R (2008) Active electrolocation in *Gnathonemus petersii*: Behaviour, sensory performance, and receptor systems. Journal of Physiology - Paris 102(4–6):279–290

600. von der Emde G, Schwarz S (2002) Imaging of objects through active electrolocation in *Gnathonemus petersii*. Journal of Physiology - Paris 96(5-6):431–444

601. von Frisch K (1950) Die Sonne als Kompass im Leben der Bienen. Experientia 6:210–221

602. von Frisch K (1967) The Dance Language and Orientation of Bees. Cambridge, MA: Harvard University Press

603. von Middendorff A T (1855) Die Isopiptesen Russlands. Grundlagen zur Erforschung der Zugzeiten und Zugrichtungen der Vögel Russlands. Mem Acad Sci St Petersbg Sci Nat 8:1–143

604. von Thienen W, Metzler D, Witte V (2016) How memory and motivation modulate the responses to trail pheromones in three ant species. Behavioral Ecology and Sociobiology 70(3):393–407

605. Walker M, Bitterman M (1985) Conditioned responding to magnetic fields by honeybees. Journal of Comparative Physiology A 157(1):67–71

606. Wallace H, Trueman S (1995) Dispersal of *Eucalyptus torelliana* seeds by the resin-collecting stingless bee, *Trigona carbonaria*. Oecologia 104(1):12–16

607. Walton A, Toth A (2016). Variation in individual worker honeybee behavior shows hallmarks of personality. Behavioral Ecology and Sociobiology 70(7):999–1010

608. Walton S, Hassan O, Morgan K, Brown M (2011) Modified cuckoo search: A new gradient free optimisation algorithm. Chaos, Solitons and Fractals 44(9):710–718

609. Wang C, Cheng H Z, Yao L Z (2008) Reactive Power Optimisation by Plant Growth Simulation Algorithm. In: Proceedings of the Third International Conference on Electric Utility regulation and Restructuring and Power Technologies (DRPT 2008), pp 771–774, New Jersey: IEEE Press

610. Wang G, Guo L, Wang H, Duan H, Liu L, Li J (2014) Incorporating mutation scheme into krill herd algorithm for global numerical optimisation. Neural Computing and Applications 24(3):853–871

611. Wang H, Zhou X, Sun H, Yu X, Zhao J, Zhang H, Cui L (2016) Firefly algorithm with adaptive control parameters. Soft Computing doi:10.1007/s00500-016-2104-3 (published online 3 March 2016)

612. Ward P and Zahavi A (1973) The importance of certain assemblages of birds as 'information centres' for food finding. Ibis 115(4):517–534

613. Webster J, Weber R (2007) Introduction to Fungi (3rd ed). Cambridge: Cambridge University Press

614. Webster M (2017) Experience and motivation shape leader follower interactions in fish shoals. Behavioral Ecology 28(1):77–84

615. Wegner N, Snodgrass O, Dewar H, Hyde J (2015) Whole-body endothermy in a mesopelagic fish, the opah, *Lampris guttatus*. Science 348(6236):786–789

616. White J, Southgate E, Thomson J, Brenner S (1986) The Structure of the Nervous System of the Nematode *Caenorhabditis elegans*. Philosophical Transactions of the Royal Society B 314(1165):1–340

617. Whitford M, Freymiller G, Clark R (2017) Avoiding the serpent's tooth: predator–prey interactions between free–ranging sidewinder rattlesnakes and desert kangaroo rats. Animal Behaviour 130:73–78

618. Widder E (2001) Marine bioluminescence. Why do so many animals in the open ocean make light? Bioscience Explained. 1(1):1–9 http://www.bioscience-explained.org/ENvol1_1/index.html

619. Wiener J, Shettleworth S, Bingman V, Cheng K, Healy S, Jacobs L, Jeffery K, Mallot H, Menzel R, Newcombe N (2011) Animal navigation: A synthesis. In: Menzel R, Fischer J (eds) Animal Thinking: Contemporary Issues in Comparative Cognition, pp 51–76, Cambridge, MA: MIT Press

620. Wiener N (1948) Cybernetics: or Control and Communication in the Animal and the Machine. Cambridge, MA: MIT Press

621. Wilkinson G (1992) Information transfer at evening bat colonies. Animal Behaviour 44(3):501–518

622. Wilkinson G, Boughman J (1998) Social calls coordinate foraging in greater spearnosed bats. Animal Behaviour 55(2):337–350

623. Willis K (ed) (2017) State of the World's Plants 2017. Royal Botanic Gardens, Kew, https://stateoftheworldsplants.com/2017/report/SOTWP_2017.pdf

624. Wilson E (1975) Sociobiology: The New Synthesis. Cambridge, MA: Harvard University Press

625. Wilson T, Hastings J W (1998) Bioluminescence. Annual Review of Cell and Developmental Biology 14:197–230

626. Wolf T, Schmid-Hempel P (1989) Extra loads and foraging lifespan in honeybee workers. Journal of Animal Ecology 58(3):943–954

627. Wolpert D, Macready W (1995) No Free Lunch Theorems for Search. Santa Fe Institute Working Paper 95–02–010

628. World Conservation Union (2015). IUCN Red List of Threatened Species, http://cmsdocs.s3.amazonaws.com/summarystats/2015-4_Summary_Stats_Page_Documents/2015_4_RL_Stats_Table_1.pdf (accessed 22 January 2016)

629. Wray M, Klein B, Seeley T (2012) Honeybees use social information in waggle dances more fully when foraging errors are more costly. Behavioral Ecology 23(1):125–131

630. Wright J, Stone R, Brown N (2003) Communal roosts as structured information centres in the raven, *Corvus corax*. Journal of Animal Ecology 72(6):1003–1014

631. Wynne C, Udell M (2013) Animal Cognition: Evolution Behavior and Cognition (2nd ed). New York: Palgrave Macmillian

632. Xu J X, Deng X (2012) Complex chemotaxis behaviors of *C. elegans* with speed regulation achieved by dynamic neural networks. In: Proceedings of IEEE 2012 International Joint Conference on Neural Networks, pp 1–8, New Jersey: IEEE Press

633. Yang C, Huang Q, Wang L, Du W-G, Liang W, Møller A (2018) Keeping eggs warm: thermal and developmental advantages for parasitic cuckoos of laying unusually thick-shelled eggs. The Science of Nature 105(1–2):10

634. Yang C, Tu X, Chen J (2007) Algorithm of Marriage in Honeybees Optimisation Based on the Wolf Pack Search. In: Proceedings of the IEEE International Conference on Intelligent Pervasive Computing, pp 462–467, New Jersey: IEEE Press

635. Yang X S (2005) Engineering Optimisation via Nature-Inspired Virtual Bee Algorithms. In: Mira J, Álvarez J (eds) Artificial Intelligence and Knowledge Engineering Applications: A Bioinspired Approach, First International Work-Conference on the Interplay Between Natural and Artificial Computation (IWINAC 2005), LNCS 3562, pp 317–323, Springer

636. Yang X S (2008) Nature-inspired metaheuristic algorithms. Frome: Luniver Press

637. Yang X S (2009) Firefly Algorithms for Multimodal Optimisation. In: Proceedings of the 5th Symposium on Stochastic Algorithms, Foundations and Applications (SAGA 2009), LNCS 5792, pp 169–178, Berlin: Springer

638. Yang X S (2010a) Firefly Algorithm, Stochastic Test Functions and Design Optimisation. International Journal of Bio-Inspired Computation 2(2):78–84

639. Yang X S (2010b) A New Metaheuristic Bat-Inspired Algorithm. In: Proceedings of Fourth International Workshop on Nature Inspired Cooperative Strategies for Optimisation (NICSO 2010), Studies in Computational Intelligence 284, pp 65–74, Springer

640. Yang X S (2011) Bat algorithm for multiobjective optimisation. International Journal of Bio-Inspired Computation 3(5):267–274

641. Yang X S (2012) Flower pollination algorithm for global optimisation. In: Proceedings of the 11th International Conference on Unconventional Computation and Natural Computation (UCNC 2012), LNCS

642. Yang X S (2013a) Firefly Algorithm: Recent Advances and Applications. International Journal of Swarm Intelligence 1(1):36–50

643. Yang X S (2013b) Bat Algorithm: Literature Review and Applications. International Journal of Bio-Inspired Computation 5(3):141–149

644. Yang X S, Deb S (2009) Cuckoo search via Lévy Flights. In: Proceedings of the World Congress on Nature and Biologically Inspired Computing (NaBIC 2009), pp 210–214, New Jersey: IEEE Press

645. Yang X S, Deb S (2010) Engineering optimisation by cuckoo search. International Journal of Mathematical Modelling and Numerical Optimisation 1(4):330–343

646. Yang X S, Deb S (2013) Multiobjective cuckoo search for design optimisation. Computers and Operations Research 40(6):1616–1624

647. Yang X S, Deb S (2014) Cuckoo search: recent advances and applications. Neural Computing and Applications 24(1):169–174

648. Yang X S, Gandomi A (2012) Bat Algorithm: A Novel Approach for Global Engineering Optimisation. Engineering Computations 29(5):464–483

649. Yang X S, Karamanoglu M, He X S (2014) Flower pollination algorithm: A novel approach for multiobjective optimisation. Engineering Optimisation 46(9):1222–1237

650. Ydenberg R, Brown J, Stephens D (2007) Foraging: An overview. In: Stephens D, Brown J, Ydenberg R (eds) Foraging Behavior and Ecology, pp 1–28, Chicago: University of Chicago Press

651. Ye H-Y, Ye B-P, Wang D-Y (2008) Molecular control of memory in nematode *Caenorhabditis elegans*. Neuroscience Bulletin 24(1):49–55

652. Yeakle J, Dunne J (2015) Modern lessons from ancient food webs. American Scientist 103(3):188–195

653. Yip E, Powers K, Avilés L (2008) Cooperative capture of large prey solves scaling challenge faced by spider societies. Proceedings of the National Academy of Sciences 105(33):11818–11822

654. Yu J J Q, Li V O K (2013) A social spider algorithm for global optimisation. Dept. of Electrial and Electronic Engineering, The University of Hong Kong, Technical Report No. TR-2013-004

655. Yu J J Q, Li V O K (2014) Base station switching problem for green cellular networks with spider algorithm. In: Proceedings of the 2014 IEEE Congress on Evolutionary Computation (CEC 2014), pp 1–7, New Jersey: IEEE Press

656. Yu J J Q, Li V O K (2015a) A social spider algorithm for global optimisation. Applied Soft Computing 30:614–627

657. Yu J J Q, Li V O K (2015b) Parameter sensitivity analysis of social spider algorithm. In: Proceedings of the 2015 IEEE Congress on Evolutionary Computation (CEC 2015), pp 3200–3205, New Jersey: IEEE Press

658. Yuce B, Packianather M, Mastrocinque E, Pham D, Lambiase A (2013) Honeybees Inspired Optimisation Method: The Bees Algorithm. Insects 4(4):646–662

659. Zahavi A (1971) The function of pre-roost gatherings and communal roosts. Ibis 113(1):106–109

660. Zhang H, Zhu Y, Chen H (2012) Root Growth Model for Simulation of Plant Root System and Numerical Function Optimisation. In: Proceedings of the 8th International Conference, Intelligent Computing Technology (ICIC 2012), LNCS 7389, pp 641–648, Berlin: Springer

661. Zhang H, Zhu Y, Chen H (2014) Root growth model: a novel approach to numerical function optimisation and simulation of plant root system. Soft Computing 18(3):521–537

662. Zhang X, Wang Q, Adamatzky A, Chan F, Mahadevan S, Deng Y (2014) An Improved *Physarum polycephalum* Algorithm for the Shortest Path Problem. The Scientific World Journal, Vol. 2014, Article ID 487069

663. Zhang Z, Gao C, Liu Y, Qian T (2014) A universal optimisation strategy for ant colony optimisation algorithms based on the *Physarum*-inspired mathematical model. Bioinspiration and Biomimetics 9(2014):036006

664. Zhao Z, Cui Z, Zeng J, Yue X (2011) Artificial Plant Optimisation Algorithm for Constrained Optimisation Problems. In: Proceedings of the 2011 Second International Conference on Innovations in Bio-inspired Computing and Applications, pp 120–123, New Jersey: IEEE Press

665. Zhu Z, van Belzen J, Hong T, Kunihiro T, Ysebaert T, Herman P, Boum T (2016). Sprouting as a gardening strategy to obtain superior supplementary food: evidence from a seed-caching marine worm. Ecology 97(12):3278–3284

666. Zhu G-Y, Zhang W-B (2017) Optimal foraging algorithm for global optimisation. Applied Soft Computing 51:294–313

667. Zwaka H, Bartels R, Gora J, Franck V, Culo A, Götsch M, Menzel R (2015) Context Odor Presentation during Sleep Enhances Memory in Honeybees. Current Biology 25(21):2869–2874

Index

© Springer Nature Switzerland AG 2018
A. Brabazon, S. McGarraghy, *Foraging-Inspired Optimisation Algorithms*,
Natural Computing Series, https://doi.org/10.1007/978-3-319-59156-8

Z

Printed in the United States
By Bookmasters